APPLIED GROUNDWATER MODELING

APPLIED GROUNDWATER MODELING
Simulation of Flow and Advective Transport

MARY P. ANDERSON

Department of Geology and Geophysics
University of Wisconsin
Madison, Wisconsin

WILLIAM W. WOESSNER

Department of Geology
University of Montana
Missoula, Montana

ACADEMIC PRESS, INC.

Harcourt Brace Jovanovich, Publishers
San Diego   New York   Boston
London   Sydney   Tokyo   Toronto

Disclaimer. In this world of legal mazes where fault is often found with technically correct work, we the authors feel the need to include a disclaimer. This textbook sets forth the general principles needed to understand how numerical models of ground-water flow are developed. While we believe that the guidelines in this book will minimize errors if followed, no text can substitute for experience or a knowledge of specific circumstances. Accordingly, we cannot assume responsibility for any errors, mistakes, or misrepresentations that may arise from misuse or misapplication of the procedures contained herein.

This book is printed on acid-free paper. ∞

Academic Press, Inc.
San Diego, California 92101

United Kingdom Edition published by
Academic Press Limited
24–28 Oval Road, London NW1 7DX

Library of Congress Cataloging-in-Publication Data

Anderson, Mary P.
    Applied groundwater modeling : simulation of flow and advective
transport / Mary P. Anderson. William W. Woessner.
        p.   cm.
    Includes bibliographical references and index.
    ISBN 0-12-059485-4
    1. Groundwater flow--Simulation methods.   I. Woessner, William W.
II. Title.
    GB1197.7A53   1991
    551.49'01'1--dc20   91-19676
                                                                    CIP

PRINTED IN THE UNITED STATES OF AMERICA
91 92 93 94   9 8 7 6 5 4 3 2 1

To our parents
Peter and Dorothy; Flora
and
the memory of Warren

"...for the knowledge gained through a lifetime."
James McNeill Whistler

CONTENTS

CHAPTER 4

BOUNDARIES

CHAPTER 5

SOURCES AND SINKS

LIST OF BOXES

PREFACE

Application of large numerical simulation models to groundwater problems involves both art and science. Successful modelers understand the science behind the models they use and have considerable experience in applying models to practical problems. The *art* of modeling is learned by practicing how to apply models. In the process, one learns how to describe the problem domain, select boundary conditions, assign model parameters, and calibrate the model. In other words, one becomes a successful modeler by knowing the science and practicing the art. There are a number of textbooks that provide advanced, intermediate, and elementary treatments of the *science* involved in numerical modeling of groundwater flow and solute transport. Students and professionals attempting to learn the *art* of modeling are commonly overwhelmed by intricacies of computer programs and modeling mechanics. Some of the art of modeling can be learned from informed reading of the documentation that accompanies computer codes and from selected papers in the literature. However, a single comprehensive reference to assist those wishing to develop proficiency in the art of groundwater modeling is not available. Our motivation for writing this book is to provide such a text and thereby encourage proper use of groundwater models.

Groundwater models can be divided broadly into two categories: groundwater flow models and solute transport models. Groundwater flow models solve for the distribution of head, whereas solute transport models solve for concentration of solute as affected by advection (movement of the solute with the average groundwater flow), dispersion (spreading and mixing of the solute), and chemical reactions, which slow down or transform solutes. A transport model consists of two parts: a computer program to solve for groundwater heads and a computer program to solve for concentration of solute. Movement of solutes or contaminants solely by advection can be described by a ground-water flow model coupled to a particle tracking code, which traces flow lines thereby allowing general contaminant pathways to be identified. This book focuses on groundwater flow models and the use of particle tracking models to simulate the movement of solutes, usually contaminants. We feel that an emphasis on flow models and particle tracking is appropriate for the following reasons:

1. Application of solute transport models requires a fundamental understanding of groundwater flow models.
2. There are still unresolved problems in developing dispersion and chemical reaction theories that apply to field problems. This causes a great deal of uncertainty in applying the currently available solute transport codes.
3. Solute transport is an area of active research; new ideas and new ways of solving the governing equation appear regularly in the literature (Sudicky, 1989; Leismann and Frind, 1989). Existing solute transport codes are likely to become outdated quickly.
4. Results of a particle tracking analysis are sufficient to answer many regulatory questions involving contaminants.

We selected three nonproprietary groundwater flow codes and several particle tracking codes to illustrate examples of modeling mechanics. The book features two finite difference models, MODFLOW (McDonald and Harbaugh, 1988) and PLASM (Prickett and Lonnquist, 1971), and one finite element model, AQUIFEM-1 (Townley and Wilson, 1980). MODFLOW and PLASM were selected because they are widely used, well documented, and available for IBM-compatible personal computers at nominal cost. AQUIFEM-1 was selected as a representative example of a finite element model, and also because it is exceptionally well documented. It too is available for IBM-compatible personal computers. A new multilayered version called AQUIFEM-N also is available (Townley, 1990). A number of particle tracking codes are discussed. Two codes (PATH3D by Zheng, 1989, and MODPATH by Pollock, 1989) were developed to link directly with MODFLOW. FLOWPATH (Franz and Guiguer, 1990) is a combined steady-state flow and particle tracking code. The flow model is based on PLASM, but a number of changes were made to accommodate the particle tracking model. Addresses for acquiring these codes are given in Table 1 of this preface. Additionally, a few special-purpose codes are mentioned in Chapters 8 and 12.

Faust et al. (1981) list seven questions that should be considered prior to a modeling study to safeguard against model misuse. One of the most important

Table 1

Code Availability for IBM-Compatible Computers

Code	Source	Price[a]
Flow codes		
MODFLOW	Scientific Software Group P.O. Box 23041	$215
	Washington, D.C. 20026-3041 (703-620-9214) FAX: (703-620-6793)	$390 (compiled for 386 and 486 computers)
AQUIFEM-1	GEOCOMP Corp. 342 Sudbury Road Concord, Massachusetts 01742 (617-369-8304)	$750
AQUIFEM-N	Dr. Lloyd Townley CSIRO Western Australian Laboratories Private Bag, PO Wembley, WA 6014 Australia	$1000
PLASM (version 5.0-1989B)	Thomas A. Prickett & Associates 6 GH Baker Drive Urbana, Illinois 61801 (217-384-0615)	$150
Particle tracking codes		
PATH3D (PC version)	S.S. Papadopulos and Associates, Inc. 12250 Rockville Pike, Suite 290 Rockville, Maryland 20852 (301-468-5760)	$500
MODPATH-MODPATH PLOT	Scientific Software Group (see address above)	$500
		$750 (compiled for 386 and 486 computers)
FLOWPATH	Scientific Software Group (see address above)	$430

[a] Prices quoted are current as of January 1991.

is, does the investigator have sufficient background in both hydrology and modeling? Although this book provides some of the background needed to apply models to field problems, it is assumed that the reader has some knowledge of basic principles of hydrogeology as contained in Domenico and Schwartz (1990), Fetter (1988), or Freeze and Cherry (1979), and has a rudimentary knowledge of some of the more theoretical aspects of groundwater modeling as contained in Bear and Verruijt (1987), and Wang and Anderson (1982), for example. In particular, the reader should be familiar with the basic theory of finite differences and finite elements.

The first two chapters provide the motivation and the equations needed for modeling. The third chapter describes how to formulate a conceptual model of a flow system and translate the conceptual model into a numerical model. Chapters 4 through 7 discuss the mechanics of applying models to groundwater flow problems with special attention to features in MODFLOW, PLASM, and AQUIFEM-1. The general principles discussed, however, apply to all comprehensive numerical codes designed to simulate groundwater problems. Code-specific discussions and other specialized topics are contained in boxes throughout the book. Chapter 8 deals with model execution and the calibration, process, and Chapter 9 offers suggestions for presenting modeling assumptions and results. Chapter 10 analyzes recent insights gained through postaudits of modeling predictions. Particle tracking is discussed in Chapter 11. Computer generation of flow nets and more advanced topics including modeling of unsaturated flow, multiphase flow, solute transport with dispersion and chemical reactions, flow through fractured media, and density-dependent flow of miscible fluids are assessed briefly in Chapter 12. This book will be useful as a textbook in courses in groundwater modeling and will be helpful to consultants, industry, government agencies, and other users.

"All things are ready, if our minds be so."
—*Henry V, Act IV*

Acknowledgments

Laura Toran, Lloyd Townley, Shlomo Neuman, Chunmiao Zheng, Scott Bair, Alan Dutton, David Patterson, Randy Hunt, and Steve Wheatcraft reviewed portions of the book in draft form. We are grateful for their many valuable comments. Erik Webb, Susan Werther, Paul Dombrowski, and John Cuplin helped with the drafting of figures.

We want to acknowledge our mentors, especially Irwin Remson and Dave Stephenson, and our students, who influenced the creation of this book both directly and indirectly. We also thank our parents for life-long encouragement to pursue our interests and for moral and financial support. Our respective spouses, Charles and Jean, gave never-ending encouragement and kept the home front afloat during this work. Kristina and David and Sasha and Mitchell also contributed by enriching our lives.

1

INTRODUCTION

"It is a capital mistake to theorise before one has data. Insensibly one begins to twist facts to suit theories, instead of theories to suit facts."
—*Sherlock Holmes in Scandal in Bohemia*

What changes can be expected in groundwater levels in the aquifers beneath Perth, Australia, in the year 2020?

How will a change in stream stage affect the water table in an adjacent alluvial aquifer in Missoula, Montana?

What is the capture area for a well field that furnishes municipal water supplies to Stevens Point, Wisconsin?

What is the most likely pathway of contaminants if there is a leak in the proposed underground repository for high-level radioactive waste in Yucca Mountain, Nevada?

Groundwater hydrologists are often called upon to predict the behavior of groundwater systems by answering questions like these. Providing answers to these seemingly simple questions involves formulating a correct conceptual model, selecting parameter values to describe spatial variability within the groundwater flow system, as well as spatial and temporal trends in hydrologic stresses and past and future trends in water levels. Although some decisions can be made using best engineering or best geologic judgment, in many instances human reasoning alone is inadequate to synthesize the conglomeration of factors involved in analyzing complex groundwater problems. The best tool available to help groundwater hydrologists meet the challenge of prediction is usually a groundwater model.

Peck et al. (1988) illustrate the complexity involved in a typical groundwater problem with the following example:

> Imagine a three-layered alluvial aquifer recharged by a stream and areally distributed recharge. For simplicity, assume the aquifer is rectangular, but the boundary conditions are complex (spatially variable). Flow in the river varies annually and therefore transient conditions must be modelled. Also for simplicity, let us assume that the system may be treated as isotropic and that changes in storage of the confining beds are negligible. The system might be simulated using finite differences with perhaps 1200 cells per layer (30 by 40). Suppose the river crosses in 70 cells and there are 20 wells tapping one layer each. Finally, we wish to simulate 12 months in which the stresses (pumping, recharge, and river stage) and boundary values vary each month.

The modeling application described above is a relatively simple one, yet it requires 37,820 pieces of data. In practice it is usually possible to identify zones, comprising several cells of a finite difference grid or several elements of a finite element mesh, in which parameter values are the same. Nevertheless, the example serves to illustrate the complexity typically encountered in addressing regional scale problems. The attraction of groundwater modeling is that it combines the subtlety of human judgment with the power of a digital computer.

1.1   What Is a Model?

A *model* is any device that represents an approximation of a field situation. *Physical models* such as laboratory sand tanks simulate groundwater flow directly. A *mathematical model* simulates groundwater flow indirectly by means of a governing equation thought to represent the physical processes that occur in the system, together with equations that describe heads or flows along the boundaries of the model (boundary conditions). For time-dependent problems, an equation describing the initial distribution of heads in the system also is needed (initial conditions). Mathematical models can be solved *analytically* or *numerically*. Either type of solution may involve a computer. When assumptions used to derive an analytical solution are judged to be too simplistic and inappropriate for the problem under consideration, a numerical model may be selected. For example, analytical solutions usually assume a homogeneous porous medium. A numerical model may also be easier to apply than an analytical model if the analytical problem involves complex superposition of solutions, e.g., with image well theory. Generally speaking, the fewer the simplifying assumptions used to formulate a model, the more complex is the model.

The set of commands used to solve a mathematical model on a computer forms the computer program or *code*. The code is generic, whereas a model includes a set of boundary and initial conditions as well as a site-specific nodal grid and site-specific parameter values and hydrologic stresses. A code is written once but a new model is designed or built for each modeling application. There are two prevalent opinions about mathematical models:

1. Models are worthless because they require too many data and therefore are too expensive to assemble and run. Furthermore, they can never be proved to be correct and suffer from a lack of scientific certainty.

   One of the most insidious and nefarious properties of scientific models is their tendency to take over, and sometimes supplant, reality. They often act as blinders, limiting attention to an excessively narrow region. No application of logic can prove a model to be true, though its lack of plausibility can often be demonstrated easily. The extravagant reliance on models has contributed much to the contrived and artificial character of large portions of current research. (Chargaff, 1978)

   The scientific community and the community of environmental regulators sorely need to be told the truth about models and the current lack of scientific certainty. (Rogers, 1983)

   Ground-water computer models are, certainly, toys which provide intellectual stimulation. They can be useful tools for advancement of the ground-water profession, but I believe that they have been blown out of proportion and that this might cause irreparable damage to our profession. (Baski, 1979)

2. Models are essential in performing complex analyses and in making informed predictions.

   The model allows more effective use of the available data; more complexities can be accounted for; and the implications of the assumptions used in the analysis and of management decisions can be evaluated. (Hamilton, 1982)

   One of the most valuable and practical tools the ground-water manager can use is the computer model.... Any professional working in the field of hydrogeology should adapt to and use ground-water models to be truly efficient. (Darr, 1979)

   But all these suggested researches should be considered as particular aspects or supporting activities of the overall objective of defining numerically the hydrologic system.... (Thomas and Leopold, 1964)

   Mathematical models have significantly expanded the nation's ability to understand and manage its water resources. ...In some cases they have increased the accuracy of estimates of future events to a level far beyond "best judgment" decisions. In other areas they have made possible analyses that cannot be performed empirically or without computer assistance. Furthermore, they have made it feasible to compare quantitatively the likely effects of different resource decisions. Models are often the best available alternative for analyzing complex resource problems. (Friedman et al., 1984)

There is some truth in the opinions quoted under #1 above. Applications of groundwater models do require extensive field information for input data and for calibration. It is also true that solutions may be nonunique and therefore results will be influenced by uncertainty. But the same could be said of just about every other type of methodology used to analyze groundwater problems. A good modeling methodology will increase confidence in modeling results. Models provide a framework for synthesizing field information and for testing ideas about how the system works. They can alert the modeler to phenomena not previously considered. They may identify areas where more field information is required. Whether models act as "blinders" depends on the way in which the modeler is influenced by the results; it has nothing to do with the model itself. Although groundwater models are time-consuming to design and therefore expensive in terms of labor time, it is also true that use of a groundwater model is the best way to make an informed analysis or prediction about the consequences of a proposed action. If the question to be answered is sufficiently important, it may be that the expense of data collection and model design and execution is a small price to pay. For these reasons, the bias of this book is, of course, toward opinion #2.

1.2 Why Model?

Most groundwater modeling efforts are aimed at *predicting* the consequences of a proposed action. There are, however, two other important types of applications (Table 1.1).

Models can be used in an *interpretive* sense to gain insight into the controlling parameters in a site-specific setting or as a framework for assembling and

Table 1.1
Types of Modeling Applications

Predictive: Used to predict the future; requires calibration.
    Examples: Karanjac et al. (1977); Andersen et al. (1984); McLaughlin and Johnson (1987); Gerhart and Lazorchick (1988)

Interpretive: Used as a framework for studying system dynamics and/or organizing field data; does not necessarily require calibration.
    Examples: Remson et al. (1980); Stephens (1983); Maslia and Johnston (1984); Krabbenhoft and Anderson (1986); Jamieson and Freeze (1983).

Generic: Used to analyze flow in hypothetical hydrogeologic systems; may be useful to help frame regulatory guidelines for a specific region; does not necessarily require calibration.
    Examples: Winter (1976); Carsel et al. (1988); Hensel et al. (1990)

---

organizing field data and formulating ideas about system dynamics. For example, the Regional Aquifer-System Analysis (RASA) program of the U.S. Geological Survey (Sun, 1986; Weeks and Sun, 1987) was established to improve understanding of regional flow systems. Models are used to help establish locations and characteristics of aquifer boundaries and assess the quantity of water within the system and the amount of recharge to the aquifer.

Models can also be used to study processes in *generic* geologic settings. For example, generic models have been used to study lake-groundwater interaction (McBride and Pfannkuch, 1975; Winter, 1976). Generic modeling studies also may be helpful in formulating regional regulatory guidelines and as screening tools to identify regions suitable or unsuitable for some proposed action (Hensel et al., 1990).

1.3 Establishing the Purpose

It is essential to identify clearly the purpose of the modeling effort at the onset in order to help avoid the situation posed by Philip (1980):

> Can it be that the vast labor of characterizing these systems, combined with the vast labor of analyzing them, once they are adequately characterized is wholly disproportionate to the benefits that could conceivably follow?

Answers to the following questions will assist in determining the type and level of modeling effort needed:

1. Is the model to be constructed for prediction, system interpretation, or a generic modeling exercise?
2. What do you want to learn from the model? What questions do you want the model to answer?
3. Is a modeling exercise the best way to answer the question(s)?
4. Can an analytical model provide the answer or must a numerical model be constructed?

The responses to these questions will determine the magnitude of the modeling effort, i.e., whether the model is steady state or transient, one-, two-, or three-dimensional, analytical or numerical, and whether the solution involves a particle tracking or full solute transport analysis.

The answers to the four questions given above will be different for each application. But it should be remembered that the first step in every model application must be to establish the purpose of the model. It may be that a model isn't necessary after all and that the question at the heart of the investigation can be answered more effectively using another approach. Or it may be that a simple analytical model can provide the answer without numerical modeling. An excellent illustration of this last point was given by McLaughlin and Johnson (1987), who showed that the Theis analytical solution was adequate to provide the answer to a question involving groundwater pumping from an aquifer in New Mexico. They also demonstrated that two relatively sophisticated numerical modeling studies were essentially overkill in this particular application. Prickett (1979) also discussed model misuse in the form of overkill.

Modeling is an excellent way to help organize and synthesize field data but it is important to recognize that modeling is only one component in a hydrogeological assessment and not an end in itself. It may be clear either before or after a modeling effort that the available data are inadequate to support modeling results. In that case, professional ethics require that modeling results should not be used at all or at least should be presented with the appropriate qualifiers. Many of today's computer codes and graphics packages are easy to use so that impressive-looking results can be readily produced using few data or data of poor quality. If it is known a priori that modeling will not produce useful results, it is ethically sound to advise against a costly modeling study. Should circumstances, such as regulatory obligations, dictate that a modeling study is mandatory even when data are inadequate, then professional ethics require that the necessary field data be collected concurrently with model design. Some of the problems involved in the use of models in regulatory applications, such as uncertainty in predictions and mandates for predictions 10,000 years into the future, are discussed in a recent report by the National Research Council (1990).

1.4 Modeling Protocol

When it has been determined that a numerical model is necessary and the purpose of the modeling effort has been clearly defined, the task of model design and application begins. A protocol for modeling includes code selection and verification, model design, calibration, sensitivity analysis, and finally prediction. The steps in a suggested modeling protocol together with relevant definitions are given below and summarized in Fig. 1.1. Each of these steps builds support in demonstrating that a given site-specific model is capable of producing meaningful results, i.e., that the model is valid. The process of model validation encompasses the entire modeling protocol described below.

1. Establish the **purpose** of the model (Section 1.3). The purpose will determine what governing equation will be solved and what code will be selected.

2. Develop a **conceptual model** of the system (Chapter 3). Hydrostratigraphic units and system boundaries are identified. Field data are assembled including information on the water balance and data needed to assign values to aquifer parameters and hydrologic stresses. During this stage a field visit to the site is highly recommended. A field visit

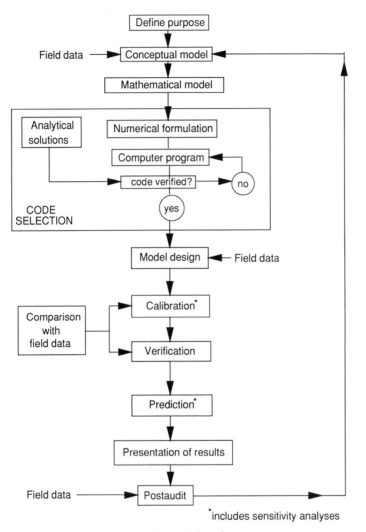

Fig. 1.1 Steps in a protocol for model application.

will help keep the modeler tied into reality and will exert a positive influence on the subjective decisions that will be made during the modeling study.

3. Select the governing equation and a **computer code** (Chapters 2 and 8). The code is the computer program that contains an algorithm to solve the mathematical model numerically. Both the governing equation and the code should be verified. Verification of the governing equation demonstrates that it accurately describes the physical processes occur-

ring in porous media. **Code verification** refers to comparison of the numerical solution generated by the model with one or more analytical solutions or with other numerical solutions. Verification of the code ensures that the computer program accurately solves the equations that constitute the mathematical model. Verification of the governing equation is the more difficult of the two verification steps. It may be possible to establish some confidence in the governing equation by comparisons with results of laboratory experiments but eventually theoretical results must be compared to field data. Hence, verification of the governing equation may be accomplished by applying a model to a number of site-specific problems, following the steps outlined in the modeling protocol.

4. **Model design** (Chapters 3–7). The conceptual model is put into a form suitable for modeling. This step includes design of the grid, selecting time steps, setting boundary and initial conditions, and preliminary selection of values for aquifer parameters and hydrologic stresses.

5. **Calibration** (Chapter 8). The purpose of calibration is to establish that the model can reproduce field-measured heads and flows. During calibration a set of values for aquifer parameters and stresses is found that approximates field-measured heads and flows. Calibration is done by trial-and-error adjustment of parameters or by using an automated parameter estimation code.

6. **Calibration sensitivity analysis** (Chapter 8). The calibrated model is influenced by **uncertainty** owing to the inability to define the exact spatial (and temporal) distribution of parameter values in the problem domain. There is also uncertainty over definition of boundary conditions and stresses. A **sensitivity analysis** is performed in order to establish the effect of uncertainty on the calibrated model.

7. **Model verification** (Chapter 8). The purpose of model verification is to establish greater confidence in the model by using the set of calibrated parameter values and stresses to reproduce a second set of field data.

8. **Prediction** (Chapter 8) quantifies the response of the system to future events. The model is run with calibrated values for parameters and stresses, except for those stresses that are expected to change in the future. Estimates of the future stresses are needed to perform the simulation. **Uncertainty** in a predictive simulation arises from uncertainty in the calibrated model and the inability to estimate accurate values for the magnitude and timing of future stresses.

9. **Predictive sensitivity analysis** (Chapter 8) is done to quantify the effect of uncertainty in parameter values on the prediction. Ranges in estimated future stresses are simulated to examine the impact on the model's prediction.

10. **Presentation of modeling design and results** (Chapter 9). Clear presentation of model design and results is essential for effective communication of the modeling effort.

11. **Postaudit** (Chapter 10). A postaudit is conducted several years after the modeling study is completed. New field data are collected to determine whether the prediction was correct. If the model's prediction was accurate, the model is validated for that particular site. Because each site is unique, a model ideally should be validated for each site-specific application.

A postaudit should occur long enough after the prediction was made to ensure that there has been adequate time for significant changes to occur. A postaudit performed too soon after the initial calibration may lead to the conclusion that the prediction came close to estimating the observed value when in fact not enough time elapsed to allow the system to move sufficiently far from the calibrated head values.

12. **Model redesign** (Chapter 10). Typically the postaudit will lead to new insights into system behavior which may lead to changes in the conceptual model or changes in model parameters.

Although few modeling studies follow all the steps in the above protocol, it represents the ideal against which the completeness of a modeling study should be measured. All modeling studies should proceed through step 6. Typically, generic and interpretive studies will not proceed beyond this. If a second set of field data does not exist, model verification (step 7) necessarily will be skipped. Model validation (step 11) has not been considered a normal part of a modeling protocol, but in view of the important information gained from the few postaudits that have been reported (Chapter 10) it is clear that model validation should be part of a modeling protocol.

The material in this book presents generally accepted procedures for model design and application. Frequent reference is made to previously published material in which the concepts under consideration were first introduced or are explained in more detail. We assume that the reader has sufficient background in groundwater hydrology to understand the processes the model represents. The reader must be aware that the applicability of the general modeling procedures discussed herein to a specific modeling problem requires a significant amount of judgment on the part of the modeler. For this reason, following the guidelines in this book will not necessarily guarantee successful completion of a modeling study unless the modeler also exercises good judgment.

1.5 Case Studies

No single case study can cover all the points relevant to modeling groundwater flow. For this reason we have selected a number of case studies to illustrate selected features of the modeling process. These examples are presented in boxes within chapters.

The reader is urged to consult the literature cited throughout this book, particularly in the figure captions, to find examples of modeling studies. Other case studies, including international examples, are presented by Bredehoeft et al. (1982). Review of this literature, together with the modeling studies the reader will eventually perform, will serve as a personal workbook of modeling case studies. The experience summarized in this workbook will help guide future modeling applications.

Problems

The problems following each chapter are intended to illustrate the main points of the chapter. Most of the problems require the use of a finite difference or finite element code and/or access to the hydrogeologic literature.

1.1 List the type of modeling (i.e., generic, interpretive, or predictive) that should be used to solve each of the following problems:
- **(a)** A planning agency wants to determine the effect of pumping a high-yield irrigation well on four existing wells used for domestic water supply and located at distances of 500, 1000, 2200, and 3000 ft from the irrigation well.
- **(b)** A hydrogeologist wants to estimate the change in hydraulic conductivity across a 500-ft-wide fault zone through an unconfined sandstone aquifer.
- **(c)** A lawyer wants to estimate seasonal fluctuations in the water table resulting from a change in timing and distribution of groundwater recharge originating from irrigation return flow. The change in recharge was brought about by recent litigation involving water rights.
- **(d)** A consulting firm wants to determine what types of aquifer heterogeneities would cause a 25% reduction in the size of the capture zone of a well designed to pump contaminated water from what was thought to be a homogeneous outwash aquifer.
- **(e)** A regulatory agency wants to estimate the probable pathway of a leachate plume emanating from a hypothetical landfill for a variety of hydrogeologic settings that are representative of a large regional area.

1.2 Review the U.S. Environmental Protection Agency's Well Head Protection Program as discussed in U.S. EPA (1987) or refer to documents that discuss other well head protection strategies. State and support your view on the types of models that would be appropriate to accomplish program goals in a specific type of hydrogeologic setting.

1.3 Describe a groundwater problem from your geographical region that could be addressed using an analytical model. What changes in modeling objectives or in hydrogeological conditions would warrant the use of a numerical model?

1.4 Step 7 in the modeling protocol presented in Section 1.4 is model verification. What would be the effects of omitting this step on the applications considered in Problem 1.1?

1.5 Review the report by Buckles and Watts (1988). Compare their modeling protocol with the one presented in Section 1.4.

2

EQUATIONS AND
NUMERICAL METHODS

"The fascinating impressiveness of rigorous mathematical analysis, with its
atmosphere of precision and elegance, should not blind us to the defects of the
premises that condition the whole process."
—*T.C. Chamberlin, 1899*

In Chapter 1, we noted that a complete mathematical description of a model
consists of a statement of the governing equation, the boundary conditions, and
the initial conditions if the problem is time dependent. In this chapter we
discuss governing equations. Boundary conditions are discussed in Chapter 4
and initial conditions are covered in Chapter 7.

2.1   Governing Equations

AQUIFER VERSUS FLOW SYSTEM VIEWPOINTS

Before we derive a governing equation for groundwater flow, we need a conceptual model of the system. There are two conceptual views of groundwater
systems—the aquifer viewpoint and the flow system viewpoint.

The *aquifer viewpoint* is based on the concept of confined and unconfined
aquifers. An *aquifer* is a unit of porous material capable of storing and transmit-
ting appreciable quantities of water to wells. A confined aquifer is overlain by a
confining bed, a unit of porous material that retards the movement of water; an
unconfined aquifer has a water table as its upper boundary. The aquifer viewpoint is especially suited to analysis of flow to pumping wells and is the basis

for many analytical solutions including those of Thiem, Theis, and Jacob. In this viewpoint, groundwater flow is assumed to be strictly horizontal through aquifers and strictly vertical through confining beds. The ability of an aquifer to transmit water is described by its *hydraulic conductivity*. In the aquifer viewpoint the hydraulic conductivity is integrated in the vertical dimension to give an average transmission characteristic known as *transmissivity*, or hydraulic conductivity times the aquifer's saturated thickness. The transmissivity of a confined aquifer is constant if the aquifer is homogeneous and of uniform thickness, but the transmissivity of an unconfined aquifer always varies spatially because saturated thickness depends on the elevation of the water table. Although assumed to be constants in the analytical solutions used in well hydraulics, hydraulic conductivity and transmissivity, of course, vary spatially in field situations because aquifers are always heterogeneous.

The aquifer viewpoint is used to simulate two-dimensional horizontal flow in confined and unconfined aquifers. Leaky confined aquifers can be simulated using a quasi three-dimensional approach whereby vertical flow through confining beds is represented by a leakage term that adds or extracts water from the aquifers overlying and underlying the confined leaky aquifer. The amount of leakage depends on the hydraulic gradient across the confining bed and the thickness and vertical hydraulic conductivity of the confining bed. Confining beds are not explicitly modeled and heads in confining beds are not calculated.

The differences between the aquifer and flow system viewpoints are illustrated in Fig. 2.1. In the aquifer viewpoint, heads in the confining beds are not of interest and are not calculated during the simulation. A numerical model of the system shown in Fig. 2.1a and based on the aquifer viewpoint would have only two layers, one for each of the aquifers (Fig. 2.1b). A one-layer model may be used to represent the system if the head in the unconfined aquifer is at steady state and will not be affected by changes in head in the confined aquifer. Attention is then focused on the confined aquifer; heads in the unconfined aquifer are not calculated but are used to calculate the hydraulic gradient across the confining bed. A numerical model of the system based on the flow system viewpoint (Fig. 2.1c) would have at least three layers and heads would be calculated in each layer. Note that in the flow system viewpoint equipotential lines pass through all geologic units, both aquifers and confining beds.

A general form of the governing equation for the aquifer viewpoint is

$$\frac{\partial}{\partial x}\left(T_x \frac{\partial h}{\partial x}\right) + \frac{\partial}{\partial y}\left(T_y \frac{\partial h}{\partial y}\right) = S \frac{\partial h}{\partial t} - R + L \tag{2.1}$$

where

$$L = -K_z' \frac{h_{source} - h}{b'}$$

The terms on the left-hand side of Eqn. 2.1 represent horizontal flow through the aquifer where h is head and T_x and T_y are components of transmissivity.

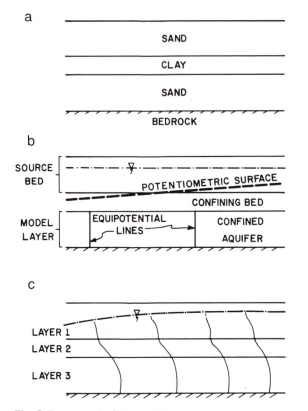

Fig. 2.1 Conceptual viewpoints.
(a) The geologic system.
(b) The aquifer viewpoint. In this example, the aquifer viewpoint focuses on the confined aquifer. The vertical hydraulic conductivity and thickness of the confining bed and the assigned head in the overlying unconfined source bed are used to calculate the leakage of water into or out of the confined aquifer. The head distribution is calculated only for the confined aquifer.
(c) The flow system viewpoint. In this viewpoint, hydraulic properties are assigned to each geologic unit and heads are calculated in all three layers.

The placement of T_x and T_y within the partial derivatives allows for spatial variation of these parameters (heterogeneity). The x and y subscripts on T indicate that the transmissivity in each of these directions may be different; i.e., the aquifer may be anisotropic. S is the storage coefficient; R is a sink/source term which is defined to be intrinsically positive to represent recharge. If withdrawal of water occurs, $R = -W$ where W is the withdrawal rate. The last term on the right-hand side (L) represents leakage through a confining bed where K_z' is the vertical hydraulic conductivity of the confining bed and b' is the thickness of the confining bed; h_{source} is the head in the source reservoir on the other side of the confining bed.

When Eqn. 2.1 is applied to an unconfined aquifer, the Dupuit assumptions are used: (1) flow lines are horizontal and equipotential lines are vertical and (2) the horizontal hydraulic gradient is equal to the slope of the free surface and is invariant with depth. It is understood that $T_x = K_x h$ and $T_y = K_y h$ where h is the elevation of the water table above the bottom of the aquifer, i.e., the saturated thickness; h may vary in both time and space. S is understood to be the specific yield. The leakage term is typically zero unless there is leakage to or from a unit located below the unconfined aquifer.

Note that in Eqn. 2.1 there is no component of transmissivity in the z direction. This is because transmissivity is a vertically averaged parameter. It is a two-dimensional areal concept that is used to describe horizontal flow in an aquifer or model layer. Therefore, transmissivity in the z direction is undefined. Hydraulic conductivity, however, describes transmission properties at a point in an aquifer and therefore has components in all three directions.

In the *flow system viewpoint,* one is not concerned with identifying aquifers and confining beds *per se* but in constructing the three-dimensional distribution of heads, hydraulic conductivities, and storage properties everywhere in the system. The flow system viewpoint allows for both vertical and horizontal components of flow throughout the system and thereby allows treatment of flow in two-dimensional profile or in three dimensions. A general form of the governing equation is

$$\frac{\partial}{\partial x}\left(K_x \frac{\partial h}{\partial x}\right) + \frac{\partial}{\partial y}\left(K_y \frac{\partial h}{\partial y}\right) + \frac{\partial}{\partial z}\left(K_z \frac{\partial h}{\partial z}\right) = S_s \frac{\partial h}{\partial t} - R^* \tag{2.2}$$

where K_x, K_y, and K_z are components of the hydraulic conductivity tensor. S_s is specific storage; R^* is a general sink/source term that is intrinsically positive and defines the volume of inflow to the system per unit volume of aquifer per unit of time. To simulate outflow $R^* = -W^*$.

Boussinesq Equation

In the discussion above, we observed that when Eqn. 2.1 is used to simulate unconfined aquifers, it is standard practice to assume that $T_x = K_x h$ and $T_y = K_y h$ where h is the saturated thickness. When $K_x h$ and $K_y h$ are substituted into Eqn. 2.1, the resulting equation is the nonlinear Boussinesq equation:

$$\frac{\partial}{\partial x}\left(K_x h \frac{\partial h}{\partial x}\right) + \frac{\partial}{\partial y}\left(K_y h \frac{\partial h}{\partial y}\right) = S_y \frac{\partial h}{\partial t} - R \tag{2.3}$$

where L in Eqn. 2.1 is equal to zero and the storage coefficient (S) is equal to specific yield (S_y). By realizing that

$$\frac{\partial h^2}{\partial x} = 2h \frac{\partial h}{\partial x}$$

$$\frac{\partial h^2}{\partial y} = 2h \frac{\partial h}{\partial y} \tag{2.4}$$

the equation can be rewritten:

$$\frac{\partial}{\partial x}\left(K_x\frac{\partial h^2}{\partial x}\right) + \frac{\partial}{\partial y}\left(K_y\frac{\partial h^2}{\partial y}\right) = 2S_y\frac{\partial h}{\partial t} - 2R \qquad (2.5)$$

Equation 2.5 is nonlinear because h appears to the second power on the left-hand side of the equation and to the first power on the right-hand side. A few investigators (e.g., Hornberger et al., 1970; Lin, 1972; Zucker et al., 1973; Gambolati et al., 1984) have solved this equation with numerical techniques specifically designed to handle nonlinear equations. There is, however, a relatively easy way to deal with the nonlinearity in Eqn. 2.3, which does not involve specialized solution techniques. By using the current (known) value of the saturated thickness, the equation can be effectively linearized in a numerical model. This is the approach used in MODFLOW, PLASM, and AQUIFEM-1.

DERIVATION OF THE GOVERNING EQUATIONS

Equation 2.2 is derived by mathematically combining a water balance equation with Darcy's law. The derivation is traditionally done by referring to a cube of porous material that is large enough to be representative of the properties of the porous medium and yet is small enough so that the change of head within the volume is relatively small. This cube of porous material is known as a representative elementary volume or REV. Its volume is equal to $\Delta x\, \Delta y\, \Delta z$ (Fig. 2.2). The flow of water through the REV is expressed in terms of the discharge rate ($\mathbf{q}$),

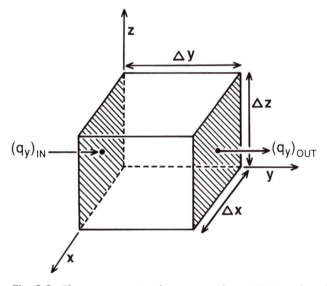

Fig. 2.2 The representative elementary volume (REV) used in the derivation of the governing equation. The components of flow along the y coordinate axis are shown.

where **q** is a vector whose magnitude can be expressed in terms of three components: q_x, q_y, and q_z. Formally speaking:

$$\mathbf{q} = q_x \mathbf{i}_x + q_y \mathbf{i}_y + q_z \mathbf{i}_z \tag{2.6}$$

where $\mathbf{i}_x$, $\mathbf{i}_y$, $\mathbf{i}_z$ are unit vectors along the x, y, and z axes.

The water balance equation (or conservation of mass) states that

$$\text{outflow} - \text{inflow} = \text{change in storage} \tag{2.7}$$

Consider flow along the y axis of the REV in Fig. 2.2. Influx to the REV occurs through the face $\Delta x\, \Delta z$ and is equal to $(q_y)_{\text{IN}}$. Flux out is equal to $(q_y)_{\text{OUT}}$. The volumetric outflow rate minus volumetric inflow rate along the y axis is

$$[(q_y)_{\text{OUT}} - (q_y)_{\text{IN}}]\, \Delta x\, \Delta z \tag{2.8}$$

We can also write this as

$$\frac{(q_y)_{\text{OUT}} - (q_y)_{\text{IN}}}{\Delta y} (\Delta x\, \Delta y\, \Delta z) \tag{2.9}$$

or dropping the IN and OUT subscripts, the change in flow rate through the REV along the y axis is

$$\frac{\partial q_y}{\partial y} (\Delta x\, \Delta y\, \Delta z) \tag{2.10}$$

Similar expressions can be written for the change in flow rate along the x and z axes. The total change in flow rate is equal to the change in storage and is expressed as

$$\left(\frac{\partial q_x}{\partial x} + \frac{\partial q_y}{\partial y} + \frac{\partial q_z}{\partial z}\right) \Delta x\, \Delta y\, \Delta z = \text{change in storage} \tag{2.11}$$

We must also allow for the possibility of a sink (e.g., a pumping well) or source of water (e.g., an injection well or some other source of recharge) within the REV. The volumetric inflow rate is represented by $R^*\, \Delta x\, \Delta y\, \Delta z$. We defined R^* to be intrinsically positive when it is a source of water; therefore, it is subtracted from the left-hand side of Eqn. 2.11. (Notice the minus sign in front of the INFLOW term in Eqn. 2.7.) The result is

$$\left(\frac{\partial q_x}{\partial x} + \frac{\partial q_y}{\partial y} + \frac{\partial q_z}{\partial z} - R^*\right) \Delta x\, \Delta y\, \Delta z = \text{change in storage} \tag{2.12}$$

Now consider the right-hand side of Eqn. 2.7. Change in storage is represented by *specific storage* (S_s), which is defined to be the volume of water released from storage per unit change in head (h) per unit volume of aquifer:

$$S_s = -\frac{\Delta V}{\Delta h\, \Delta x\, \Delta y\, \Delta z} \tag{2.13}$$

The convention used in Eqn. 2.13 is that ΔV is intrinsically positive when Δh is negative, or in other words, water is released from storage when head decreases. The rate of change in storage in the REV is

$$\frac{\Delta V}{\Delta t} = -S_s \frac{\Delta h}{\Delta t} \Delta x \, \Delta y \, \Delta z \tag{2.14}$$

Combining Eqns. 2.12 and 2.14 and dividing through by $\Delta x \, \Delta y \, \Delta z$ yields the final form of the water balance equation:

$$\frac{\partial q_x}{\partial x} + \frac{\partial q_y}{\partial y} + \frac{\partial q_z}{\partial z} = -S_s \frac{\partial h}{\partial t} + R^* \tag{2.15}$$

This equation is of little use, however, because we cannot measure $\mathbf{q}$ directly. Darcy's law is used to define the relation between $\mathbf{q}$ and h; head is a variable that we can measure directly. Darcy's law in three dimensions is written as follows:

$$q_x = -K_x \frac{\partial h}{\partial x}$$

$$q_y = -K_y \frac{\partial h}{\partial y} \tag{2.16}$$

$$q_z = -K_z \frac{\partial h}{\partial z}$$

When Eqns. 2.16 are substituted into Eqn. 2.15, the result is Eqn. 2.2. Equation 2.1 is derived from 2.2 by setting $\partial h/\partial z$ equal to zero, multiplying through by b, adding the leakage term, and setting $bS_s = S$ and $bR^* = R$.

In Eqns. 2.1 and 2.2 it is assumed that K_x, K_y, and K_z (or T_x, T_y) are colinear to the x, y, and z axes (Fig. 2.3a). If the geology is such that it is not possible to align the principal directions of the hydraulic conductivity tensor with a rectilinear coordinate system (Fig. 2.3b), a modified form of the governing equation that utilizes all of the components of the hydraulic conductivity tensor ($\overline{K}$) is required (Fig. 2.3b). The hydraulic conductivity tensor is written

$$\overline{K} = \begin{bmatrix} K_{xx} & K_{xy} & K_{xz} \\ K_{yx} & K_{yy} & K_{yz} \\ K_{zx} & K_{zy} & K_{zz} \end{bmatrix} \tag{2.17}$$

The components of the tensor may be measured during a pumping test (Quinones-Aponte, 1989; Maslia and Randolph, 1987) but when the principal directions are known, coordinate rotations are done so that the off-diagonal components of the tensor go to zero within cells or elements. This is accomplished by defining a global coordinate system for the entire problem domain and local coordinate systems for each cell or element in the grid. In the local systems, the off-diagonal components of the hydraulic conductivity tensor are

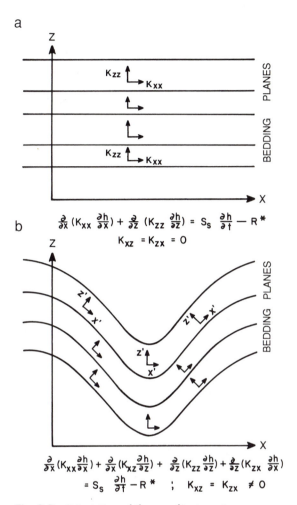

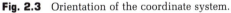

Fig. 2.3 Orientation of the coordinate system.
(a) The x–z coordinate system is aligned with the principal directions of the hydraulic conductivity tensor. The governing equation for two-dimensional transient flow is shown.
(b) A global coordinate system (x–z) is defined. Local coordinates (x′–z′) are aligned with the principal directions of the local hydraulic conductivity tensor. The governing equation for two-dimensional transient flow in the global coordinate system is shown.

equal to zero. By means of coordinate rotation it is possible to derive equations relating the principal components of hydraulic conductivity defined in the local coordinate system to the components of the hydraulic conductivity tensor defined in the global coordinate system. Details are given by Wang and Anderson (1982, Appendix A) for two-dimensional problems and Bear (1972) for three-dimensional problems.

When the principal components of the hydraulic conductivity tensor are colinear with a rectilinear coordinate system, the off-diagonal components of the tensor are zero and the second subscript on the principal components may be dropped as in Eqn. 2.2. Equations 2.1 and 2.2 also assume that the density of water is constant. Density-dependent flow is a complication in problems involving saline water, immiscible contaminants, and certain miscible contaminants. If density effects are important, a different governing equation and a special-purpose numerical code are needed. Density-dependent flow is discussed in Chapter 12 (Section 12.7).

2.2   Numerical Methods

When simplified, Eqns. 2.1 and 2.2 can be solved analytically. The simplifications usually involve assumptions of homogeneity and one- or two-dimensional flow. Except for applications to well hydraulics, analytical solutions for flow are not widely used in practical applications. Numerical solutions are much more versatile and, with the widespread availability of computers, are now easier to use than some of the more complex analytical solutions. This book deals exclusively with numerical solutions.

The following five numerical methods are used in groundwater modeling: finite differences, finite elements, integrated finite differences, the boundary integral equation method, and analytic elements. The boundary integral equation method (Liggett and Liu, 1983; Liggett, 1987) and analytic elements (Strack, 1987, 1988) are relatively new techniques and are not yet widely used. Integrated finite difference (IFD) techniques are closely related to the finite element method. Applications of IFD were reported by Narasimhan and Witherspoon (1978) for the TRUST model and by Fogg (1986) for the TERZAGHI model. Finite differences and finite elements are more commonly used to solve flow problems. The discussion in this book focuses on these two numerical methods.

A computer program or code solves a set of algebraic equations generated by approximating the partial differential equations (governing equation, boundary conditions, and initial conditions) that form the mathematical model. Approximating techniques such as the finite difference and finite element methods operate on the mathematical model and change it into a form that can be solved quickly by a computer. The set of algebraic equations produced in this way can be expressed as a matrix equation. Numerical methods are used to solve the matrix equation (Box 2.1). Thus, the solution process has two steps: (1) application of finite differences or finite elements to the original mathematical model and (2) solving the resulting matrix equation.

Table 2.1 gives the names of the three flow codes featured in this book along with the relevant approximation method and solution technique. MODFLOW and PLASM are widely used codes. No finite element code has gained such wide acceptance. AQUIFEM-1 was selected as a representative finite element code for discussion here because of its versatility and its exceptionally good user's manual. All three codes are in the public domain; the source code for each is for sale for a nominal charge (see Table 1 in the Preface). Summary tables for other codes can be found in van der Heijde et al. (1988). Most of the guidelines given in this book are applicable to all finite difference or all finite element models. Methods specific to a particular code are discussed in boxes in selected chapters.

The choice between a finite difference and a finite element model depends on the problem to be solved and on the preference of the user. Finite differences are easy to understand and program. In general, fewer input data are needed to construct a finite difference grid. Finite elements are better able to approximate irregularly shaped boundaries than standard finite differences. (Integrated finite differences, however, *can* handle irregular boundaries as well as finite elements.) It is easier to adjust the size of individual elements as well as the location of boundaries with the finite element method, making it much easier to test the effect of nodal spacing on the solution (Section 8.2). Finite elements are

Table 2.1
Groundwater Flow Codes

Name*	Type	Solution technique
MODFLOW	3D finite differences	SIP (strongly implicit procedure); SSOR (slice successive over-relaxation)
PLASM	2D finite differences	IADIP (iterative alternating direction implicit procedure)
AQUIFEM-1	2D finite elements; linear triangular elements	Direct solution using Crout's method

* See the Preface for reference citations and addresses to obtain each code listed.

also better able to handle internal boundaries such as fault zones and can simulate point sources and sinks, seepage faces, and moving water tables better than finite differences. For other problems, selection of an approximation method generally is based on the preference of the user. Along these lines, Gray (1984) observed that "discussions which argue that one method is better than the other in fact often reflect that the author has merely utilized and programmed one method better...." We agree with Gray (1984) that "each method has special features which may be desirable for a particular application."

In fact, Pinder and Gray (1976) and Wang and Anderson (1977), among others, demonstrated that the finite difference method is a special case of the finite element method. For problems having a mesh of regularly spaced nodal points, the finite element method yields the standard finite difference five-point star equation. That is, the same set of algebraic equations results from

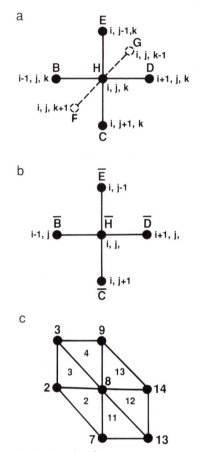

Fig. 2.4 Computational molecules.
(a) Three-dimensional finite difference molecule.
(b) Two-dimensional finite difference molecule.
(c) Patch of six finite elements around node 8. Both node numbers (large type) and element numbers (small type) are shown. The patch is taken from the two-dimensional finite element grid shown in Fig. 3.10d.

both the finite element and finite difference methods. There is a fundamental difference in philosophy, however, between the two methods. Finite difference methods compute a value for the head at the node which also is the average head for the cell that surrounds the node. No assumption is made about the form of the variation of head from one node to the next. Finite elements, on the other hand, precisely define the variation of head within an element by means of interpolation (basis) functions. Heads are calculated at the nodes for convenience, but head is defined everywhere by means of basis functions.

The general form of the finite difference expression for Eqn. 2.2 is written for the computational molecule shown in Fig. 2.4a as follows:

$$Bh_{i-1,j,k} + Ch_{i,j+1,k} + Dh_{i+1,j,k} + Eh_{i,j-1,k} + Fh_{i,j,k+1} + Gh_{i,j,k-1}$$
$$+ Hh_{i,j,k} = RHS_{i,j,k} \tag{2.18}$$

The equation for the head at the node i, j, k ($h_{i,j,k}$) involves the head at the node itself as well as heads at the six surrounding nodes. Each head is multiplied by a coefficient ($B, C, D, E, F, G,$ or H) that is a function of the hydraulic conductivity between the nodes. Coefficient H is also a function of the storage term. The term $RHS_{i,j,k}$ includes storage and recharge terms on the right-hand side of the equation. A form of Eqn. 2.18 is used in MODFLOW. The two-dimensional finite difference equation, as used in PLASM for example, is written for the computational molecule shown in Fig. 2.4b:

$$\overline{B}h_{i-1,j} + \overline{C}h_{i,j+1} + \overline{D}h_{i+1,j} + \overline{E}h_{i,j-1} + \overline{H}h_{i,j} = RHS_{i,j} \tag{2.19}$$

where again $\overline{B}, \overline{C}, \overline{D}, \overline{E},$ and $\overline{H}$ are coefficients. Both Eqns. 2.18 and 2.19 can be written as matrix equations of the form $[A]\{h\} = \{f\}$, where $[A]$ is the coefficient matrix, $\{h\}$ is the array of unknown heads, and $\{f\}$ is the array of terms on the right-hand side (RHS) of the equation.

AQUIFEM-1 uses a two-dimensional finite element approximation based on the Galerkin finite element method that yields a matrix equation of the form $[G]\{h\} = \{f\}$, where $[G]$ is the coefficient matrix, also known as the conductance matrix. The matrix equation is assembled by calculating contributions to each term in the coefficient matrix from a patch of elements like the one shown in Fig. 2.4c. Additional details regarding the implementation of the finite element method in AQUIFEM-1 may be found in Wilson et al. (1979).

Details of the theory behind finite differences and finite elements as well as details of the techniques used to solve the resulting matrix equations are beyond the scope of this book. The reader is referred to standard textbooks such as Istok (1989), Bear and Verruijt (1987), Kinzelbach (1986), Huyakorn and Pinder (1983), Wang and Anderson (1982), Pinder and Gray (1977), and Remson et al. (1971).

Box 2.1
Spreadsheet Modeling

The set of algebraic equations that result when approximating a groundwater flow model using the method of finite differences is normally solved using a combination of matrix and iterative solution techniques. A simple iterative solution of these equations can be obtained with the aid of a spreadsheet (Olsthoorn, 1985;

Ousey, 1986).

In spreadsheet modeling, each entry in the spreadsheet is a finite difference cell. The technique can be simply illustrated by considering the two-dimensional steady-state governing equation for homogeneous, isotropic aquifers (the Laplace equation):

$$\frac{\partial^2 h}{\partial x^2} + \frac{\partial^2 h}{\partial z^2} = 0 \qquad (1)$$

The finite difference equation for a regular grid where Δx and Δz are constant and equal is

$$\frac{h_{i+1,j} + h_{i-1,j} + h_{i,j+1} + h_{i,j-1}}{4} = h_{i,j} \qquad (2)$$

That is, the head in each finite difference cell is equal to the average of the heads in the four neighboring cells. An equation of this type is entered for each cell in the spreadsheet. The spreadsheet is set up to solve the equations repeatedly using an iteration option. Specified head boundary conditions are handled by entering the boundary head values into the appropriate cells in the spreadsheet. Spreadsheet models implicitly assume mesh-centered boundary conditions for both specified head and flux boundaries (Chapter 4, Box 4.2).

For example, consider the problem illustrated in Fig. 1, which represents a cross section through a regional flow system. The water table forms the top boundary. The side boundaries represent regional groundwater divides and the no-flow boundary at the bottom of the system represents impermeable bedrock. The spreadsheet is set up using six rows and eleven columns. The equation for each cell in the spreadsheet is written following Eqn. 2 (Fig. 2a). Notice that equations in the first and last columns and the last row have been modified for no-flow boundary conditions. The solution, shown in Fig. 2b, may be compared with the solution generated by PLASM modified for mesh-centered boundaries as described in Box 4.2 (Fig. 2a).

Spreadsheet modeling is a useful pedagogical tool because it is instructional to write the equation associated with each finite difference cell. However, from an operational standpoint it is doubtful that spreadsheet solutions offer any advantages over standard computer codes. The equations one needs to enter into the spreadsheet become increasingly complex when sources, sinks, and transient conditions are represented. Olsthoorn (1985) outlined the equations necessary to use spreadsheets to simulate multilayer aquifers, three-dimensional flow, and unconfined conditions. The time required to set up and test a complex spreadsheet model is likely to be equal to or greater than the time needed to set up and run a standard flow code. Moreover, the standard flow codes listed in Table 2.1 are versatile, readily available at nominal cost, contain options for computing boundary fluxes and other water balance terms, and are well tested and accepted by the modeling community.

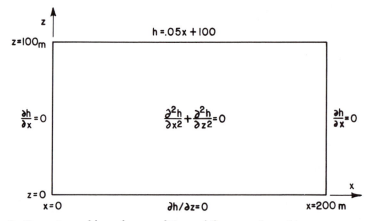

Fig. 1 Geometry and boundary conditions of the example problem representing a cross section through a regional groundwater flow system.

a

	A	B	C	D	E	F	G	H	I	J	K
	100.00	101.00	102.00	103.00	104.00	105.00	106.00	107.00	108.00	109.00	110.00
2	(2.*B2+A1+A3)/4.	(B1+B3+A2+C2)/4.	(C1+C3+B2+D2)/4.	(D1+D3+C2+E2)/4.	(E1+E3+D2+F2)/4.	(F1+F3+E2+G2)/4.	(G1+G3+F2+H2)/4.	(H1+H3+G2+I2)/4.	(I1+I3+H2+J2)/4.	(J1+J3+I2+K2)/4.	(2.*J2+K1+K3)/4.
3	(2.*B3+A2+A4)/4.	(B2+B4+A3+C3)/4.	(C2+C4+B3+D3)/4.	(D2+D4+C3+E3)/4.	(E2+E4+D3+F3)/4.	(F2+F4+E3+G3)/4.	(G2+G4+F3+H3)/4.	(H2+H4+G3+I3)/4.	(I2+I4+H3+J3)/4.	(J2+J4+I3+K3)/4.	(2.*J3+K2+K4)/4.
4	(2.*B4+A3+A5)/4.	(B3+B5+A4+C4)/4.	(C3+C5+B4+D4)/4.	(D3+D5+C4+E4)/4.	(E3+E5+D4+F4)/4.	(F3+F5+E4+G4)/4.	(G3+G5+F4+H4)/4.	(H3+H5+G4+I4)/4.	(I3+I5+H4+J4)/4.	(J3+J5+I4+K4)/4.	(2.*J4+K3+K5)/4.
5	(2.*B5+A4+A6)/4.	(B4+B6+A5+C5)/4.	(C4+C6+B5+D5)/4.	(D4+D6+C5+E5)/4.	(E4+E6+D5+F5)/4.	(F4+F6+E5+G5)/4.	(G4+G6+F5+H5)/4.	(H4+H6+G5+I5)/4.	(I4+I6+H5+J5)/4.	(J4+J6+I5+K5)/4.	(2.*J5+K4+K6)/4.
6	(2.*A5+2.*B6)/4.	(2.*B5+A6+C6)/4.	(2.*C5+B6+D6)/4.	(2.*D5+C6+E6)/4.	(2.*E5+D6+F6)/4.	(2.*F5+E6+G6)/4.	(2.*G5+F6+H6)/4.	(2.*H5+G6+I6)/4.	(2.*I5+H6+J6)/4.	(2.*J5+I6+K6)/4.	(2.*K5+2.*J6)/4.

b

	A	B	C	D	E	F	G	H	I	J	K
1	100.00	101.00	102.00	103.00	104.00	105.00	106.00	107.00	108.00	109.00	110.00
2	101.60	101.96	102.60	103.35	104.17	105.00	105.83	106.65	107.40	108.04	108.40
3	102.47	102.65	103.08	103.65	104.31	105.00	105.69	106.35	106.92	107.35	107.53
4	102.98	103.10	103.41	103.87	104.41	105.00	105.59	106.13	106.59	106.90	107.02
5	103.25	103.35	103.61	104.00	104.48	105.00	105.52	106.00	106.39	106.65	106.75
6	103.34	103.43	103.67	104.04	104.50	105.00	105.50	105.96	106.33	106.57	106.66

Fig. 2

(a) Finite difference equations for each spreadsheet cell (rows 1–6 and columns A–K) with specified head boundary values (in meters) in the first row. The nodal spacing is 20 m. (b) Solution generated by the spreadsheet MathPlan (WordPerfect Corporation). Heads are in meters.

Problems

2.1 Figure P2.1 is a schematic representation of the groundwater flow system for Long Island, New York. Prepare schematic diagrams of this hydrogeologic setting that illustrate a model representing

(a) the aquifer viewpoint

(b) the flow system viewpoint

2.2 Show how the Laplace equation

$$\frac{\partial^2 h}{\partial x^2} + \frac{\partial^2 h}{\partial y^2} + \frac{\partial^2 h}{\partial z^2} = 0$$

may be generated from Eqn. 2.1. State the assumptions used to reduce Eqn. 2.1 to the Laplace equation.

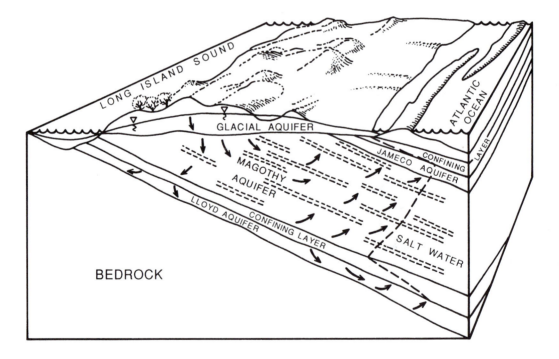

Fig. P2.1   Groundwater flow system for Long Island, New York (adapted from Cohen et al., 1968).

2.3 Equations 2.1 and 2.2 assume that the principal components of the hydraulic conductivity tensor, K_x, K_y, and K_z, are aligned with the principal or global coordinate axes, x, y, and z. In many geologic situations this can be accomplished by proper alignment of the coordinate axes when designing the grid of the numerical model. When it is not possible to align the global coordinate system with the principal components of the hydraulic conductivity tensor, a different form of the governing equation must be used (Fig. 2.3). It is then necessary to calculate the off-diagonal components of the tensor, K_{xy}, K_{xz}, K_{yx}, K_{yz}, K_{zx}, K_{zy}.

Consult Appendix A of Wang and Anderson (1982) and Bear (1972) and derive expressions for the off-diagonal components for a

(a) two-dimensional flow system

(b) three-dimensional flow system.

2.4 Examine the documentation accompanying a code such as PLASM, MODFLOW, or AQUIFEM-1, in which the numerical method used to solve the set of n equations with n unknowns is described (Table 2.1). Outline the basic steps in the numerical solution technique and review the computer code that executes the solution routine.

3

THE CONCEPTUAL MODEL AND GRID DESIGN

"Everything should be made as simple as possible, but not simpler."
—*Albert Einstein*

3.1   Building the Conceptual Model

The first step in the modeling protocol discussed in Chapter 1 is to establish the purpose of the model; the second step is to formulate a conceptual model of the system. A conceptual model is a pictoral representation of the groundwater flow system, frequently in the form of a block diagram or a cross section (Fig. 3.1). The nature of the conceptual model will determine the dimensions of the numerical model and the design of the grid.

The purpose of building a conceptual model is to simplify the field problem and organize the associated field data so that the system can be analyzed more readily. Simplification is necessary because a complete reconstruction of the field system is not feasible. The data requirements for a groundwater flow model are listed in Table 3.1. These data should be assembled when formulating the conceptual model. In theory, the closer the conceptual model approximates the field situation, the more accurate is the numerical model. However, in practice it is desirable to strive for *parsimony*, by which it is implied that the conceptual model has been simplified as much as possible yet retains enough complexity so that it adequately reproduces system behavior. It is critical that the conceptual model be a valid representation of the important hydrogeologic conditions; failure of numerical models to make accurate predictions can often be attributed to errors in the conceptual model (Chapter 10).

The first step in formulating the conceptual model is to define the area of interest, i.e., to identify the boundaries of the model. Numerical models require boundary conditions, such that the head or flux must be specified along the boundaries of the system (Chapter 4). Whenever possible the natural hydrogeologic boundaries of the system should be used as the boundaries of the model. However, for some problems it may be necessary to restrict the problem domain to less than that encompassed by the natural aquifer boundaries. In either case, the true hydrogeologic boundaries of the system should be identified when formulating the conceptual model. There are three steps in building a conceptual model: (1) defining hydrostratigraphic units; (2) preparing a water budget; (3) defining the flow system.

DEFINING HYDROSTRATIGRAPHIC UNITS

Geologic information including geologic maps and cross sections, well logs, and borings, are combined with information on hydrogeologic properties to

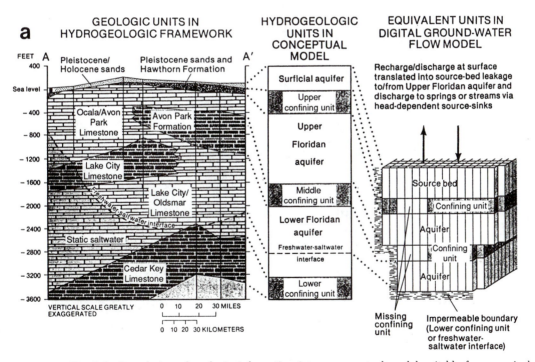

Fig. 3.1 Translation of geologic information into a conceptual model suitable for numerical modeling.
(a) Floridan aquifer system, west to east cross section for central Florida (Bush and Johnston, 1986).
(b) Snake River Plain aquifer system, southwest to northeast cross section (Lindholm, 1986).
(c) Sand and gravel aquifer in Pensacola, Florida (Franks, 1988).
(d) Glacial-drift river-valley aquifer in Rhode Island (Morrissey, 1989).

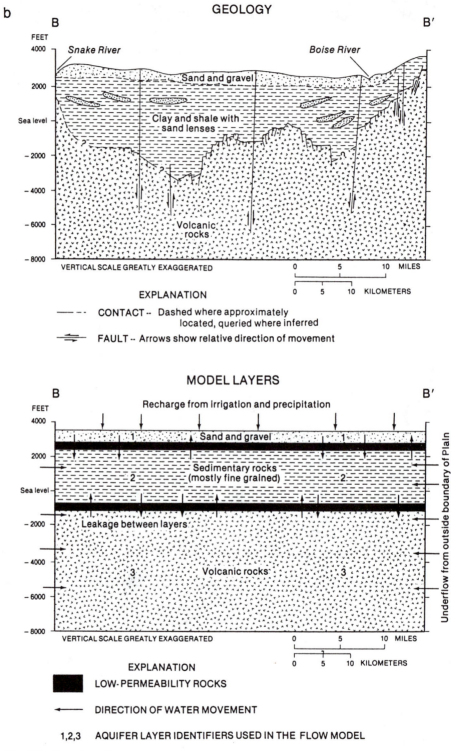

b

GEOLOGY

B   B′

FEET
4000

Snake River   *Boise River*

Sand and gravel

2000

Sea level

Clay and shale with sand lenses

−2000

−4000

Volcanic rocks

−6000

−8000

VERTICAL SCALE GREATLY EXAGGERATED

0   5   10   MILES

0   5   10   KILOMETERS

EXPLANATION

— — - -   CONTACT -- Dashed where approximately located, queried where inferred

⇌   FAULT -- Arrows show relative direction of movement

MODEL LAYERS

B   B′

FEET
4000

Recharge from irrigation and precipitation

1   Sand and gravel   1

2000

2   Sedimentary rocks (mostly fine grained)   2

Sea level

−2000

Leakage between layers

−4000

3   Volcanic rocks   3

−6000

−8000

VERTICAL SCALE GREATLY EXAGGERATED

Underflow from outside boundary of Plain

0   5   10   MILES

0   5   10   KILOMETERS

EXPLANATION

▆   LOW-PERMEABILITY ROCKS

⟵   DIRECTION OF WATER MOVEMENT

1,2,3   AQUIFER LAYER IDENTIFIERS USED IN THE FLOW MODEL

Fig. 3.1 *(continued)*

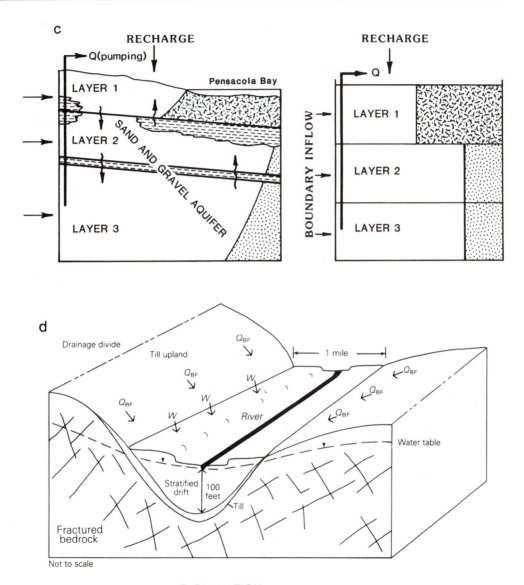

EXPLANATION

W, Aquifer recharge from precipitation

Q_{BF}, Boundary flux, recharge from till upland areas

Fig. 3.1 (*continued*)

Table 3.1
Data Requirements for a Groundwater Flow Model

A. *Physical framework*
   1. Geologic map and cross sections showing the areal and vertical extent and boundaries of the system.
   2. Topographic map showing surface water bodies and divides.
   3. Contour maps showing the elevation of the base of the aquifers and confining beds.
   4. Isopach maps showing the thickness of aquifers and confining beds.
   5. Maps showing the extent and thickness of stream and lake sediments.

B. *Hydrogeologic framework*
   1. Water table and potentiometric maps for all aquifers.
   2. Hydrographs of groundwater head and surface water levels and discharge rates.
   3. Maps and cross sections showing the hydraulic conductivity and/or transmissivity distribution.
   4. Maps and cross sections showing the storage properties of the aquifers and confining beds.
   5. Hydraulic conductivity values and their distribution for stream and lake sediments.
   6. Spatial and temporal distribution of rates of evapotranspiration, groundwater recharge; surface water–groundwater interaction, groundwater pumping, and natural groundwater discharge.

Adapted from Moore, 1979.

define hydrostratigraphic units for the conceptual model. In modeling regional flow systems, aquifers and confining beds are defined using the concept of the hydrostratigraphic unit, which was introduced by Maxey (1964) and reassessed by Seaber (1988). Simply stated, hydrostratigraphic units comprise geologic units of similar hydrogeologic properties. Several geologic formations may be combined into a single hydrostratigraphic unit or a geologic formation may be subdivided into aquifers and confining units. For example, in Table 3.2 the Tertiary age Chickahominy Formation and the Piney Point Formation are combined into a single aquifer, whereas the Cretaceous Potomac Formation is subdivided into the Upper, Middle, and Lower Potomac aquifers and two confining beds.

Geologists rely on stratigraphic information and an understanding of the depositional history to help in reconstructing the environment of deposition. Facies models are helpful in this analysis. A facies is a unit of material of similar physical properties that was deposited in the same geologic environment. Facies models are conceptual models of the expected distribution of facies (Fig. 3.2). Knowledge of the depositional history of an area might be helpful in predicting the occurrence of sediment types when geologic information is sparse (Anderson, 1989). Geologic facies can then be used to define hydrostratigraphic units.

Special techniques may be necessary when defining hydrostratigraphic units consisting of thick sand sequences such as those typical of the Gulf Coastal Plain. Subdivision of these sedimentary units into model layers on the

Table 3.2
Lithostratigraphy and Hydrostratigraphy of the Virginia Coastal Plain (Meng and Harsh, 1988)

PERIOD	EPOCH	AGE	STRATIGRAPHIC FORMATION	VIRGINIA RASA HYDROGEOLOGIC UNIT
QUATERNARY	HOLOCENE	POST-GLACIAL	Holocene deposits	Columbia aquifer
	PLEISTOCENE	WISCONSIN TO NEBRASKAN	Pleistocene undifferentiated deposits	
TERTIARY	PLIOCENE	PIACENZIAN	Bacons Castle Formation (Oaks and Coch, 1973)	Yorktown confining unit
		ZANCLEAN	Chesapeake Group: Yorktown Formation	Yorktown-Eastover aquifer
	MIOCENE	MESSINIAN	Chesapeake Group: Eastover Formation	
		TORTONIAN		St. Marys confining unit
		SERRAVALLIAN	Chesapeake Group: St. Marys Formation	
			Chesapeake Group: Choptank Formation	St. Marys-Choptank aquifer
		LANGHIAN	Chesapeake Group: Calvert Formation	Calvert confining unit
		BURDIGALIAN		
		AQUITANIAN	Chesapeake Group: Old Church Formation	Chickahominy-Piney Point aquifer
	OLIGOCENE	CHICKASAWHAYAN[1]	Not present in study area	
		VICKSBURGIAN[1]		
	EOCENE	JACKSONIAN[1]	Pamunkey Group: Chickahominy Formation	Chickahominy-Piney Point aquifer
		CLAIBORNIAN[1]	Pamunkey Group: Piney Point Formation	
	EOCENE / PALEOCENE	SABINIAN[1]	Pamunkey Group: Nanjemoy Formation	Nanjemoy-Marlboro clay confining unit
			Pamunkey Group: Marlboro clay	
	PALEOCENE		Pamunkey Group: Aquia Formation	Aquia aquifer
		MIDWAYAN[1]	Pamunkey Group: Brightseat Formation	Brightseat confining unit
				Brightseat aquifer
CRETACEOUS	LATE CRETACEOUS	MAASTRICHTIAN	Undifferentiated sediments	Upper Potomac confining unit
		CAMPANIAN		
		SANTONIAN		
		CONIACIAN		
		TURONIAN		
		CENOMANIAN	Potomac Formation	Upper Potomac aquifer
	EARLY CRETACEOUS	ALBIAN		Middle Potomac confining unit
				Middle Potomac aquifer
		APTIAN		Lower Potomac confining unit
		BARREMIAN		Lower Potomac aquifer
		HAUTERIVIAN		
		VALANGINIAN		
		BERRIASIAN		

[1]Commonly used ages in Atlantic Coastal Plain province

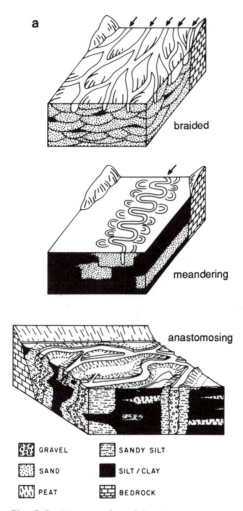

Fig. 3.2 Conceptual models of geologic facies (Anderson, 1989).
(a) Alluvial environments of deposition.
(b) Outwash plain, showing the proximal, medial, and distal facies proceeding outward from the ice margin.
(c) Lodgement and basal till overlain by a supraglacial sediment complex.

basis of stratigraphic boundaries is sometimes inappropriate, as is evident from the electric log shown in Fig. 3.3a. Formation boundaries may be ignored and hydrostratigraphic units defined directly from electric logs. When there are thick sequences of interbedded sand and clay (Fig. 3.3b), model layers may be defined using regional head data to identify units of similar hydrogeologic properties (Weiss and Williamson, 1985). Figure 3.3c shows four clusters of

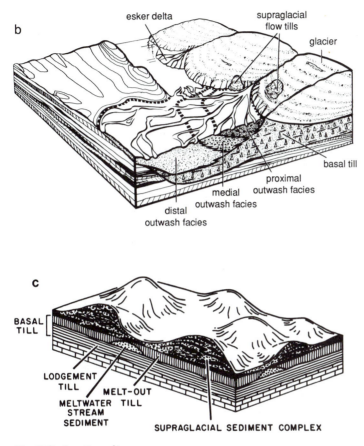

Fig. 3.2 (continued)

head data that were used to define the four layers shown. Simulated heads should fall roughly in the middle of each cluster of observed heads.

The concept of a hydrostratigraphic unit is most useful for simulating geologic systems at a regional scale. At smaller scales, one would like to group adjacent geologic units of similar hydrogeologic properties into locally defined hydrostratigraphic units. In practice, however, this is difficult to do because detailed site-specific information on stratigraphy and hydraulic conductivity is required to define hydrostratigraphic units at this scale. Such detailed information is seldom available. Facies models are of limited use here because they are idealized summaries of environments of deposition, representing a synthesis of the features of many similar environments. They do not represent the features of any one site. For example, the facies model for a till complex (Fig. 3.2b) indicates the presence of meltwater stream sediment in this environment. Such

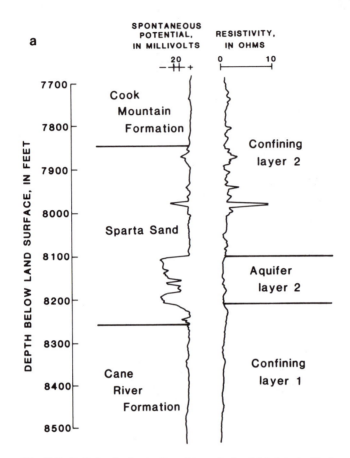

Fig. 3.3 Defining hydrostratigraphic units for thick interbedded sandy aquifers (Weiss and Williamson, 1985).

(a) Electric log for Gulf Coastal Plain sediments with lithostratigraphic units noted on the left and the interpretation of the hydrostratigraphy on the right.

(b) Schematic diagram of an intermixed sand and clay layer (aquifer layer 2) with over- and underlying confining units. The aquifer layer was assumed to behave as a homogeneous anisotropic unit.

(c) Use of head data to define hydrostratigraphic units. The water levels in wells penetrating the thick sandy sequence between 1800 to 2600 ft, from which water is produced, were lower than heads at shallower and deeper depths. This was interpreted to reflect the presence of a zone of higher hydraulic conductivity that is over- and underlain by less permeable material. A layer representing this zone (layer 4) was included in the conceptual model, based on the measured head differences.

units will form preferential flow paths and hence are important features of the hydrogeology. The facies model, however, cannot predict the precise location or even the density of meltwater stream sediment within any specific till complex.

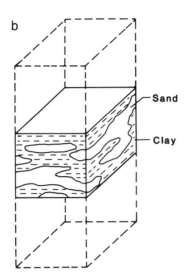

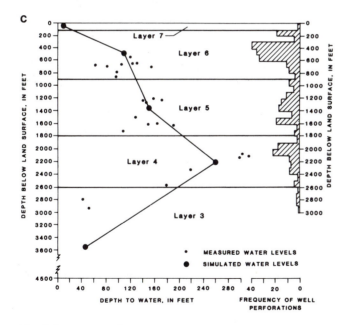

Fig. 3.3 (continued)

PREPARING THE WATER BUDGET

The sources of water to the system as well as the expected flow directions and exit points should be part of the conceptual model (Fig. 3.1b–d). The field-estimated inflows may include groundwater recharge from precipitation, over-land flow, or recharge from surface water bodies. Outflows may include springflow, baseflow to streams, evapotranspiration, and pumping. Underflow may occur as either inflow or outflow. A water budget should be prepared from the field data to summarize the magnitudes of these flows and changes in storage. During model calibration the field-estimated water budget will be compared with the water budget computed by the model (Chapter 8, Section 8.2).

DEFINING THE FLOW SYSTEM

The hydrostratigraphy forms the framework of the conceptual model. Hydro-logic information is used to conceptualize the movement of groundwater through the system. Hydrologic information on precipitation, evaporation, and surface water runoff, as well as head data and geochemical information are used in this analysis. Water level measurements are used to estimate the gen-eral direction of groundwater flow, the location of recharge and discharge areas, and the connection between aquifers and surface water systems. Definition of the flow system may be based solely on physical hydrologic data, but it is advisable to use geochemical data whenever possible to strengthen the concep-tual model. Water chemistry data can be used to infer flow directions (Swen-son, 1968; Bredehoeft et al., 1983), identify sources and amounts of recharge (Knott and Olimpio, 1986), estimate groundwater flow rates (Krabbenhoft et al., 1990), and define local, intermediate, and regional flow systems (Lee and Strickland, 1988). Chemical analyses typically include concentrations of major cations (Ca^{+2}, Mg^{+2}, Na^{+}) and anions (SO_4^{-2}, HCO_3^{-}, Cl^{-}), temperature, and pH. Depending on the purpose of the chemical study, analyses may also in-clude trace metals, stable and radiogenic isotopes, and organic compounds.

3.2　Types of Models

There are several ways to classify groundwater flow models. Models can be either transient or steady state, confined or unconfined, and consider one, two, or three spatial dimensions. In setting up the grid of a numerical model, the classification that is most relevant is one based on spatial dimension. We can classify models in terms of spatial dimension as two-dimensional areal, two-dimensional profile, quasi three-dimensional, and full three-dimensional. Two-dimensional areal and quasi three-dimensional models assume the aqui-fer viewpoint, while two-dimensional profile and full three-dimensional models use the flow system viewpoint. Particle tracking codes to simulate the advective transport of contaminants (Chapter 11) can be used with any of these models. If unsaturated flow, immiscible flow, density effects, dispersion, or flow through fractures is an important feature of the conceptual model, it will be necessary to solve a different governing equation than the general saturated flow equation discussed in Chapter 2 and use the special-purpose codes discussed in Chapter 12.

TWO-DIMENSIONAL AREAL MODELS

Two-dimensional areal simulations may consider four different types of aquifers. These are confined aquifers, leaky confined aquifers, unconfined aquifers, and mixed aquifers.

Confined Aquifers

When simulating confined aquifers, transmissivity and storage coefficient are specified for each node, cell, or element. Variation in transmissivity may represent changes in either hydraulic conductivity or aquifer thickness (Fig. 3.4). In a two-dimensional areal model, anisotropy in transmissivity is represented by the difference between transmissivity in the x and y directions (T_x and T_y). Input to the model may consist of the two transmissivity arrays or the T_x array

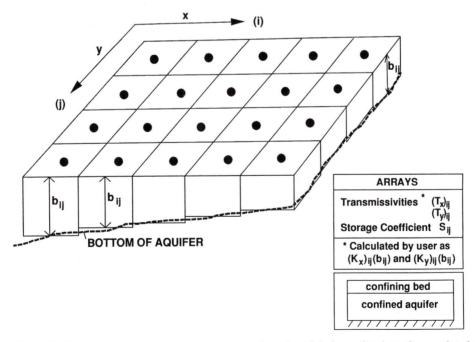

Fig. 3.4 Block-centered grid for a two-dimensional areal model of a confined aquifer, overlain by a confining bed and underlain by impermeable material.

Table 3.3
Ranges of Values for Hydraulic Conductivity

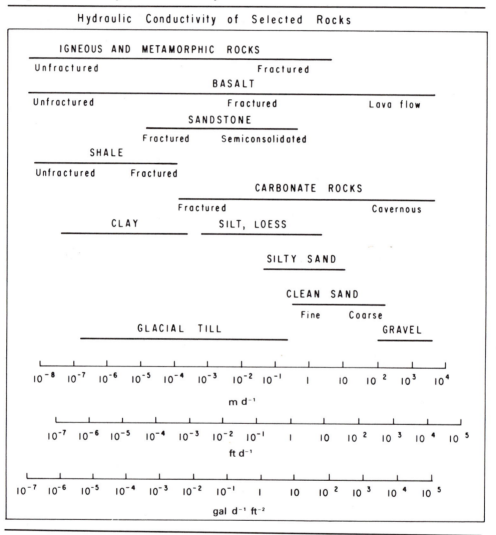

From Heath, 1983.

and an anisotropy factor to compute the T_y array from the T_x array. Transmissivity can be estimated from literature values of hydraulic conductivity (Table 3.3) and estimates of aquifer thickness. Values for transmissivity and storage coefficient are generally obtained from pumping test results. Storage coefficient can be calculated from estimates of specific storage (Table 3.4).

Table 3.4
Ranges of Values of Specific Storage (S_s)

Material	Specific storage (S_s) (m^{-1})
Plastic clay	2.0×10^{-2}–2.6×10^{-3}
Stiff clay	2.6×10^{-3}–1.3×10^{-3}
Medium-hard clay	1.3×10^{-3}–9.2×10^{-4}
Loose sand	1.0×10^{-3}–4.9×10^{-4}
Dense sand	2.0×10^{-4}–1.3×10^{-4}
Dense sandy gravel	1.0×10^{-4}–4.9×10^{-5}
Rock, fissured, jointed	6.9×10^{-5}–3.3×10^{-6}
Rock, sound	Less than 3.3×10^{-6}

Adapted from Domenico, 1972.

Leaky Confined Aquifers

In a leaky confined system, the confining bed and adjacent aquifer that supplies leakage to the confined aquifer are not explicitly represented in the model but are simulated by means of a leakage term (see Eqn. 2.1). The leakage term is a function of the *leakance*, which is the ratio of the vertical hydraulic conductivity (K'_z) of the confining bed to the thickness of the confining bed (b')

$$\text{Leakance} = K'_z/b' \qquad (3.1)$$

The system is understood to be configured as depicted in Fig. 3.5. Arrays specify leakance values and the distribution of head in the source aquifer. Transmissivity and storage coefficient of the leaky confined aquifer are also needed.

The source of water to the leaky confined aquifer may be another confined

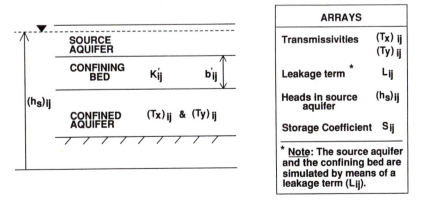

Fig. 3.5 Schematic view in cross section of a two-dimensional areal model of a leaky confined aquifer.

aquifer or an unconfined aquifer or even a surface water body; the model generally assumes that the head in the source reservoir is invariant with time. In other words, hydrologic stresses to the leaky confined aquifer do not exert any effect on the head in the source aquifer or reservoir.

Another assumption used in simulating leaky confined aquifers in two-dimensional areal models is that there is no release of water from storage within the confining bed. The transient release of water from the confining bed can be assumed to occur within the period $t_s = S'_s b'/2K'_z$ (Trescott et al., 1976; Frind, 1979) where S'_s and K'_z are the specific storage and vertical hydraulic conductivity, respectively, of the confining bed and b' is the thickness of the confining bed. Prior to t_s, the solution may be affected by errors owing to the neglect of transient leakage. Trescott et al. (1976) included a correction term to the leakage term (L) in Eqn. 2.1 to allow for transient release of storage from the confining bed. These effects can also be simulated by including the confining bed in a full three-dimensional simulation.

Unconfined Aquifers

Most modeling applications involving unconfined aquifers use the Dupuit assumptions (Chapter 2), which ensure horizontal flow by requiring that there is no change in head with depth. Use of the Dupuit assumptions in effect turns a three-dimensional problem into a two-dimensional areal problem and a two-dimensional profile problem into a one-dimensional problem. The model calculates the elevation of the water table for each node.

Simulations involving an unconfined aquifer require arrays specifying hydraulic conductivity, specific yield, and the elevation of the datum (Fig. 3.6). Because the simulation is usually done in two-dimensional areal view, hydraulic conductivity values should be vertically averaged when using point data (Section 3.4, Eqn. 3.4a) or obtained from pumping tests. Specific yield can also be obtained from pumping tests but these are not always reliable. Specific yield ranges from 0.1 to 0.4 (Table 3.5). Given this small range of possible values for specific yield, it is common to select a value within this range and then test the sensitivity of the model to specific yield during sensitivity analyses (Chapter 8).

When it is not appropriate to use the Dupuit assumptions, a profile model or a full three-dimensional model must be used instead of a two-dimensional areal model.

Mixed Aquifer Systems

A mixed aquifer system consists of some combination of the above three aquifer types. An aquifer may vary spatially from unconfined to confined conditions (Fig. 3.7). Or an aquifer may undergo conversion from confined to unconfined conditions as a result of pumping. All arrays required by the aquifer types shown in Figs. 3.4–3.6 are used in a mixed aquifer simulation.

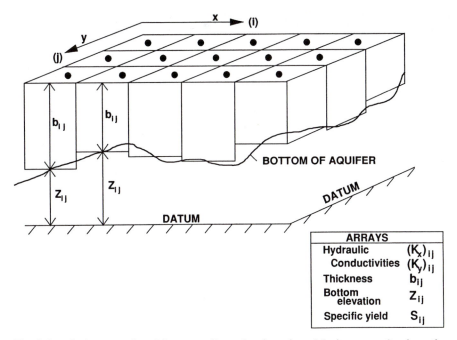

Fig. 3.6 Block-centered grid for a two-dimensional areal model of an unconfined aquifer.

Table 3.5
Ranges of Values of Specific Yield (S_y)

Material	No. of analyses	Range	Arithmetic mean
Sedimentary materials			
Sandstone (fine)	47	0.02–0.40	0.21
Sandstone (medium)	10	0.12–0.41	0.27
Siltstone	13	0.01–0.33	0.12
Sand (fine)	287	0.01–0.46	0.33
Sand (medium)	297	0.16–0.46	0.32
Sand (coarse)	143	0.18–0.43	0.30
Gravel (fine)	33	0.13–0.40	0.28
Gravel (medium)	13	0.17–0.44	0.24
Gravel (coarse)	9	0.13–0.25	0.21
Silt	299	0.01–0.39	0.20
Clay	27	0.01–0.18	0.06
Limestone	32	0–0.36	0.14
Wind-laid materials			
Loess	5	0.14–0.22	0.18
Eolian sand	14	0.32–0.47	0.38
Rock			
Schist	11	0.22–0.33	0.26
Turr	90	0.02–0.47	0.21

From Morris and Johnson, 1967.

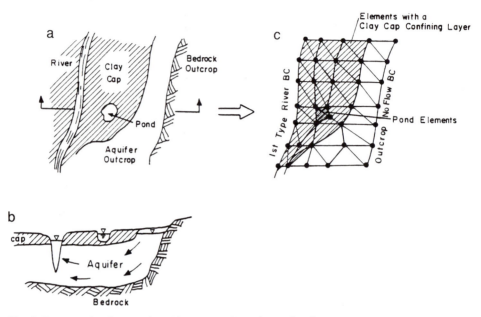

Fig. 3.7 Example of a mixed aquifer system (Townley and Wilson, 1980).
(a) Plan view.
(b) Cross section showing the transition from unconfined to confined conditions.
(c) Finite element grid.

QUASI THREE-DIMENSIONAL MODELS

A quasi three-dimensional model simulates a sequence of aquifers with intervening confining layers (Fig. 3.8). Like two-dimensional areal models of leaky confined aquifers (Fig. 3.5) confining layers are not explicitly represented in a quasi three-dimensional model, nor are heads in the confining beds calculated. The effect of a confining bed is simulated by means of a leakage term ($L_{i,j}$) representing vertical flow between two aquifers. The leakage term is a function of the leakance (Eqn. 3.1) and the head difference across the confining bed (Eqn. 2.1). Release of water from storage within the confining bed typically is not considered in this approach. In a quasi three-dimensional model the head in the unit overlying the top confining bed, usually an unconfined aquifer, can be calculated directly by the model. Codes designed to treat quasi three-dimensional conditions are discussed by Aral (1989), Walton (1989), and Hemker and van Elburg (1987).

Ignoring horizontal flow in the confining beds causes less than a 5% difference in heads in modeled layers when the contrast in hydraulic conductivity between the aquifer and confining beds is at least two orders of magnitude

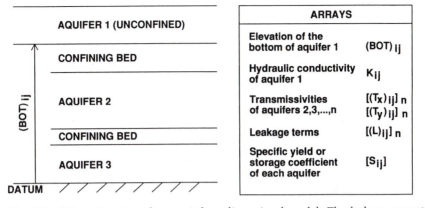

Fig. 3.8 Schematic view of a quasi three-dimensional model. The leakage properties of the confining beds, $(L_{i,j})_n$, are used to connect aquifers 1, 2, and 3. The confining beds are not represented as model layers, nor are storage properties of confining beds included in the model.

(Neuman and Witherspoon, 1969). When there is less than two orders of magnitude difference between the hydraulic conductivity of the confining beds and the aquifers, a full three-dimensional model may be preferred.

PROFILE AND FULL THREE-DIMENSIONAL MODELS

Two-dimensional profile models and full three-dimensional models assume the flow system viewpoint. Full three-dimensional models have essentially the same array requirements as two-dimensional areal models except that parameter arrays must be specified for each layer of the model (Fig. 3.9). Profile models are a special class of models; data input can be structured similarly to two-dimensional areal confined models or to full three-dimensional models. Details on profile modeling are given in Chapter 6.

Profile or full three-dimensional models are used to simulate unconfined aquifers when vertical head gradients are important. In these models, the water table (and seepage face, if present) forms part of the boundary. Both finite difference and finite element models are able to simulate aquifers in profile, but movement of the water table and seepage face is more easily handled with a finite element model. Some finite element codes are designed to move nodes in designated deformable elements, thereby allowing for adjustment of the water table and the seepage face during the solution. Treatment of the water table and seepage face as a boundary condition is discussed in more detail in Chapter 4.

Full three-dimensional models may be used to represent transient release of water from storage within confining beds by including the confining bed as a layer with an assigned value of specific storage. Release of water from storage within interbeds accompanied by compaction of the interbeds is included as an option in some codes (Leake, 1990).

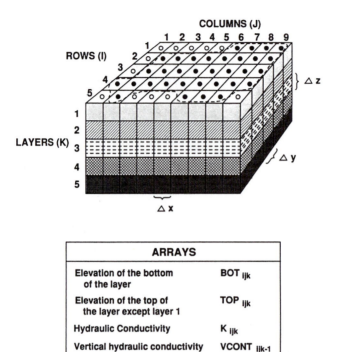

ARRAYS	
Elevation of the bottom of the layer	BOT $_{ijk}$
Elevation of the top of the layer except layer 1	TOP $_{ijk}$
Hydraulic Conductivity	K $_{ijk}$
Vertical hydraulic conductivity terms or leakage terms between layers	VCONT $_{ijk-1}$
Storativity	S $_{ijk}$

Fig. 3.9 Schematic diagram of a full three-dimensional model (adapted from McDonald and Harbaugh, 1988). All hydrostratigraphic units are represented by one or more layers

3.3   Laying Out the Grid

In a numerical model, the continuous problem domain is replaced by a discretized domain consisting of an array of nodes and associated finite difference blocks (cells) or finite elements. The nodal grid forms the framework of the numerical model. The conceptual model and the selection of model type as discussed in Sections 3.1 and 3.2 will determine the overall dimensions of the grid. The selection of either a finite difference or finite element code influences the structure of the grid.

TYPES OF GRIDS

There are two types of *finite difference grids*: the block-centered grid (Fig. 3.10b) and the mesh-centered grid (Fig. 3.10c). The difference between them lies mainly in the way in which flux boundaries are handled. In the block-centered approach flux boundaries always are located at the edge of the block. In a mesh-centered grid, the boundary coincides with a node. Details on the way in which boundaries are represented in a finite difference model are presented in Chapter 4.

In large general computer codes, the finite difference mathematics for boundaries are more easily treated with the block-centered approach. Consequently most codes, including MODFLOW, use this type of grid. PLASM uses a block-centered approach that allows the user to switch from block-centered to mesh-centered treatment of boundaries (Chapter 4, Box 4.2).

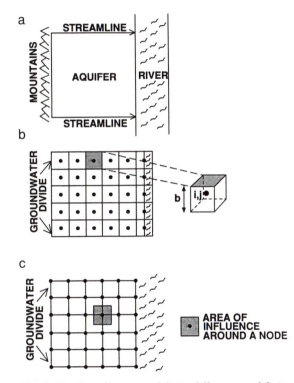

Fig. 3.10 Two-dimensional finite difference and finite element grids.
(a) Problem domain. No-flow boundaries are designated at the mountain range and along the streamlines. The river fully penetrates the aquifer and may be represented by a constant-head boundary.
(b) Block-centered finite difference grid. No-flow boundaries are located at the edge of blocks. The specified-head river boundary is located on nodes. The grid is larger than the problem domain.
(c) Mesh-centered finite differences. Both no-flow and constant-head boundaries fall directly on the nodes.
(d) Triangular finite elements. Node numbers are shown; element numbers are circled. Both no-flow and specified-head boundaries fall directly on the nodes.
(e) Quadrilateral finite elements. Node numbers are shown; element numbers are circled. Both no-flow and specified-head boundaries fall directly on the nodes.

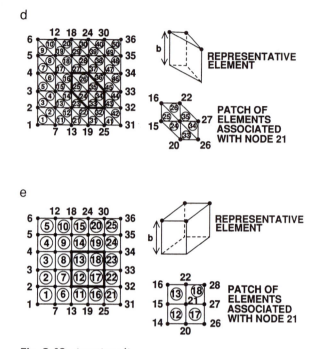

Fig. 3.10   (*continued*)

Finite elements allow more flexibility in designing a grid. Two-dimensional elements are either triangles or quadrilaterals (Figs. 3.10d,e and 3.11). Three-dimensional elements are tetrahedrons, hexahedrons, or prisms (Fig. 3.12). The nature of the interpolation (basis) function used to define heads within the element determines whether the element is linear, quadratic, or cubic. The most commonly used element is the linear element. AQUIFEM-1 uses linear triangular elements exclusively. Some codes (e.g., FREESURF, by Neuman, 1976) allow a mix of triangles and quadrilaterals.

DEFINING MODEL LAYERS

The selection of model type will determine whether one or more than one model layer is required. If only one layer is needed, this layer typically represents a single hydrostratigraphic unit. The conceptual model is used to help decide how many layers are needed.

When quasi three-dimensional models are used to represent regional flow systems, hydrogeologic units are assumed to be horizontal or, in other words, have zero dip. Most geologic units slope at some angle to the horizontal (Fig. 3.13a). In many cases the slope (dip) is quite small (1 or 2 degrees) and the units can be represented by horizontal model layers (Fig. 3.13b) for modeling pur-

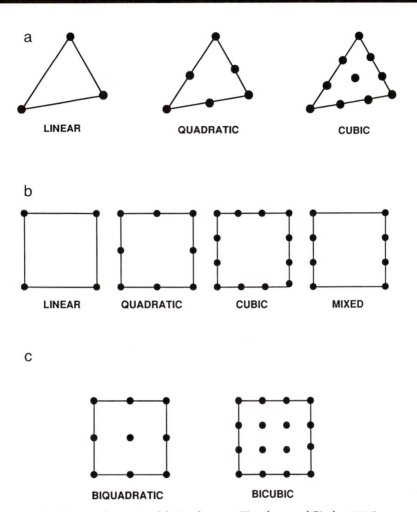

Fig. 3.11 Two-dimensional finite elements (Huyakorn and Pinder, 1983).
(a) Triangular elements.
(b) Quadrilateral elements (Serendipity family).
(c) Quadrilateral elements (Lagrange family).

poses. Horizontal model layers that represent dipping units are shown in Fig. 3.14. Pinching out of aquifers or confining beds in quasi three-dimensional simulations can be handled as illustrated in Fig. 3.15, where changes in lithology within the layer are accommodated by changing the transmissivity of the layer or the leakance of the confining bed. If necessary, profile and full three-dimensional models may be used to simulate dipping units as shown in Fig. 3.16.

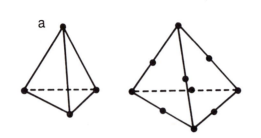
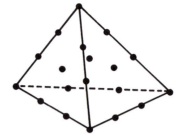
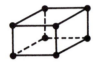
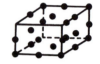
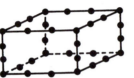

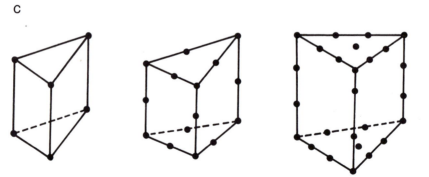
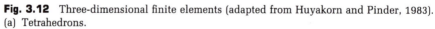

Fig. 3.12 Three-dimensional finite elements (adapted from Huyakorn and Pinder, 1983).
(a) Tetrahedrons.
(b) Hexahedrons.
(c) Prisms.

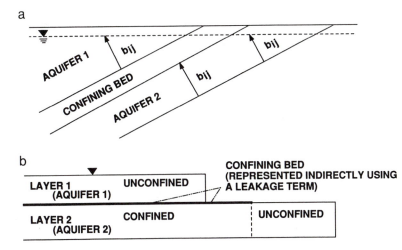

Fig. 3.13 Definition of model layers in cross section.
(a) Dipping geologic units are identified as hydrostratigraphic units.
(b) Representation of the dipping units in (a) as layers in a quasi three-dimensional model. The units are modeled as horizontal layers, so that in the model the coordinate system is aligned with the principal directions of the hydraulic conductivity tensor. Aquifer 1 is represented as layer 1 and is unconfined. Aquifer 2 is represented as confined, except where it crops out. The confining bed is not modeled explicitly but is represented by a leakage term that is a function of the head difference across the confining bed and the vertical hydraulic conductivity and thickness of the confining bed.

ORIENTING THE GRID

The grid should be drawn on an overlay of a map of the area to be modeled. The horizontal plane of the grid should be aligned so that the x and y coordinate axes are colinear with K_x and K_y (Fig. 3.17a). The vertical axis of the model, when present, should be aligned with K_z (Fig. 3.17b). However, it is not always practical to do this. In Fig. 3.13, for example, model layers are horizontal to the surface although the bedding is at an angle to the surface; hence, the z axis is not parallel to K_z. Likewise the vertical axes for the grids shown in Fig. 3.16 are not perpendicular to bedding planes. In such cases it is typically assumed that the angle between the dip of the beds and the horizontal axis is small so that K_z can be assumed to be nearly colinear with the vertical axis. If it is not possible to align the grid with the principal directions of the hydraulic conductivity tensor and if the anisotropy effects are determined to be important, the governing equation should be written to include the off-diagonal components of the hydraulic conductivity tensor (Fig. 2.3). Some finite element codes allow for these conditions.

In a finite difference model it is also important to orient the grid to mini-

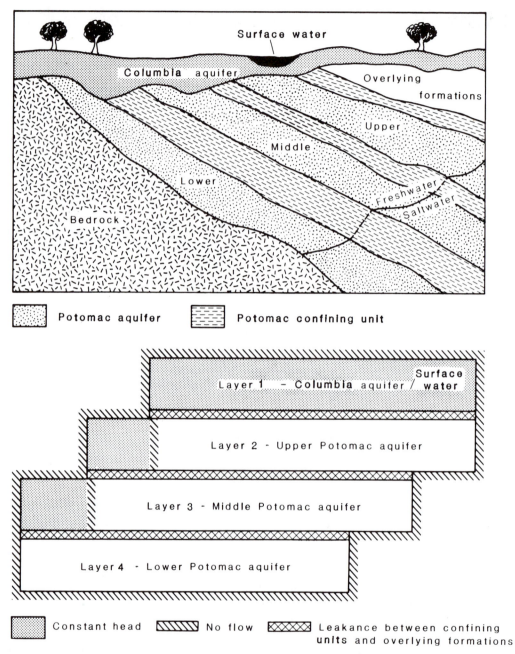

Fig. 3.14 Representation of dipping hydrostratigraphic units in Delaware using a quasi three-dimensional model (Phillips, 1987). Groundwater flow is downward from the unconfined Columbia aquifer (layer 1) to the three units of the Potomac aquifer. Leakage of water from the unconfined aquifer is computed from assigned specified head boundary nodes in layer 1 and leakance values assigned to represent the transmission characteristics of material between layers 1 and 2. The contact of the other two layers with the unconfined system is simulated by the specified heads and leakance properties assigned to the two shaded rectangular areas on the left side of layer 2 and layer 3. These areas have no-flow boundaries on three sides but are connected to the layers by leakance terms associated with the fourth side.

a

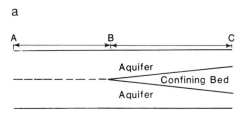

Case 1--Confining Bed Pinchout

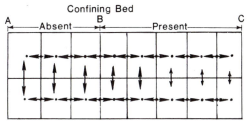

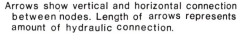

Arrows show vertical and horizontal connection between nodes. Length of arrows represents amount of hydraulic connection.

b

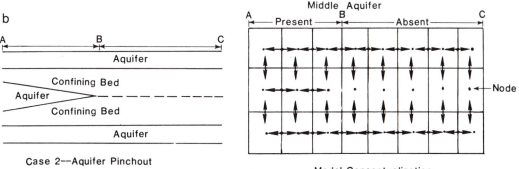

Case 2--Aquifer Pinchout

Physical Configuration

Model Conceptualization

Fig. 3.15 Representation of pinchouts (Leahy, 1982).
(a) Pinchout of a confining bed as simulated in a quasi three-dimensional model. The confining bed is represented by lower leakance terms when the confining bed is present (B–C). Where no confining strata are present (A–B), leakance terms between layers are calculated based on the harmonic mean of aquifer hydraulic conductivities and thickness (Box 3.1, Eqn. 1).
(b) Pinchout of an aquifer as simulated in a full three-dimensional model. Confining units and the aquifer are represented by model layers. The pinchout of the aquifer is represented by changing hydrogeologic properties in layer 2 to reflect the change from aquifer to confining bed.

mize the number of nodes that fall outside the boundaries of the modeled area. These nodes are called *inactive nodes*, whereas nodes that fall within the modeled area are *active nodes*. Inactive nodes are not part of the solution but still use up storage space in the arrays needed by the code. The problem of inactive nodes arises because finite difference grids are rectangular while the area to be modeled frequently is not (e.g., Fig. 3.17a).

Finite element grids do not have inactive nodes because the elements are fitted exactly to the boundary (e.g., Fig. 3.17b). When it is necessary to simulate interaction between the groundwater system and the boundaries, it may be critical to approximate the boundaries as closely as possible. An example of such a simulation is one involving the calculation of flow into (or out of) a

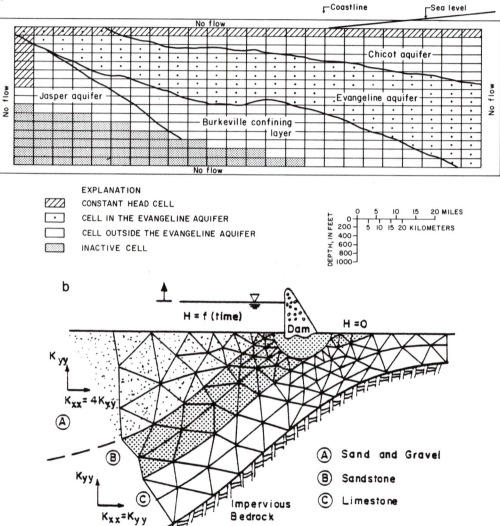

Fig. 3.16 Representation of dipping hydrostratigraphic units in profile models.

(a) A west-to-east finite difference model of the coastal aquifer system in Texas was developed to examine the boundary effects of the Chicot aquifer and Bunkerville confining layer on flow in the Evangeline aquifer (Groschen, 1985).

(b) The design of a finite element mesh to account for dipping beds and boundary conditions for a dam seepage problem. The detailed grid near the base of the dam is not shown (Townley and Wilson, 1980).

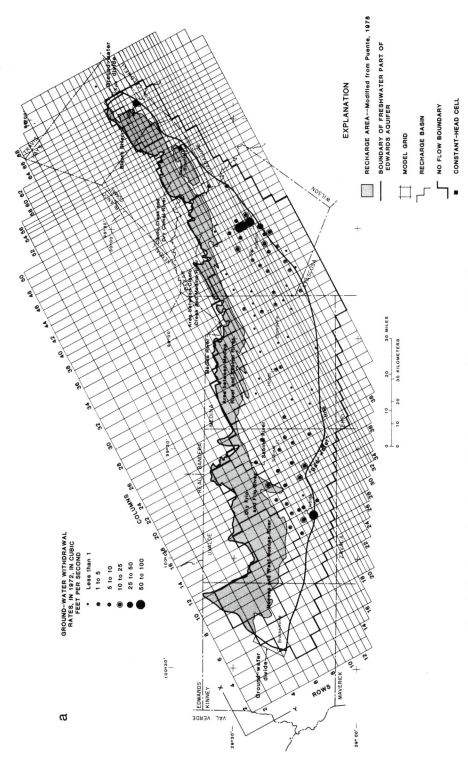

Fig. 3.17 Orientation of grids with features or conditions controlling flow.

(a) Orientation of a finite difference grid to align with northeast-southwest trending faults in the Edwards aquifer, Texas (Maclay and Land, 1988).

(b) Orientation of local coordinates within a finite element grid to the stratification of geologic units shown in profile. The detailed grid near the sheet pile is not shown (Townley and Wilson, 1980).

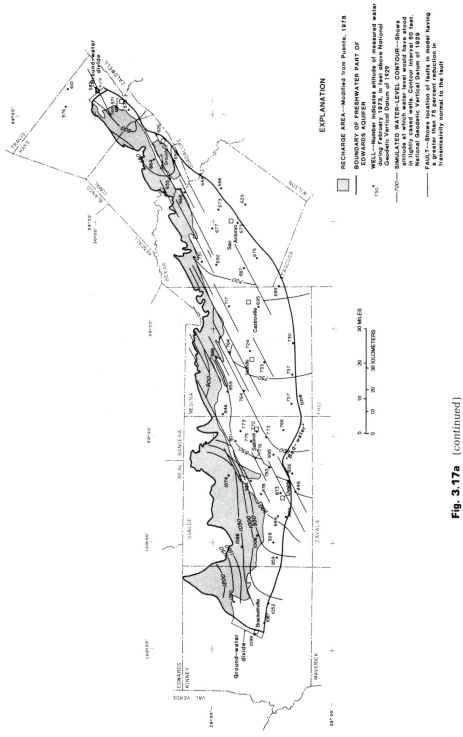

Fig. 3.17a (continued)

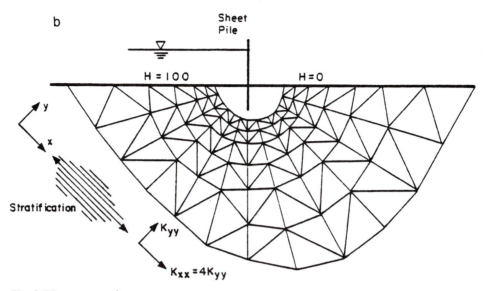

Fig. 3.17 *(continued)*

surface water body (Fig. 3.18a,b). Irregular boundaries can be simulated with special finite difference approximations (Remson et al., 1971; Rushton and Redshaw, 1979) but these formulas are not included in most finite difference codes. If exact representation of boundaries is important, it is preferable to use a finite element code rather than attempt to modify a finite difference code.

When fitting the grid to the boundaries, care should be taken that the node falls directly on the boundary when using a finite element or a mesh-centered finite difference code (Fig. 3.10c,d,e). In a finite difference block-centered grid, the grid is designed so that flux boundaries fall on the edge of the blocks and specified head boundaries fall on the node (Fig. 3.10b). If necessary, block-centered codes can be modified to move specified head boundaries to the edge of the blocks (Huyakorn and Pinder, 1983, pp. 346–348).

When concern focuses on the interior of the grid, the boundaries may be set far from the area of interest so that imposed stresses to the interior part of the system do not reach the boundaries. In such a simulation, fitting the grid to the exact shape of the boundary is not crucial to the modeling effort. For example, Fig. 3.19a shows a finite difference grid used to simulate the effects of pumping from municipal water supply wells in Madison, Wisconsin. The wells pump from a sandstone aquifer that is part of a large regional aquifer system extending into eastern Wisconsin, Illinois, and Iowa. Therefore, when boundaries are placed sufficiently far from the center of the grid, the effects of pumping do not reach the boundaries within the time period used in the transient simulation.

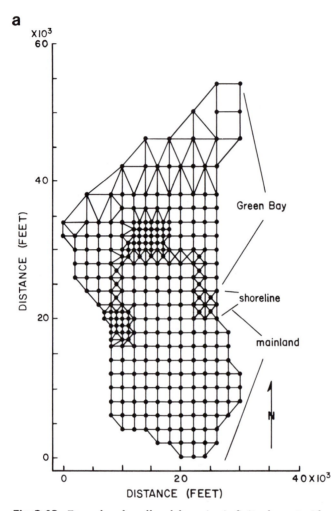

a

Fig. 3.18 Examples of small nodal spacing in finite element grids near the boundary with surface water bodies.

(a) Finite element grid for a peninsula in Wisconsin, constructed with a mix of triangular and quadrilateral elements (Bradbury, 1982). Fine nodal spacing is used to represent the shoreline area where groundwater discharges to Green Bay.

(b) A grid showing the use of small nodal spacing near a barrier wall designed to protect a nuclear reactor located in northwestern Italy from a high water table when the Po River floods. The Po River forms the southern boundary and the river bank is also represented by a fine grid. The northern and southern boundaries are specified head and the eastern and western boundaries are no-flow boundaries (Gambolati, Toffolo, and Uliana, Water Resources Research, 20(7), pp. 903–913, 1984, copyright by the American Geophysical Union).

b

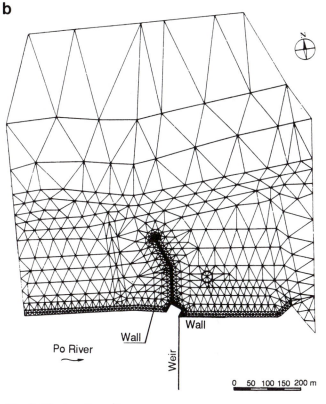

Fig. 3.18 (*continued*)

SPATIAL SCALES

Selecting the size of the nodal spacing is a critical step in grid design. The size of the nodal spacing in the horizontal dimension is a function of the expected curvature in the water table or potentiometric surface. Finer nodal spacing will be required to define highly curved surfaces. Similarly, the change in head in the vertical direction will influence the selection of the vertical nodal spacing.

A secondary consideration in selecting the nodal spacing is the variability in aquifer properties. Model layers typically correspond to hydrostratigraphic units. However, if there are significant vertical head gradients, two or more layers should be used to represent a single hydrostratigraphic unit. Variation of aquifer properties in the horizontal dimension will usually occur over a longer length interval than variations in the vertical dimension. Finally, the variability in areal recharge, pumping, and recharge or discharge to rivers should be con-

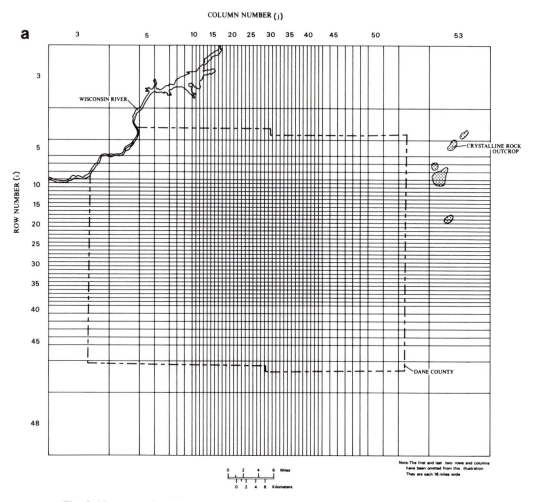

Fig. 3.19 Examples of irregular finite difference grids.

(a) The grid for a model of Dane County, Wisconsin (McLeod, 1975). Fine grid spacing in the interior of the grid is needed to represent municipal pumping. No-flow boundaries are located in columns 1 and 55 and in rows 1 and 50, which are not shown in the figure.

(b) A grid designed to accommodate closely spaced nodes near production wells and simulate flow and infiltration from the Susquehanna River to a shallow aquifer in southern New York State (Yager, 1986). Details of the grid used in the central portion of the modeled area are shown in the inset.

(c) A grid designed to provide finer nodal spacing in areas where pumping stresses are expected to occur for the Black Mesa area in northeastern Arizona (Brown and Eychaner, 1988). All boundaries are no flow.

b

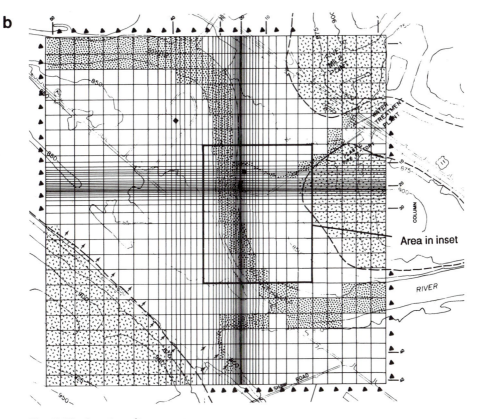

Fig. 3.19 (*continued*)

sidered. Smaller grid spacing will be needed to represent river nodes and pumping nodes. It is conceivable that areal recharge may change significantly from one cell or element to another, but typically very little information is available on field-measured recharge rates and a constant value is usually assumed over large portions of the grid.

The overall size of the modeled area will also affect the selection of nodal spacing. A grid with a small number of nodes is preferred in order to minimize data handling and computer storage and computation time. Yet, it is desirable to use a large number of nodes to represent the system accurately. The need to select meaningful boundaries may require modeling a large area. A compromise between accuracy and practicality is necessary. One way to resolve the trade-off between number of nodes and required level of detail is to use a technique sometimes called telescopic mesh refinement (Ward et al., 1987) whereby a coarse grid is used to model a large problem domain bounded by the physical limits of the aquifer system. The solution is used to define subregional

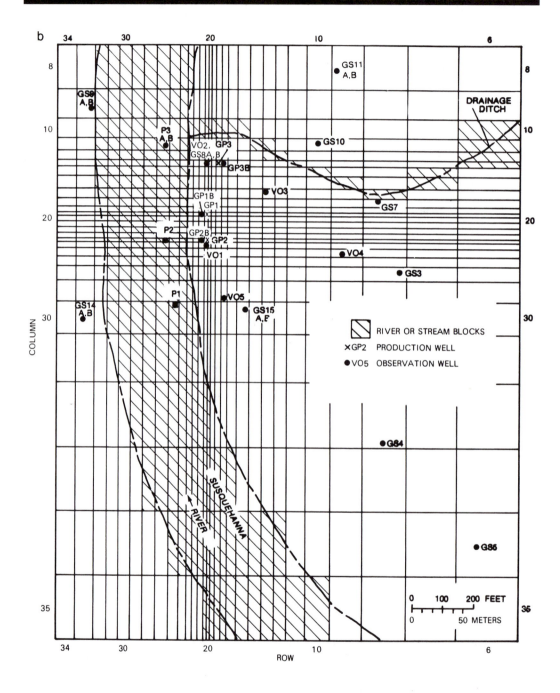

Fig. 3.19 (continued)

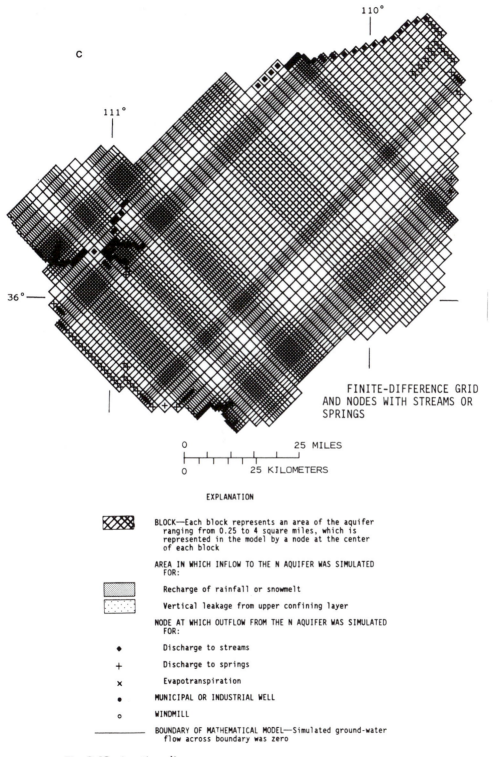

C

110°

111°

36°

FINITE-DIFFERENCE GRID
AND NODES WITH STREAMS OR
SPRINGS

0   25 MILES
0   25 KILOMETERS

EXPLANATION

BLOCK—Each block represents an area of the aquifer
ranging from 0.25 to 4 square miles, which is
represented in the model by a node at the center
of each block

AREA IN WHICH INFLOW TO THE N AQUIFER WAS SIMULATED
FOR:

Recharge of rainfall or snowmelt

Vertical leakage from upper confining layer

NODE AT WHICH OUTFLOW FROM THE N AQUIFER WAS SIMULATED
FOR:

◆   Discharge to streams

+   Discharge to springs

×   Evapotranspiration

•   MUNICIPAL OR INDUSTRIAL WELL

○   WINDMILL

_____ BOUNDARY OF MATHEMATICAL MODEL—Simulated ground-water
flow across boundary was zero

Fig. 3.19 (continued)

boundaries, which then define a new smaller problem domain. The telescoping continues until the grid is small enough to obtain the desired detail (Section 4.2 and Box 4.3).

Finite Difference Grids

Nodes are labeled using an (i, j, k) indexing convention (Fig. 3.9), to refer to the position within a row, column, and layer. A two-dimensional finite difference grid is generated by specifying arrays of values for Δx and Δy for the horizontal plane. In a three-dimensional simulation an array of Δz values may also be specified, although MODFLOW uses a different approach for discretizing the vertical dimension (Box 3.1).

Nodes may be regularly spaced so that Δx, Δy, and Δz are all constants but not necessarily equal to each other. More commonly, however, an irregular grid is necessary in order to use small grid spacing in one area of the problem domain. The grid is designed so that nodes are closely spaced where steep hydraulic gradients are expected (e.g., near pumping wells), or to define narrow rivers or other hydrologically significant features. Examples of finite difference grids with irregularly spaced nodes are shown in Fig. 3.19.

When small grid spacing is used in the interior of the grid, it may be necessary to increase the nodal spacing as the grid is expanded out to the boundaries. For finite difference models the rule of thumb is to expand the grid by increasing nodal spacing no more than 1.5 times the previous nodal spacing. For example, if the smallest nodal space is one meter, the next space should be no more than 1.5 meters. The next space beyond it should be no more than 2.25 meters, and so on. A factor of two may be used for a few rows and columns but it is advisable not to expand the entire grid using a factor of two. The limit on grid expansion arises because the finite difference expression for the second derivative has a larger error when derived for irregular grid spacing. The finite difference expression for an irregular grid is correct to the first order, whereas for a regular grid the finite difference expression is correct to the second order. The difference in truncation error can be demonstrated theoretically by means of a Taylor series expansion (Remson et al., 1971).

A heuristic explanation of the cause of the larger error when dealing with irregular nodal spacing follows. Consider the one-dimensional system shown in Fig. 3.20. For an irregular grid Δx is not a constant and the second derivative in the neighborhood of (i) is the difference in the first derivatives as calculated at points $i - 1/2$ and $i + 1/2$ (Fig. 3.20). Using the notation defined in Fig. 3.20, the second derivative may be approximated as follows:

$$\left.\frac{d^2h}{dx^2}\right|_i = \frac{d}{dx}\left(\left.\frac{dh}{dx}\right|_{i+1/2} - \left.\frac{dh}{dx}\right|_{i-1/2}\right) \tag{3.2a}$$

$$\left.\frac{d^2h}{dx^2}\right|_i = \frac{1}{\Delta x_i}\left(\frac{h_{i+1} - h_i}{\Delta x_{i+1/2}} - \frac{h_i - h_{i-1}}{\Delta x_{i-1/2}}\right) \tag{3.2b}$$

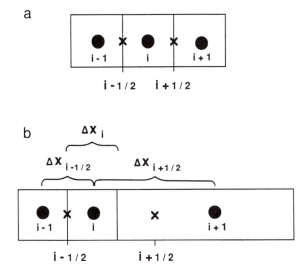

Fig. 3.20 Finite difference grids where $i + 1/2$ represents the location halfway between i and $i + 1$; $i - 1/2$ is the location halfway between $i - 1$ and i. $\Delta x_{i+1/2}$ is the distance between nodes i and $i + 1$; $\Delta x_{i-1/2}$ is the distance between nodes $i - 1$ and i and Δx_i is the length of the cell.
(a) Regular grid for which $\Delta x_{i-1/2} = \Delta x_i = \Delta x_{i+1/2}$.
(b) Irregular grid for which $\Delta x_{i-1/2} \neq \Delta x_i \neq \Delta x_{i+1/2}$.

Note that first derivatives are calculated at $i + 1/2$ and $i - 1/2$. When Δx is a constant (Fig. 3.20a), the points $i + 1/2$ and $i- 1/2$ coincide with the edges of the nodal block so that the nodal point (i) is centered between $i - 1/2$ and $i + 1/2$. However, for the irregular grid shown in Fig. 3.20b, the location of $i + 1/2$ does not coincide with the edge of the nodal block that contains node i and nodal point (i) is no longer centered between $i - 1/2$ and $i + 1/2$. The finite difference solution calculates the head at the point halfway between $i - 1/2$ and $i + 1/2$. When the node is not centered between $i - 1/2$ and $i + 1/2$, the location for which the head is calculated does not coincide with the location of the node, thereby introducing some error into the solution. Note that for illustration purposes, the grid shown in Fig. 3.20b was expanded using a factor of four rather than the recommended factor of 1.5.

Finite Element Grids

A disadvantage of finite element models is that the input of data required to define the grid is more laborious than for finite difference models. Finite element models require that each node and element be numbered (Fig. 3.10d,e) and that the coordinate location (x, y, z) of each node and the node numbers

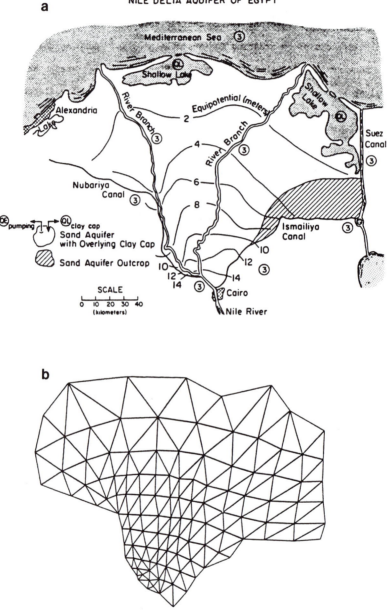

Fig. 3.21 Irregular finite element grid of the Nile delta (Townley and Wilson, 1980).
(a) Map of the Nile delta with boundary conditions indicated by circled numbers: 2 for a type two or flux boundary and 3 for a mixed boundary (see Section 4.1).
(b) Triangular elements are shaped to fit the irregular boundary.
(c) Nodal numbering is shown for a truncated version of the grid. Note that the nodes are numbered consecutively along the shortest grid direction. The bandwidth is 35.

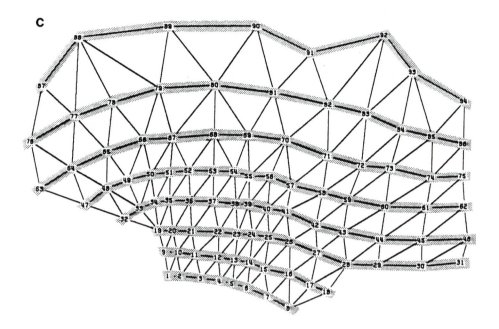

Fig. 3.21 (*continued*)

affiliated with each element be input to the model. Numbering of nodes is done systematically from top to bottom (or bottom to top) and from left to right and sequentially across the shortest dimension of the problem domain (Fig. 3.21). The finite element method treats each element separately and then assembles equations for all elements into a global matrix equation. Systematic numbering across the shortest dimension of the grid reduces the bandwidth of the coefficient matrix and thereby reduces computer storage requirements and computation time. The semibandwidth (SBW) of a symmetric matrix (i.e., entries reflect across the diagonal) is the maximum number of columns between the diagonal and the last nonzero entry, inclusive, along any row of the matrix. The bandwidth is equal to 2 * (SBW − 1). The semibandwidth may be calculated from

$$SBW = R + 1 \qquad (3.3)$$

where R is the maximum difference in any two node numbers that define a single element within the grid (Istok, 1989). For example, in the grid shown in Fig. 3.21c, R is 17, the semibandwidth is 18, and the bandwidth is 35.

In designing a finite element grid for isotropic materials, each element should be constructed so that its *aspect ratio* (the ratio of maximum to minimum element dimensions) is close to unity. This requirement is similar to the factor of 1.5 used in expanding finite difference grids and is necessary to minimize numerical errors. For example, numerical errors can be minimized by exclusive use of equilateral triangular elements. Experience has shown that aspect ratios greater than five should be avoided. Furthermore, a transition region should be used to change element sizes gradually (Fig. 3.18). When dealing with anisotropic materials, the shape of the elements should be considered in the equivalent transformed isotropic domain and designed so that the aspect ratio in the isotropic domain does not exceed five. For details on transforming from an anisotropic to an isotropic domain see Freeze and Cherry (1979, pp. 174–178). Because design of a finite element grid can be quite time consuming, the use of a preprocessor to assist in grid generation and optimal node numbering is highly recommended for complex problems (e.g., Tucciarelli, 1989).

3.4 Assigning Parameter Values

DATA NEEDS

Data needed for groundwater flow models are summarized in Table 3.1. The data can be grouped into two general categories. Data under category A, the physical framework, define the geometry of the system including the thickness and areal extent of each hydrostratigraphic unit. Hydrogeologic data include information on heads and fluxes (items B.1 and B.2 in Table 3.1), which are needed to formulate the conceptual model and check model calibration (Chapter 8). Hydrogeologic data also define aquifer properties and hydrologic stresses (items B.3–B.6 in Table 3.1). In addition, the distribution of effective porosity is required when calculating average linear velocities from head data for input to particle tracking codes (Chapter 11).

Obtaining the information necessary for modeling is not an easy task. Some data may be obtained from existing reports, but in most cases additional on-site field work will be required. A discussion of the field techniques for acquiring these data is beyond the scope of this book. Only a brief overview of methodologies will be presented below. Transmissivity and storage coefficient are typically obtained from pumping test results (Walton, 1987). For modeling at a local scale, values of hydraulic conductivity can be determined by pumping tests if volume-averaged values are desired or by slug tests if point values are required (Bouwer, 1989; Bouwer and Rice, 1976; Hvorslev, 1951). For unconsolidated sand-size sediment, hydraulic conductivity may also be obtained from laboratory grain size analyses (Masch and Denny, 1966) or from laboratory permeability tests using permeameters. Permeameter results must be used with caution because hydraulic conductivity values obtained from permeameter tests typically are several orders of magnitude smaller than values measured *in situ* (e.g., Tanaka and Hollowell, 1966; Herzog and Morse, 1984; White, 1988). The discrepancy is caused by rearrangement of grains during repacking the sample into the permeameter. Furthermore, large-scale features such as fractures, gravel lenses, or other types of bedding that may impart transmission characteristics to the hydrogeologic unit as whole are not captured in a sample the size of a laboratory column. Hence, even intact core samples taken from the field for laboratory analyses tend to yield lower hydraulic conductivity values than are measured in the field.

Laboratory measurements of specific yield suffer from the same scale problems as hydraulic conductivity measurements. Field measurements of specific yield from pumping tests are also uncertain, as are estimates of effective porosity measured by tracer experiments (DeMarsily, 1986). While hydraulic conductivity ranges over 13 orders of magnitude, specific yield and effective porosity vary mainly within 1 order of magnitude. Consequently, there is less uncertainty associated with specific yield and porosity estimates relative to hydraulic conductivity. In the absence of site-specific field or laboratory measurements, initial estimates for aquifer properties may be taken from tables like Table 3.3, 3.4, and 3.5. Also see Mercer et al. (1982) for a compilation of parameter values used in modeling.

When simulating anisotropic media, information is required on the three principal components of the hydraulic conductivity tensor, K_x, K_y, and K_z. Anisotropy is sometimes quantified by an anisotropy ratio. Horizontal anisotropy is represented by the ratio between K_x and K_y, while vertical anisotropy is represented by the ratio between K_x and K_z. Horizontal anisotropy may be caused by fracture sets or by sedimentary structures such as imbrication. Vertical anisotropy is caused mainly by bedding planes and laminae within a sequence of sediment layers but may also be affected by fractures and sedimentary structures. Horizontal anisotropy may be estimated from field measurements (Quinones-Aponte, 1989; Neuman et al., 1984).

At a small enough scale, isotropic hydrogeologic units may be identified in the field. However, for most groundwater problems it is difficult, if not impossible, to model geologic units at this scale. When the thickness of the model layer ($B_{i,j}$) is much larger than the thickness of the isotropic layer (b_{ijk}), assuming this thickness can even be identified from information about bedding, it is possible to calculate hydrologically equivalent horizontal and vertical hydraulic conductivities for the model layer using the following equations:

$$(K_x)_{i,j} = \sum_{k=1}^{m} \frac{K_{i,j,k} b_{i,j,k}}{B_{i,j}} \tag{3.4a}$$

$$(K_z)_{i,j} = \frac{B_{i,j}}{\displaystyle\sum_{k=1}^{m} b_{i,j,k}/K_{i,j,k}} \tag{3.4b}$$

$$B_{i,j} = \sum_{k=1}^{m} b_{i,j,k}$$

If sufficient stratigraphic information were available it would be possible to use Eqns. 3.4a and 3.4b to calculate vertical anisotropy for each hydrogeologic unit (model layer) prior to calibration (e.g., Guswa and Le Blanc, 1985). Vertical anisotropy can also be estimated directly from pumping tests (Neuman, 1975; Weeks, 1969). The anisotropy ratio also may be estimated during calibration. It is advisable, however, to use Eqns. 3.4 to obtain an order of magnitude estimate of a realistic anisotropy ratio for the specific geologic situation being considered. Vertical anisotropy ratios ranging from 1 to 1000 are common in model application (Fig. 3.22a). Horizontal anisotropy ratios are typically less.

The vertical hydraulic conductivity of hydrostratigraphic units controls the flow of water between layers. If structures within a hydrostratigraphic unit are aligned to create anisotropy that is at an angle to the top of the unit, e.g., cross-bedding, one could correct for that effect. Hearne (1985) presented an equation attributed to Theis for correcting for the effects of layering in dipping stratigraphic units that are simulated as horizontal layers in a model.

$$\frac{K_x}{K_y} = \frac{R}{1 - (1 - R)\cos^2 A} \tag{3.5}$$

where K_x = the horizontal hydraulic conductivity in the direction of the dip (L/T)

          K_y = the horizontal hydraulic conductivity in the direction of the strike (L/T)

          R = the ratio of cross-bed to in-bed hydraulic conductivity

          A = the angle of dip

In practice, vertical anisotropy is often unknown and is estimated during calibration.

The thickness and vertical hydraulic conductivity of stream and lake sediments are needed for leakage calculations. These values can be estimated from field measurements (Lapham, 1989; Barwell and Lee, 1981) or during model calibration.

Hydrologic stresses include pumping, recharge, and evapotranspiration. Of these, pumping rates may be the easiest to estimate. Recharge is one of the most difficult parameters to estimate. Further discussion of recharge is reserved for Chapter 5 (Section 5.2). Likewise, field information for estimating evapotranspiration (ET) is likely to be sparse. Field measurements using lysimeters and studies of the vegetation may be helpful in estimating ET rate or areal patterns of ET (Thomas et al., 1989).

Fig. 3.22 Zonation of aquifer parameters.
(a) Zonation of vertical anisotropy ratio (K_x/K_z) for a portion of Cape Cod, Massachusetts (Guswa and Le Blanc, 1985). Five zones are identified with ratios ranging from 10 to 1000. Vertical anisotropy was calculated using information from well logs and Eqn. 3.4b.
(b) Finite difference grid zoned by geologic facies in Owens Valley, California (Danskin, 1988). Each facies is assigned a different hydraulic conductivity value giving five zones. Impermeable rock in the interior of the grid forms impermeable interior boundaries (Section 4.4).

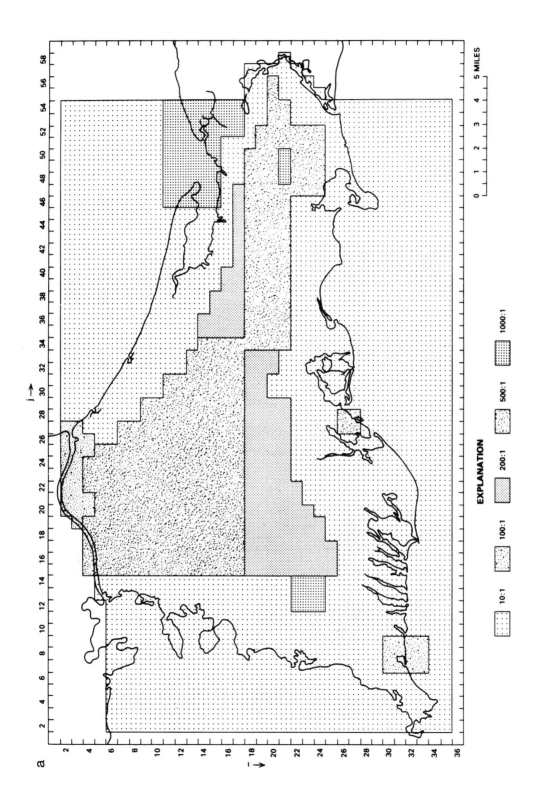

EXPLANATION

10:1	100:1	200:1	500:1	1000:1	

0 1 2 3 4 5 MILES

a

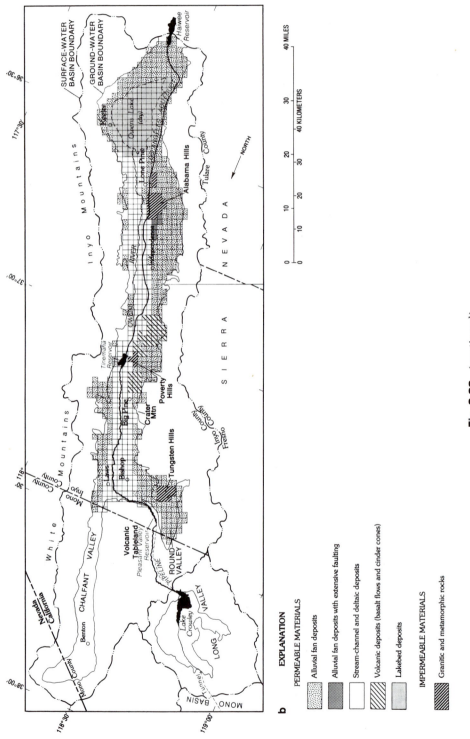

EXPLANATION

PERMEABLE MATERIALS

Alluvial fan deposits

Alluvial fan deposits with extensive faulting

Stream-channel and deltaic deposits

Volcanic deposits (basalt flows and cinder cones)

Lakebed deposits

IMPERMEABLE MATERIALS

Granitic and metamorphic rocks

Fig. 3.22 (continued)

TRANSFERRING FIELD DATA TO THE GRID

The first consideration in translating field data to the grid is to match parameter values to the scale of the model. For example, profile and full three-dimensional models require point measurements of hydraulic conductivity. Ideally, these data are obtained from point measurements of hydraulic conductivity in the field. Two-dimensional areal models and quasi three-dimensional models require vertically averaged values that may be obtained indirectly by averaging point measurements (e.g., Eqn. 3.4a) or directly from pumping tests in wells that fully penetrate the aquifer.

When the field data are determined to be compatible with the scale of the model, aquifer properties may be assigned to each hydrostratigraphic unit identified in the conceptual model. The grid is divided into zones so that certain sets of nodes have similar aquifer properties based on the areal extent of the hydrostratigraphic units (Fig. 3.22). Thickness of each hydrostratigraphic unit is also assigned to each node. When hydrostratigraphic units are defined at a local scale, interfingering of two or more types of material may occur within a single cell or element (Fig. 3.3b). In this case, average properties for the cell or element are computed. The *geometric mean* is used if the pattern of heterogeneity is random, whereas the *arithmetic mean* is used if layering is present.

A finite difference model calculates the head at the node. This head value is also the average head for the finite difference cell. In a block-centered grid, aquifer properties and hydraulic stresses are typically assigned to the block surrounding the node (Figs. 3.10b and 3.23). Other conventions for defining areas of influence in block-centered grids are possible. For example, the convention used in PLASM is discussed in Box 3.2. In a mesh-centered grid, properties are assigned to the area of influence surrounding a node as shown in Fig. 3.10c.

In finite element models, aquifer properties may be assigned either to the node or to the element. Some codes (e.g., SUTRA, by Voss, 1984b) assign some properties to the element, some to the node, and some to a cell or area of influence around the node. AQUIFEM-1 (Box 3.3) assigns properties to either the node or the element, allowing the user to select a preference. When linear triangular elements are used exclusively, it may be easier to assign properties to the nodes because there are always fewer nodes than elements in this type of grid. Note that the grid shown in Fig. 3.10d, for example, contains 50 elements but only 36 nodes. When aquifer properties vary sharply, parameter assignment should be done for the elements. Furthermore, boundaries between two types of porous media should always coincide with element boundaries (Fig. 3.16b).

KRIGING

Assigning parameter values to the grid is difficult because the model requires values for each node, cell, or element and field data are typically sparse. Inter-

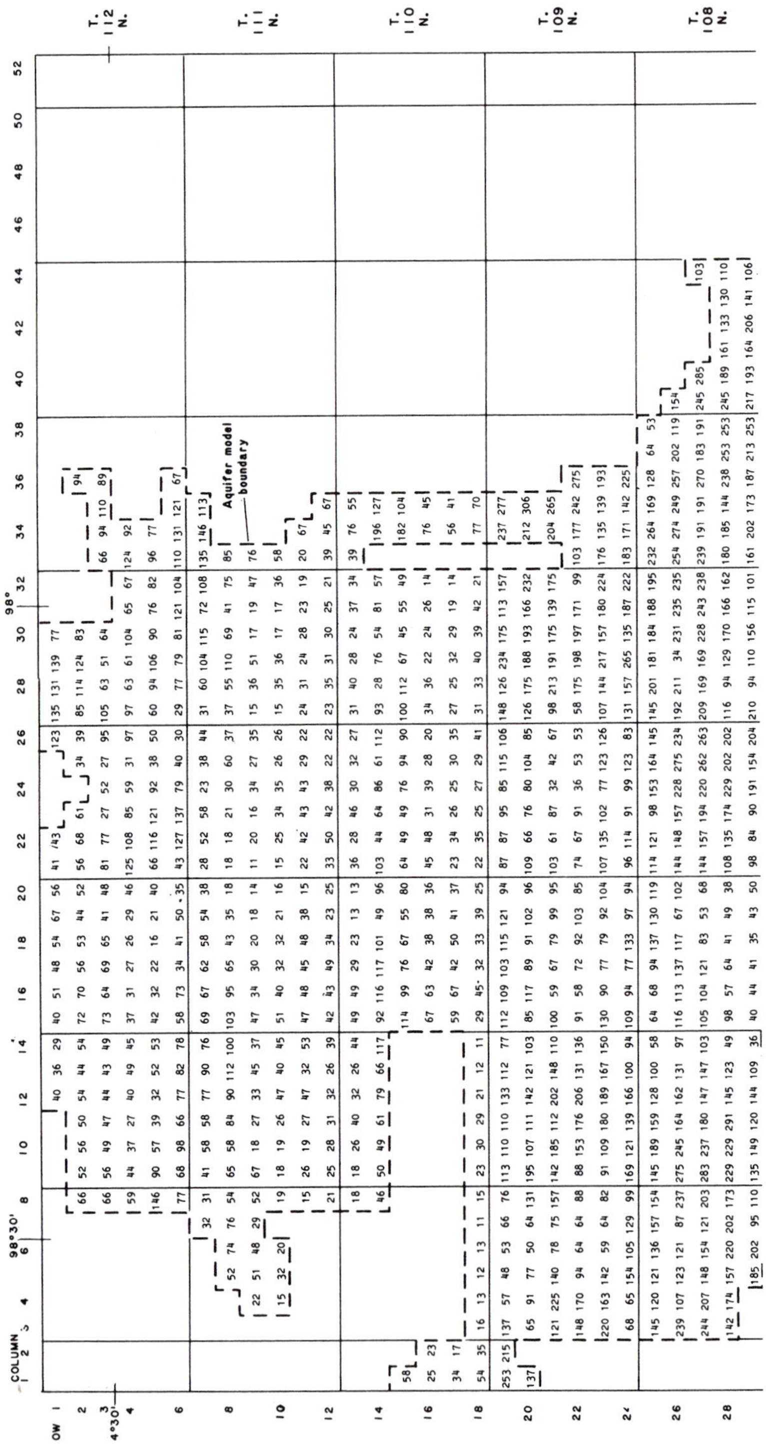
Aquifer model boundary
COLUMN
ROW
98°30'
98°
44°30'
T. 112 N.
T. 111 N.
T. 110 N.
T. 109 N.
T. 108 N.

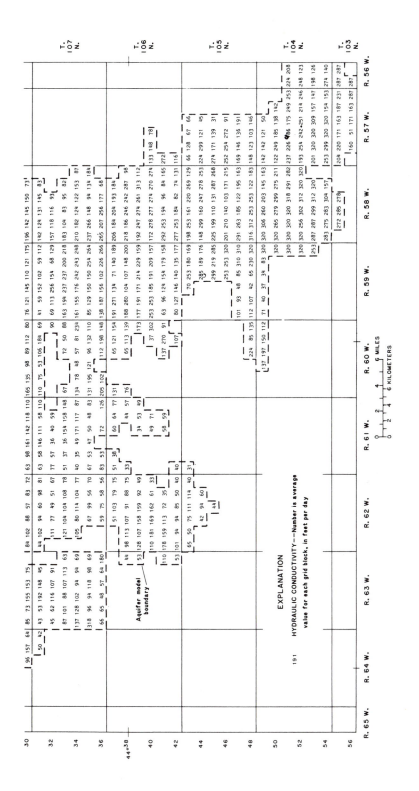

Fig. 3.23 Assignment of hydraulic conductivity (ft/day) to cells in a finite difference grid (Emmons, 1988). Average hydraulic conductivity values were calculated from composite geologic logs developed for each one square mile section. These logs were used to assign hydraulic conductivity values to each cell depending on grain size.

polation of measured data points can help in defining the spatial variability over the problem domain. Kriging is the interpolation method used most frequently for this purpose. However, other interpolation methods may also be used (e.g., Williams and Williamson, 1989) or parameter values may be assigned to nodes using hydrogeologic judgment.

Kriging is a statistical interpolation method that chooses the best linear unbiased estimate (BLUE) for the variable in question. The variable is assumed to be a random function whose spatial correlation (structure) is defined by a variogram (Fig. 3.24), which is a measure of the change in the variable with changes in distance. Higher correlation between measurement points is expected for small separation distances. Kriging differs from other interpolation methods because it considers the spatial structure of the variable and provides an estimate of the interpolation error in the form of the standard deviation of the kriged values. Such error estimates are needed when assigning plausible ranges of parameter values prior to model calibration. Kriging also preserves the field value at measurement points, unlike some other interpolation schemes such as least-squares fitting of a polynomial.

Kriging was first used in applications to ore deposits (Journel and Huijbregts, 1978). DeMarsily (1986) provides a good overview of kriging applied to groundwater problems and gives selected applications. Applications of kriging of transmissivity data were also reported by Clifton and Neuman (1982), Aboufirassi and Marino (1984), and Pucci and Murashige (1987), among others. Kriging can also be used to interpolate heads (Pucci and Murashige, 1987; DeMarsily, 1986; Dunlap and Spinazola, 1984; Neuman and Jacobson, 1984) and produce kriged contour maps based on measured heads that can be compared with simulated heads during model calibration. A number of software packages are available to aid in geostatistical analysis including kriging. GEO-

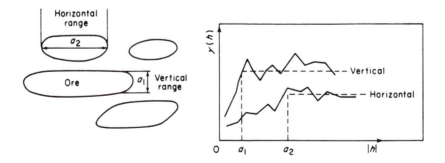

Fig. 3.24 Example variograms, where h is separation distance and γ is the variance of the separation distance or the variogram function (Journel and Huijbregts, 1978). Variograms and kriging were first used in mining for ore bodies. σ_1 and σ_2 define the sills of the variograms, which represent the horizontal and vertical dimensions of the ore body. In hydrogeological applications the variogram is used to quantify the structure in the aquifer caused by the arrangement of heterogeneities.

EASE (Geostatistical Environmental Assessment Software) is a public domain code prepared by the U.S. Environmental Protection Agency that is designed for variogram estimation and kriging. GEOKRIG is another kriging software package; it is available from Scientific Software Group (see Table 1 in the preface for the address). SURFER (Golden Software) also uses kriging to produce contour maps.

Whatever method is used to assign parameter values to nodal points, care should be taken that the resulting distributions make sense in hydrogeologic terms and that the values fall within reasonable ranges for the relevant geologic setting.

Box 3.1
Introduction to MODFLOW

MODFLOW is a block-centered finite difference code that can simulate all of the aquifer types discussed in Section 3.2. The sophistication present in MODFLOW means that the input assembly is complex, and readers are referred to the user's manual (McDonald and Harbaugh, 1988) for details. Preprocessors are available to help with data assemble and post-processors can assist in viewing the output (Rumbaugh and Duffield, 1989). Several features of the model are highlighted below.

Specifying the Vertical Grid Spacing

MODFLOW views a three-dimensional system as a sequence of layers of porous material (Fig. 3.9). The horizontal grid is generated in the usual way by specifying grid dimensions in the x and y directions. As with all finite difference grids, the horizontal grid must be the same for each layer. The model does not require input of a Δz array. Instead Δz is specified indirectly. The user may input *layer transmissivities*, which are equal to the hydraulic conductivity of the layer times the layer thickness (Δz). Alternatively, the user may input hydraulic conductivity arrays for each layer and arrays giving the elevation of the top and bottom of the layer. MODFLOW then calculates transmissivity for the layer after first computing layer thickness from the top and bottom elevations.

Transmissivity at each (i, j) location within a layer may vary owing to spatial variations in aquifer thickness and/or hydraulic conductivity. This means that in effect Δz varies spatially within a layer. This procedure allows greater flexibility in fitting hydrostratigraphic units into a finite difference grid (Fig. 1). However, it distorts the layers (Fig. 2), thereby introducing error into the finite difference approximation. According to McDonald and Harbaugh (1988), the error is generally small.

Types of Model Layers

Layers may be designated as always confined, always unconfined, or capable of being either confined or unconfined (convertible). If the layer is confined, the user inputs the transmissivity and storage coefficient of the layer. The top layer in the system is typically designated to be unconfined and the user inputs the hydraulic conductivity, specific yield, and the bottom of the layer. MODFLOW calculates the transmissivity of the layer by multiplying hydraulic conductivity by the saturated thickness of the layer. Heads in the layer are calculated under the Dupuit assumptions. After each iteration, the saturated thickness in the layer is updated and new transmissivities are calculated. MODFLOW allows the water table to rise to infinity in the top unconfined layer. That is, the top layer is assumed to be infinitely thick.

If the layer is designated to be convertible, hydraulic conductivities and the elevations of the top and bottom of the aquifer are input and

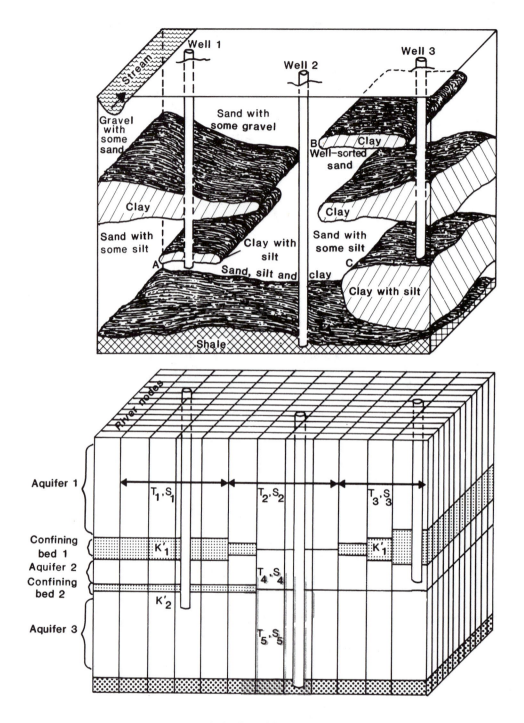

Fig. 1 Fitting layers to irregularly shaped hydrostratigraphic units (Peters, 1987).

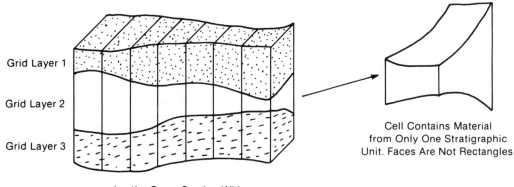

**Aquifer Cross Section With
Deformed Grid Superimposed**

Fig. 2 Distortion of hydrostratigraphic layers is preserved in the block-centered grids used in MODFLOW by adjusting layer transmissivities (McDonald and Harbaugh, 1988).

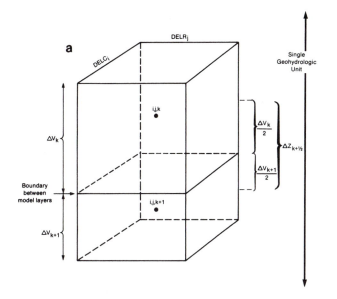

Fig. 3 Grid configurations assumed in the VCONT calculation (McDonald and Harbaugh, 1988). (a) This configuration is used when VCONT represents vertical anisotropy in hydraulic conductivity within a single hydrostratigraphic unit. (b) This configuration is used to represent differences in the vertical hydraulic conductivity between two hydrogeologic units. (c) This configuration is used in quasi three-dimensional modeling where the semiconfining layer is not represented explicitly in the model.

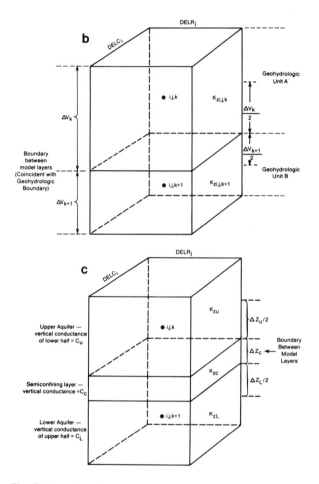

Fig. 3 *(continued)*

MODFLOW calculates layer transmissivities. After each iteration the model checks to determine whether the head in the layer is above or below the elevation of the top of the layer. If the head in the layer is higher than the elevation of the top of the layer, the layer is assumed to be confined. If the head in the layer is less than the elevation of the top of the layer, the layer is assumed to be unconfined.

The VCONT Arrays

For MODFLOW simulations involving more than one layer, the user must calculate a vertical transmission or leakage term, known as VCONT, for each nodal block in the grid, except for blocks in the bottom layer. A VCONT array is not required for the bottom layer because the model assumes that the bottom layer is underlain by impermeable material and VCONT is zero. VCONT is a function of the vertical hydraulic conductivity between layers and the thickness of the layers. Several options are available for computing VCONT depending on whether a quasi or a full three-dimensional simulation is performed. The most general formula, which is used for full three-dimensional simulations, is

$$\text{VCONT}_{i,j,k+1/2} = \frac{2}{\dfrac{\Delta v_k}{(K_z)_{i,j,k}} + \dfrac{\Delta v_{k+1}}{(K_z)_{i,j,k+1}}} \qquad (1)$$

Refer to Fig. 3a,b for definition of terms. Equation 1 is the harmonic mean of $(K_z)_{i,j,k}/\Delta v_k$ and $(K_z)_{i,j,k+1}/\Delta v_{k+1}$ and is used when each layer represents a different hydrostratigraphic unit as in Fig. 3b. Equation 1 is also used when two or more layers represent a single hydrostratigraphic unit (Fig. 3a). In this case, $(K_z)_{i,j,k}$ equals $(K_z)_{i,j,k+1}$ and Eqn. 1 becomes

$$\text{VCONT}_{i,j,k+1/2} = \frac{2K_z}{\Delta v_k + \Delta v_{k+1}} \qquad (2)$$

The formula for quasi three-dimensional simulations is

$$\text{VCONT}_{i,j,k+1/2} = \frac{2}{\dfrac{\Delta z_u}{(K_z)_u} + \dfrac{2\Delta z_c}{(K_z)_c} + \dfrac{\Delta z_L}{(K_z)_L}} \qquad (3)$$

Refer to Fig. 3c for definition of terms. When $(K_z)_c \ll (K_z)_u$ and $(K_z)_L$ this formula simplifies to

$$\text{VCONT}_{i,j,k+1/2} = \frac{(K_z)_c}{\Delta z_c} \qquad (4)$$

which is simply the definition of leakance (Eqn. 3.1).

When the top layer is unconfined, the layer has no top. Typically the average water-table elevation expected to occur during the simulation is used to define the top of the cell and to calculate the VCONT value between layers 1 and 2.

Internodal Transmissivities

The VCONT arrays quantify vertical transmission characteristics *between the nodes* in each layer, i.e., in the neighborhood of $i, j, k + 1/2$. Vertical internodal values are required because the finite difference equations used in MODFLOW require the transmission characteristics of the volume of the aquifer located *between* nodes. The finite difference equations also require the horizontal transmission properties *between* nodes. However, the user is asked to specify transmissivities (T_x and T_y) or hydraulic conductivities (K_x and K_y) for each cell or block around the node. The model converts block transmissivities to transmissivites between the nodes by using a weighted harmonic mean to compute internodal transmissivities. For example, consider the one-dimensional array of nodal points shown in Fig. 3.20. The transmissivity between nodes i and $i + 1$ would be calculated by taking the harmonic mean of $T_i/\Delta x_i$ and $T_{i+1}/\Delta x_{i+1}$:

$$\frac{T_{i+1/2}}{\Delta x_{i+1/2}} = \frac{2T_i T_{i+1}}{\Delta x_i T_{i+1} + \Delta x_{i+1} T_i} \qquad (5)$$

The harmonic mean gives the exact flow rate between adjacent blocks at steady state when transmissivity changes abruptly at the cell boundary. It also allows convenient simulation of no-flow boundaries because the internodal transmissivity is equal to zero when T_{i+1} (or T_{i-1}) is equal to zero. Appel (1976) discusses other ways of computing internodal averages. For example, the following formula may be used to calculate an internodal average when transmissivity is a linear function between nodal points

$$T_{i+1/2} = (T_{i+1} - T_i)/\ln(T_{i+1}/T_i) \qquad (6)$$

Box 3.2
Introduction to PLASM

PLASM is the Prickett-Lonnquist Aquifer Simulation Model, first documented in Prickett and Lonnquist (1971). It was one of the first readily available, well-documented groundwater flow models. Since its first publication, the code has been revised for incorporation as a component of a solute transport code (Prickett et al., 1981) and updated to a user-friendly format for an IBM-compatible microcomputer (see the table in the preface).

PLASM is a finite difference code that can simulate two-dimensional problems in both areal and profile orientation. It is a block-centered model but can accommodate mesh-centered flux boundary conditions (see Box 4.2). PLASM requires the user to assign a storage parameter (either storage coefficient or specific yield) to the area around the node and to specify transmissivities for the area *between* nodes according to the scheme illustrated in Fig. 1.

In the original PLASM model (Prickett and Lonnquist, 1971) the user specified transmissivity (or hydraulic conductivity) under the assumption that the grid was regular, i.e., that Δx and Δy are constants. If this was not the case, the user corrected the transmissivity (or hydraulic conductivity) arrays manually by multiplying T_x (or K_x) by $2\Delta y_{i,j}/(\Delta x_{i,j} + \Delta x_{i+1,j})$ and

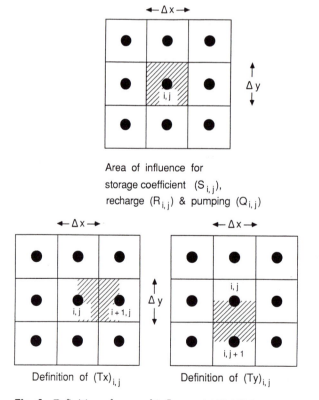

Area of influence for
storage coefficient $(S_{i,j})$,
recharge $(R_{i,j})$ & pumping $(Q_{i,j})$

Definition of $(T_x)_{i,j}$ Definition of $(T_y)_{i,j}$

Fig. 1 Definition of areas of influence in PLASM.

T_y (or K_y) by $2\Delta x_{i,j}/(\Delta y_{i,j} + \Delta y_{i,j+1})$. Furthermore, the original version of PLASM required a user-supplied storage factor equal to the storage coefficient (or specific yield) multiplied by Δx Δy. In the 1981 version incorporated in the random walk solute transport code (Prickett et al., 1981) and in the version for IBM-compatible microcomputers the input was changed so that the user now specifies transmissivity (or hydraulic conductivity), storage coefficient (or specific yield), *and* arrays of Δx and Δy. The model internally performs the corrections required for irregular nodal spacing. The user may select inconsistent units of gallons, feet, and days, or consistent units of cubic meters, meters, and days.

In addition to the three versions of PLASM mentioned above, there are documented versions produced by persons other than the original author. For example, PLASM was modified by Potter and Gburek (1987) to accommodate discharge of water through a seepage face and distribution of the water to an overland flow routing procedure. Additionally, changes were made in data entry to facilitate the description of boundary conditions, aquifer geometry, and irregular grid spacing. Changes were also made in the numerical solution technique and presentation of output. Walton (1989) modified PLASM to accommodate quasi three-dimensional simulations involving multiple stacks of aquifers and confining beds where each stack consists of a confined aquifer overlain by a confining bed. The entire sequence is topped by an unconfined aquifer. He also added subroutines to simulate partially penetrating wells, effects of well casing storage, multiaquifer production wells, and flowing wells. Walton's modification of PLASM is renamed GWFL3D.

Box 3.3
Introduction to AQUIFEM-1

AQUIFEM-1 is a two-dimensional finite element code that uses triangular elements (Townley and Wilson, 1980). The "1" indicates that the code normally simulates horizontal flow in *one* aquifer layer. The aquifer can be unconfined, confined, or leaky confined. AQUIFEM-1 also can simulate two-dimensional flow in cross section. AQUIFEM-N (Townley, 1990) is a relatively new extension of AQUIFEM-1 that can simulate quasi three-dimensional flow in a sequence of aquifers and can adjust the water table boundary in a profile model by using deforming elements (Chapter 4). The reader should be aware that AQUIFEM-1 and AQUIFEM-N are unrelated to the codes AQUIFEM (Pinder and Voss, 1979) and AQUIFEM-SALT (Voss, 1984a).

AQUIFEM-1 has the same simulation capabilities as PLASM, but because it is a finite element model it has more flexibility in grid design and in simulating sink/sources terms. The user has the option of assigning aquifer properties and sink/source terms either to nodes or to elements. The grid can be fitted to irregular boundaries. All boundary conditions can change with time, provided the user specifies the changes at the beginning of the simulation (Box 4.1). AQUIFEM-1 is well suited to simulating geologic faults or other internal boundaries. Faults may be impermeable barriers to flow, as in Fig. 1a, or may be permeable avenues of broken and crushed rock as in Fig. 1b. An impermeable fault can be simulated by using elements with nonconnecting nodes as in Fig. 2a; the shaded element does not exist in the model. Impermeable faults can also be simulated using low-transmissivity elements as shown in Fig. 2b. A permeable fault would be simulated using high-transmissivity elements. Partially penetrating rivers in the interior of the model can also be simulated more realistically by AQUIFEM-1 than by finite difference meth-

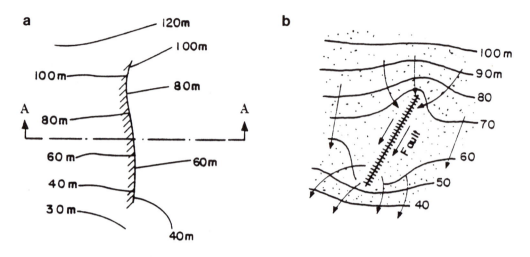

Fig. 1 Faults (Townley and Wilson, 1980).
(a) Plan view with equipotential lines for an impermeable fault.
(b) Plan view with equipotential lines showing funnelling of water along a permeable fault.

ods (Box 4.1). Point sinks (wells) and sources can be simulated effectively as nodes. The code then calculates the head at the point allowing direct computation of well drawdown. In contrast, finite difference models approximate point sources and sinks as blocks of dimensions Δx by Δy and the code cannot compute the head in the well directly. See Chapter 5 for more discussion of sinks and sources.

A useful feature of AQUIFEM-1 is that it allows the user to assign *external* node numbers to each node. The sequence of external node numbers is unimportant, so that if a node is accidently omitted when numbering, it can be

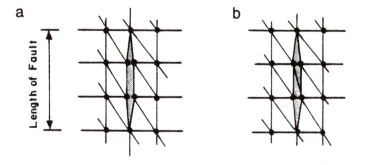

Fig. 2 Representation of faults with a finite element grid (Townley and Wilson, 1980).
(a) Use of nonconnected nodes to represent an impermeable fault.
(b) Use of small elements to represent a fault. Transmissivity values assigned to the elements determine whether it acts as a no-flow or a permeable fault.

added at the end. In the grid shown in Fig. 3, for example, node 66 is out of sequence, occurring between nodes 24 and 25. When entering node numbers into the data file, external node numbers must be entered in sequence as they occur in the mesh. For example, in the data file for the grid in Fig. 3, node 66 would be entered after node 24 and before node 25. AQUIFEM-1 then renumbers the nodes internally according to the order in which they are read, giving them *internal* node numbers. The internal node numbers define the bandwidth of the matrix used in the numerical solution. Nodes should be entered into the data file sequentially along the shortest problem domain in order to minimize the bandwidth of the coefficient matrix. The code also requires that the (x, y) coordinates of each node be given. For example, node 66 in Fig. 3 is located at $(x, y) = (80, 0)$. The

data entry would be

$$66 \qquad 80 \qquad 0$$

Element input consists of the element number together with the numbers of the nodes that define the element. The number of the element is unimportant, as is the order in which elements are arranged in the data file. This is because finite element models keep track of the location of the node; each element is linked in the data file to the nodes it contains. Node numbers for each element are read into the data file in counterclockwise fashion. For example, element 96 in Fig. 3 is defined by nodes 62, 63, and 57, in that order. Similarly, element 41 is defined by nodes 66, 30, and 25. The data entry for element 41 would be

$$41 \qquad 66 \qquad 30 \qquad 25$$

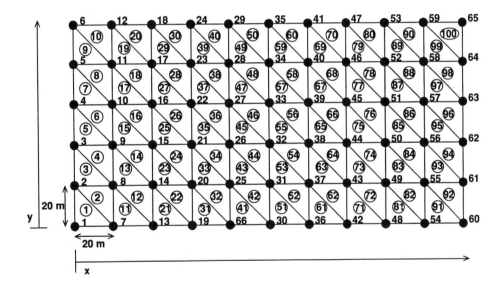

Fig. 3 Example of a finite element grid with 66 nodes and 100 elements. Note that node 66 is out of sequence between nodes 24 and 25.

Box 3.4
Parameter Assignment: A Case Study _____

Patrick et al. (1989) designed a quasi three-dimensional model to study the effect of pumping a confined aquifer beneath Anchorage, Alaska. Well logs (Fig. 1) indicate that the geology consists of unconsolidated layers and lenses of gravel, sand, silty sand, and silty clay. Three hydrostratigraphic units were identified forming a confined aquifer, confining bed, and upper unconfined aquifer. Model layers were used to represent the two aquifers. The confining bed was represented by a leakage term.

Thicknesses of the units, as estimated from well logs, were used to construct isopach maps. Hydraulic properties of each unit were calculated from lithologic data available from over 200 well logs and around 50 aquifer or well performance tests (Fig. 1). Information on the local hydrogeology was used to interpolate between wells. Maps of relative permeability of geologic materials were also used to guide interpolation where field measurements were absent. The results of this analysis were maps showing the zonation of hydraulic conductivity in the unconfined aquifer (Fig. 2a), the zonation of vertical hydraulic conductivity in the confining layer (Fig. 2b), and the zonation of transmissivity in the confined aquifer (Fig. 2c).

Areal recharge to the unconfined aquifer was also zoned by calculating direct recharge as precipitation minus evapotranspiration. Recharge rates ranged from 3 in/yr in the lowland areas to 11 in/yr near the mountains (Fig. 3). Leakage from streams and seepage to wetlands were simulated using head-dependent conditions (Box 4.1). Boundary conditions and locations of head-dependent nodes are shown in Fig. 4.8b (Chapter 4).

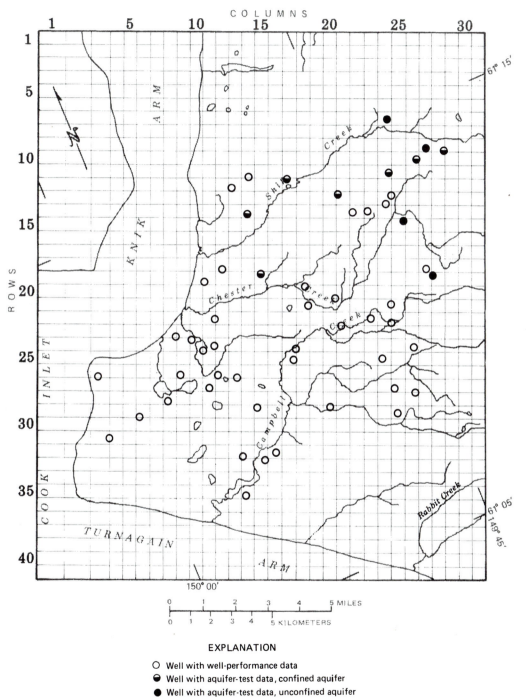

Fig. 1 Locations of wells used to estimate aquifer properties (Patrick et al., 1989).

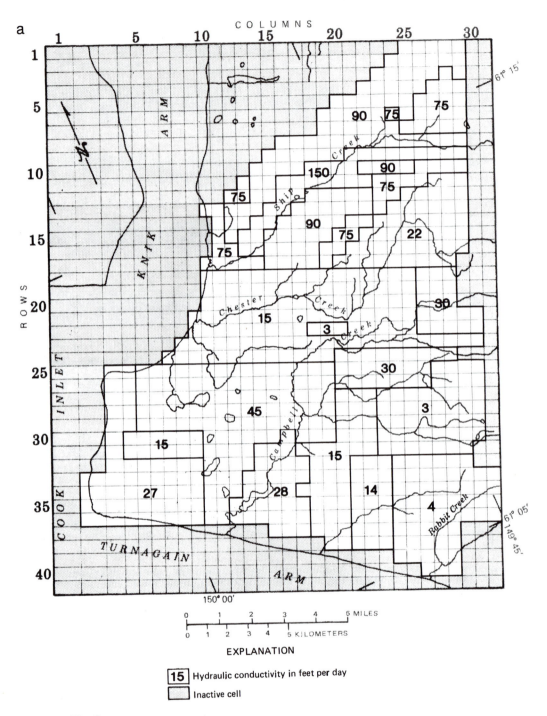

Fig. 2

(a) Zonation of hydraulic conductivity in layer 1, which represents the unconfined aquifer.
(b) Zonation of vertical hydraulic conductivity in the confining bed.
(c) Zonation of transmissivity in layer 2, which represents the confined aquifer. (Parts a–c from Patrick et al., 1989.)

b

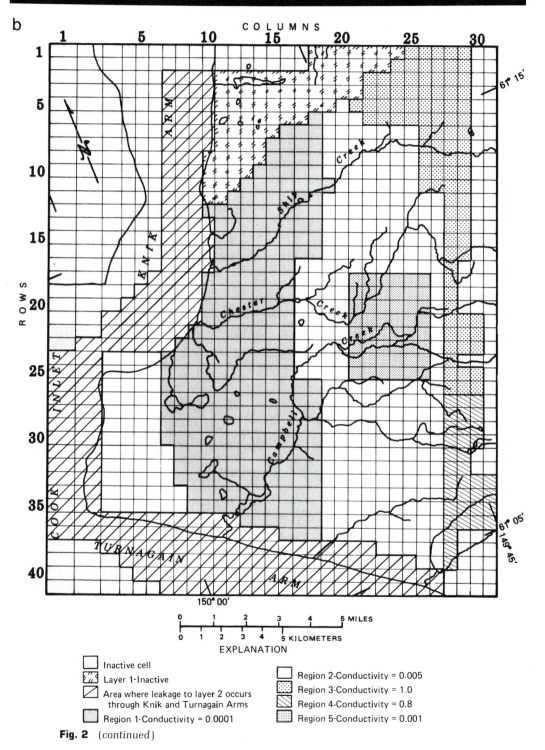

Fig. 2 (continued)

EXPLANATION

- Inactive cell
- Layer 1-Inactive
- Area where leakage to layer 2 occurs through Knik and Turnagain Arms
- Region 1-Conductivity = 0.0001
- Region 2-Conductivity = 0.005
- Region 3-Conductivity = 1.0
- Region 4-Conductivity = 0.8
- Region 5-Conductivity = 0.001

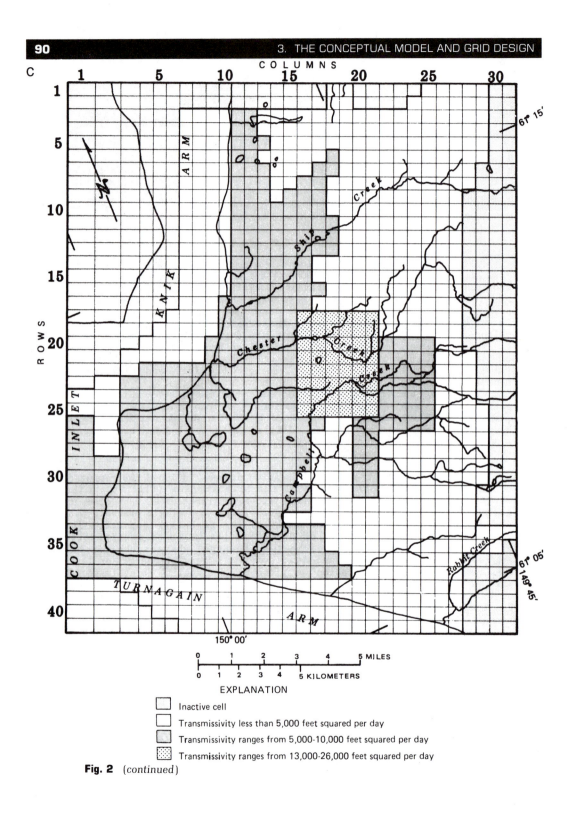

EXPLANATION

☐ Inactive cell

☐ Transmissivity less than 5,000 feet squared per day

▨ Transmissivity ranges from 5,000-10,000 feet squared per day

▨ Transmissivity ranges from 13,000-26,000 feet squared per day

Fig. 2 (continued)

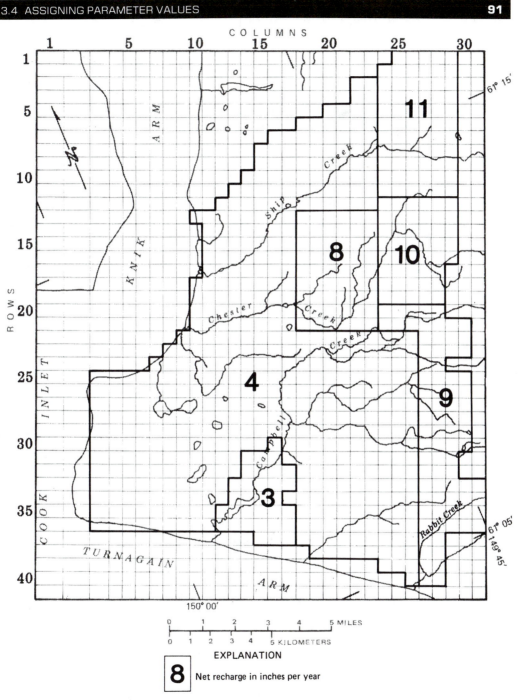

Fig. 3 Zonation of areal recharge to the unconfined aquifer (Patrick et al., 1989).

Problems

3.1 Consider the stratigraphy shown in Table P3.1. Using the descriptions of water-yielding characteristics, define hydrostratigraphic units for this setting. Draw a hydrostratigraphic column showing the new units and their thicknesses. Use Tables 3.3 and 3.4 to assign approximate hydraulic conductivity and storage values to each hydrostratigraphic unit.

3.2 Read the report by Franks (1988). Critique his conceptual model of the hydrogeologic system. Comment on the use of parsimony.

3.3 An assumption necessary to develop a two-dimensional or quasi three-dimensional model with leakage through a confining bed is that there is no transient release of water from storage within the confining bed (Section 3.2). The time to reach steady-state conditions in the confining bed can be estimated by calculating t_s (Section 3.2).

    (a) Using Fig. 3.5 as a conceptual model, select a hydraulic conductivity and specific storage for a 5-m-thick clay confining bed and calculate the time, t_s, at which steady-state conditions will occur in the confining bed after pumping is initiated in the underlying aquifer.

    (b) Graph t_s vs. b' for values of b' between 1 and 100 m for selected values of the ratio S_s'/K'. Under what set of circumstances is a steady-state model of flow through the confining bed warranted?

3.4 Winter (1976) used an anisotropy ratio of horizontal to vertical hydraulic conductivity (K_x/K_z) of 1000 in profile simulations of hypothetical lake systems set in glacial deposits. Consult the geologic literature to find information on glacial stratigraphy. Then use Eqn. 3.4 and Table 3.3 to demonstrate that a ratio of this magnitude is reasonable.

3.5 The river valley shown in Fig. P3.1 is filled with 30 m of fluvial sand and gravel underlain by 200 m of clayey alluvial-fan deposits, which overlies limestone bedrock. The climate is humid and the river gains water from groundwater discharge as it flows from Black Canyon across the XYZ fault. A well field designed to tap the unconfined alluvial sand and gravel is proposed east of the river.

    (a) Develop a conceptual model of the groundwater system including the boundaries of the flow system.

    (b) Make three copies of Fig. P3.1 and design a mesh-centered grid, a block-centered grid, and a triangular finite element grid for the region. Comment on the use of square or rectangular nodes vs. triangular elements.

3.6 In order to solve a groundwater model, n equations, one for each of the unknown heads, are solved. Thus, in general, a problem constructed with a grid having a large number of nodes requires more computer memory and computational time than a smaller grid.

Table P3.1

System	Series (group)	Geologic unit	Thickness (feet)	Physical characteristics	Hydrologic characteristics
Quaternary	Holocene	Alluvium	0–80	Stream-laid deposits ranging from clay and silt to sand and gravel that occur in the principal stream valleys.	Well yields range from 50–1,000 gallons per minute. Hydraulic and storage properties vary within short distances.
	Pleistocene	Dune sand	0–75	Fine to medium quartzose sand with small amounts of silt, clay, and coarse sand deposited into dunes by the wind.	Lies above the water table and does not yield water to wells. In areas where dune sand occurs, infiltration is rapid, and areas of appreciable ground-water recharge are delineated.
		Loess	0–45	Silt with minor amounts of very fine sand and clay deposited as windblown dust.	Lies above the water table and does not yield water to wells. In areas where loess occurs, specific retention is appreciable, and areas of minimal ground-water discharge are delineated.
Quaternary and Tertiary	Pleistocene-Miocene	Undifferentiated deposits	0–550	Composite of Quaternary-age sand, gravel, silt, clay, and caliche that overlie the Ogallala Formation, where present; composite of stream-laid and windblown deposits	The sand and gravel of the undifferentiated deposits and the Ogallala Formation are the principal water-bearing deposits in the area. Well yields range from 100–3,100 gallons per minute, and water in the aquifer is unconfined to semiconfined.
Tertiary	Miocene	Ogallala Formation	0–500	Poorly sorted clay, silt, sand, and gravel generally calcareous; when cemented by calcium carbonate, forms caliche or mortar beds.	
Cretaceous	Upper Cretaceous (Colorado Group)	Niobrara Chalk	0–250	Upper part, Smoky Hilk Chalk Member is yellow to orange-yellow chalk and light- to dark-gray chalky shale; basal part, Fort Hays Limestone Member is white to yellow, massive, chalky limestone that contains thin beds of dark-gray, chalky shale.	Areas of secondary porosity may yield water to wells. Where secondary porosity values are small, areas may yield small quantities of water to stock and domestic wells.
		Carlile Shale	0–330	Upper part consists of a dark-gray to blue-black, noncalcareous to slightly calcareous shale that is locally interbedded with calcareous, silty, very fine grained sandstone. Lower part consists of very calcareous, dark-gray shale and thin, gray interbedded limestone layers.	Sandstone in upper part may yield small quantities of water, 5–10 gallons per minute, to wells.

(continued)

Table P3.1 (continued)

System	Series (group)	Geologic unit	Thickness (feet)	Physical characteristics	Hydrologic characteristics
Cretaceous		Greenhorn Limestone	0–200	Chalky, light yellow-brown shale with thin-bedded limestone. Dark-gray, calcareous shale and light-gray, thin-bedded limestone; contains layers of bentonitic shale.	Not known to yield water to wells in southwest Kansas.
		Graneros Shale	0–130	Dark-gray, calcareous shale interbedded with black, calcareous shale; contains thin beds of bentonitic shale. Also contains thin-bedded, gray limestone and fine-grained silty sandstone.	Not known to yield water to wells.
	Lower Cretaceous	Dakota Sandstone	0–400	Brown, yellow, white, and gray fine- to medium-grained sandstones; interbedded with gray sandy shale, and varicolored clays; contains lignite, pyrite, and siderite. Generally has an upper shaley sandstone, middle sandy shale, and lower sandstone. Marine and fluvial-deltaic deposits	In some areas, the sandstones may yield more than 1,000 gallons per minute to wells. Water in the aquifer generally is confined where the Niobrara-Graneros confining unit is present. The principal aquifer in parts of Hodgeman and northern Ford Counties.
		Kiowa Shale	0–300	Dark-gray to black shale, interbedded with light-yellow-brown to gray, fine-grained sandstone.	Does not yield water to wells in southwest Kansas.
		Cheyenne Sandstone	0–200+	Gray, brown, and white, very fine to medium-grained sandstone with interbedded dark-gray shale.	Upper part (Cheyenne Sandstone) may yield up to 1,000 gallons per minute to wells in southeast Colorado; undeveloped in southwest Kansas. In some areas contains mineralized water. Lower part (undifferentiated Jurassic rocks) may yield water to wells in parts of southwest Kansas, but water is generally mineralized.
Jurassic	Upper and Middle Jurassic	Undifferentiated rocks	0–150+	Dark-gray shale interbedded with grayish-green and blue-green calcareous shale; contains very fine to medium-grained, silty sandstone and thin limestone beds at the base.	Water is generally highly mineralized.
	Upper Permian	Big Basin Formation	0–160	Brick-red to maroon siltstone and shale; contains very fine grained sandstone.	In Morton County, wells developed in solution cavities yield 300–1,000 gallons per minute of sulfate water.
		Day Creek Dolomite	0–80	White to pink anhydrite and gypsum with interbedded dark-red shale.	
	Lower Permian (Nippewalla Group)	Whitehorse Sandstone	160–350	Red to maroon, fine-grained, silty sandstone, siltstone, and shale.	Not known to yield water to wells in southwest Kansas; may contain highly mineralized water.

			Thickness (ft)	Character	Water supply
Permian		Dog Creek Shale	15–60	Maroon, silty shale, siltstone, very fine grained sandstone, and thin layers of dolomite and gypsum.	Not known to yield water to wells. Water is probably highly mineralized.
	Lower Permian (Nippewalla Group)	Blaine Formation	20–150	Generally consists of four gypsum and anhydrite beds separated by red shale. Bedded halite is present in some areas. Locally, a marker bed on geophysical logs.	Not known to yield water to wells; contains highly mineralized water.
		Flowerpot Shale	140–340	Gypsiferous shale and silty shale with thin beds of sandstone and siltstone; locally contains up to 250 feet of bedded gypsum and halite.	
		Cedar Hills Sandstone	77–180	Saliferous and gypsiferous, amber to pink, fine- to coarse-grained, shaley sandstone.	Generally contains highly mineralized water in southwest Kansas. In some areas it is used for brine disposal.
		Salt Plains Formation and Harper Sandstone	300–500	Upper unit (Salt Plains Formation)—reddish-brown siltstones, thin sandy siltstones, and very fine grained sandstone. Lower unit (Harper Sandstone)—brownish-red siltstones and silty shales with a few silty sandstones. In the subsurface, may contain bedded halite, anhydrite, and gypsum.	Not known to yield water to wells in southwest Kansas. May contain highly mineralized water. Oilfield-brine disposal zone. Contains highly mineralized water.
	Lower Permian (Sumner Group)	Stone Corral Formation	25–100	Dolomite, anhydrite, gypsum, and halite; gray to mottled with interbedded red shale. Distinctive marker bed on geophysical logs.	Not known to yield water to wells in southwest Kansas.

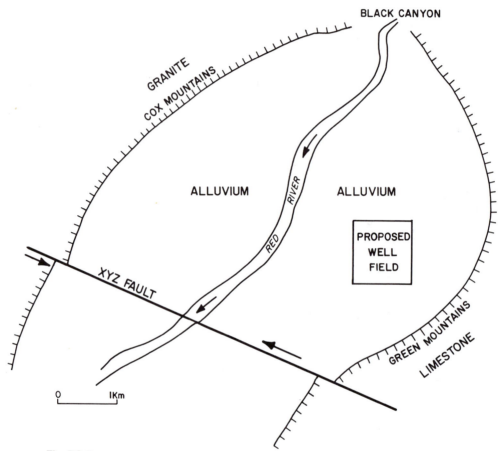

Fig. P3.1

(a) Franks (1988) used a variable grid to focus on the area of interest. How
    many nodes, including boundary nodes, did he use? How many active
    nodes did he use? (Note that there are three layers.)

(b) If he had used a uniform grid with 60 by 60m nodal spacing for the entire
    model, how many nodes, including boundary nodes, would be required?
    How many active nodes would be required?

BOUNDARIES

*"...if one advances confidently in the direction of his dreams ... he will meet
with success unexpected in his common hours. He will put something behind,
will pass an invisible boundary..."*
—*Henry David Thoreau*

4.1 Types of Boundaries

Mathematical models consist of a governing equation (Chapter 2), boundary
conditions (this chapter), and initial conditions (Chapter 7). Boundary conditions are mathematical statements specifying the dependent variable (head) or
the derivative of the dependent variable (flux) at the boundaries of the problem
domain. Some typical boundaries for two- and three-dimensional problem do-
mains are shown in Fig. 4.1.

Correct selection of boundary conditions is a critical step in model design.
In steady-state simulations, the boundaries largely determine the flow pattern.
Boundary conditions influence transient solutions when the effects of the transient stress reach the boundary. In this case, the boundaries must be selected so
that the simulated effect is realistic. According to Franke et al. (1987), setting
boundary conditions is the step in model design that is most subject to serious
error.

Physical boundaries of groundwater flow systems are formed by the physical presence of an impermeable body of rock or a large body of surface water.
Other boundaries form as a result of hydrologic conditions. These invisible
boundaries are *hydraulic boundaries* that include groundwater divides and
streamlines. For example, the groundwater divide and the streamline boundaries in Fig. 3.10 are hydraulic boundaries but the river is a physical boundary.
In Fig. 4.2, two regional flow systems are bounded by physical boundaries:

a

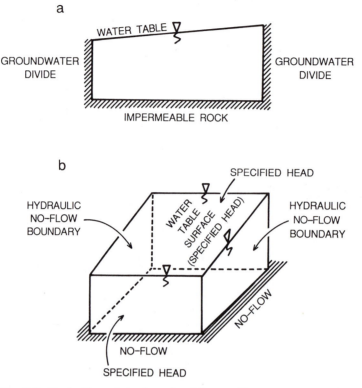

Fig. 4.1   Typical boundaries in regional flow problems.
(a) Two-dimensional problem in profile view.
(b) Three-dimensional problem domain.

impermeable rock at the bottom, the water table at the top, and rivers along part of the side boundaries (ABC in the inset). Groundwater divides form hydraulic boundaries whose locations are influenced by the presence of physical features—topographic lows that hold major rivers and a topographic high. On the left-hand side of the diagram, shallow local flow systems are separated from an intermediate flow system by *dividing streamlines* that form no-flow hydraulic boundaries. The intermediate flow system is separated from the regional flow system by another dividing streamline. All hydraulic boundaries, including those that coincide with physical features, are transitory features that may shift location or disappear altogether if hydrologic conditions change.

Hydrogeologic boundaries are represented by the following three types of mathematical conditions:

Type 1. *Specified head boundaries (Dirichlet conditions)* for which head is given.

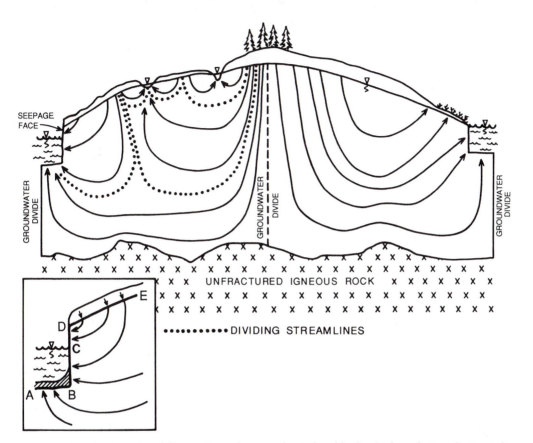

Fig. 4.2 Regional flow systems showing physical and hydraulic boundaries. The inset shows a close-up of several types of boundaries. The river boundary (ABC) could be treated as a specified head boundary, a specified flow boundary, or a head-dependent flow boundary. The seepage face forms a boundary along CD and the water table forms the boundary along DE. Bedrock forms a physical boundary to groundwater flow.

Type 2. *Specified flow boundaries (Neumann conditions)* for which the derivative of head (flux) across the boundary is given. A no-flow boundary condition is set by specifying flux to be zero.

Type 3. *Head-dependent flow boundaries (Cauchy or mixed boundary conditions)* for which flux across the boundary is calculated given a boundary head value. This type of boundary condition is sometimes called a mixed boundary condition because it relates boundary heads to boundary flows. There are several types of head-dependent flow boundaries.

The location of a boundary condition within the grid is dependent on whether a block-centered finite difference, mesh-centered finite difference, or finite element grid is used (Fig. 3.10).

4.2 Setting Boundaries

When selecting boundaries the modeler should visualize the probable flow pattern that will be induced by the boundaries. Does the flow pattern make sense? Where are the inflow and outflow locations? Does the simulated flow pattern agree with the general flow directions observed in the field? It is advisable to select physical boundaries whenever possible because they usually are stable features of the flow system. Impermeable rock typically forms the lower boundary of a modeled system (Figs. 4.1, 4.2, and 3.16). A two order of magnitude contrast in hydraulic conductivity may be sufficient to justify placement of an impermeable boundary. This type of contrast in hydraulic conductivity causes refraction of flow lines such that flow in the higher-conductivity layer is essentially horizontal and flow in the lower-conductivity layer is essentially vertical (Freeze and Witherspoon, 1967; Neuman and Witherspoon, 1969). If hydraulic gradients across the boundary are also low, flow out of the higher-conductivity layer will be negligible, and the boundary can be considered impermeable. If leakage across the boundary is significant, boundary fluxes or heads can be specified, if known. Otherwise, it will be necessary to simulate the lower-conductivity unit and continue down in the sequence until an impermeable lower boundary is found. Surface water bodies that fully penetrate the aquifer form ideal specified head boundaries (e.g., the Mediterranean Sea in Fig. 3.21). Termination of an aquifer at an impermeable rock unit forms a convenient physical no-flow boundary. Some fault zones and the saltwater interface in some coastal aquifers also form ideal no-flow boundaries.

It may not be possible or convenient to design a grid that includes the physical boundaries of the system if the focus of interest is far removed from the boundaries. In this case, it may be possible to identify a regional groundwater divide closer to the area of interest, which could be used as a no-flow boundary. Regional groundwater divides are typically found near topographic highs and may form beneath partially penetrating surface water bodies (Fig. 4.2). Zheng et al. (1988a,b) studied the permanence of groundwater divides beneath drainage ditches in central Wisconsin and concluded that the formation of fully penetrating divides beneath the ditches was inversely proportional to the regional slope of the water table and directly proportional to the head gradient across the streambed sediments. Although hydraulic boundaries in general are not stable features of the flow system, regional groundwater divides are likely to be more permanent than other types of hydraulic boundaries.

A flow system will usually have a mix of specified head and specified flow boundaries but occasionally the conceptual model of the problem may be formulated entirely with flux boundaries. For example, the groundwater flow system in Fig. 4.3a could be represented by a two-dimensional areal model with a no-flow boundary representing the groundwater divide along the perimeter of the basin, except for the segment labeled ED, which could be represented by a specified flux boundary to simulate underflow. Recharge to the

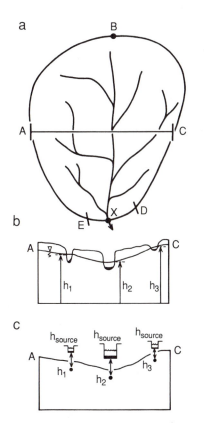

Fig. 4.3 Boundary conditions for a groundwater basin.

(a) A groundwater basin with no-flow boundaries along its perimeter, EABCD, and a specified flow boundary along ED that represents underflow from the basin. A constant-head boundary node might be specified at point X to represent the head in the river.

(b) Representation of the river system in a two-dimensional profile or full three-dimensional model by representing the river as a node within the grid. The head at the river node is specified to be equal to the stream stage.

(c) The use of leakage conditions to simulate the partially penetrating river system. The river is not represented within the grid but leakage is simulated as a head-dependent condition. River stages and the vertical hydraulic conductivity and thickness of riverbed sediments are assigned. The head in the aquifer below the stream is calculated by the model based on leakance of the riverbed sediments and the head difference between the stream and the aquifer.

model then could be entered via a source term (Chapter 5) and would represent the net flux of water into the aquifer. The numerical model for this conceptual model would have all flux boundary conditions. Although hydrogeologically defensible, exclusive use of flux boundaries generally should be avoided for the following mathematical reason. The governing equation is written in terms of derivatives, or differences in head, so that the solution will be nonunique if the boundary conditions also are specified as derivatives. Steady-state problems require at least one boundary node with a specified head in order to give the model a reference elevation from which to calculate heads. In transient solutions (Chapter 7), the initial conditions provide the reference elevation for the head solution so that the use of all flux boundaries may be justified for certain types of problems. Examples are given below under the discussion of distant boundaries.

In the problem shown in Fig. 4.3, a logical place to use a specified head node would be at point X, provided the vertically averaged groundwater head were known at that point. Alternatively, or additionally, head-dependent (leakage) conditions (Section 4.3 and Box 4.1; Section 5.3) could be used to simulate the partially penetrating rivers (Fig. 4.3c). This option requires the user to specify the river stage, h_{source}, and thereby provides reference elevations for the head solution.

If it is not possible to use physical boundaries and regional groundwater divides, it will be necessary to select other boundaries. These are typically hydraulic boundaries defined from information on the configuration of the flow system. Care must be taken when defining such boundaries to demonstrate both conceptually and numerically that the model boundaries will not cause the solution to differ significantly from the response that would occur in the field. When documenting model results (Chapter 9), the boundaries of the regional flow system should be identified as precisely as possible even if they are distant from the area of concern and are not included in the model.

In selecting boundaries that do not coincide with regional boundaries, two options are possible:

1. *Distant Boundaries.* In a transient simulation, boundaries may be arbitrarily located far from the center of the grid as long as the stresses to the system will not reach the boundaries during the simulation. That is, the assumption is that heads and flows in the vicinity of the boundaries will not change during the simulation. For example, the simulation may be terminated before the cone of depression caused by pumping at the center of the grid reaches the boundary. An analytical solution may be helpful in estimating the length of time required for the cone of depression to intercept the boundary. Another example of a transient simulation for which boundaries may not be affected by pumping stresses occurs when the well derives sufficient water from nearby sources of water such as a stream or reinjection of water into a well. When such distant boundaries are simulated, the grid is designed with small nodal spacing in the area of interest and very large nodal spacing near the boundaries (Fig. 3.19a).

2. *Hydraulic Boundaries.* Hydraulic boundaries that do not coincide with regional boundaries may be defined to create a smaller problem domain. The hydraulic boundaries may be specified head, no-flow, or specified flow boundaries. They are introduced for convenience to mimic the type

of flow desired in a portion of the larger problem domain. Boundaries defined in this way are sometimes called artificial boundaries.

Hydraulic boundaries may be defined from a water table map of the area to be modeled. In the water table contour map in Fig. 4.4a, regional flow system boundaries are physical boundaries: the two rivers, an impermeable fault, and an impermeable rock outcrop. To define a smaller problem domain, no-flow boundaries are located along flowlines at BC and AD and a specified head boundary is located along contour line AB. The new problem domain is ABCDA. The simulation must be structured so that pumping from the well shown in the figure will not affect heads or fluxes near the hydraulic boundaries.

When modeling two-dimensional seepage beneath engineering structures in profile view (Fig. 4.4b) the lateral boundaries of the model typically are hydraulic ones simulated as no-flow streamlines. Experience suggests that the distance to these lateral boundaries should be at least three times the depth of the flow system for an isotropic medium (Franke et al., 1987). In anisotropic media, the boundaries should be located a distance of $3(K_x/K_z)^{1/2}$ times the depth of the flow system.

Hydraulic boundaries can be set using a procedure termed *telescopic mesh refinement* or TMR by Ward et al. (1987). Using TMR, a grid with coarse nodal spacing is fitted to the regional boundaries, and boundary conditions for models covering successively smaller geographic areas are defined from the regional scale simulation (Fig. 4.5). Ward et al. (1987) illustrate the procedure with an application to aquifer remediation of a hazardous waste site (Box 4.3). In transient simulations, it is necessary to change conditions explicitly along the boundary with time (Miller and Voss, 1987) or run the entire sequence of models from regional to site scale for every time step. Buxton and Reilly (1986) report such an application for simulating the effects of a decrease in groundwater recharge owing to installation of sanitary sewers with ocean outfalls (Box 4.3).

Hydraulic boundaries can be used legitimately to produce a steady-state flow field for calibration purposes. However, they may or may not be acceptable for transient problems or for steady-state predictive simulations. This is because the model assumes that conditions on the boundaries do not change from their initial values unless they are explicitly changed by the modeler during the simulation. Under transient conditions in the field, heads along hydraulic boundaries might change owing to stresses imposed on the system. Because the model will not allow the boundary conditions to change in response to an applied stress, the solution in the interior portion of the grid will be in error. As an example, consider the pumping well shown in Fig. 4.4a. Pumping at this location in the field could cause drawdown at the location of the specified head boundary, AB. Pumping in the field could also change the configuration of flow, causing water to flow across BC and AD. Hence, transient effects could cause the conditions along the three hydraulic boundaries to be violated, in

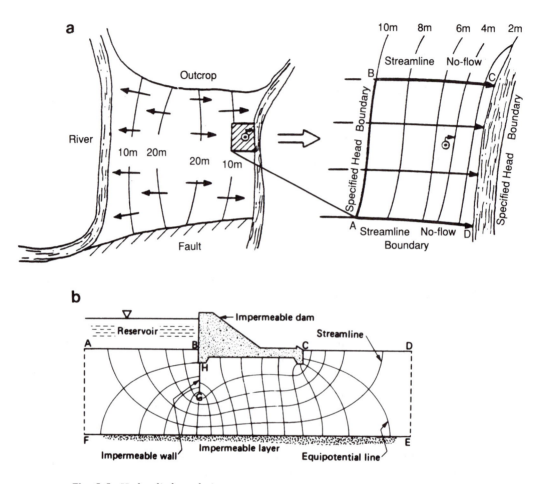

Fig. 4.4 Hydraulic boundaries.
(a) Water table contour maps showing a regional problem domain on the left with physical boundaries and a local problem domain on the right with three hydraulic boundaries (Townley and Wilson, 1980).
(b) AF and DE are no-flow streamlines used as hydraulic boundaries for a problem involving flow through a dam (Franke et al., 1987).

which case the modeling results would be unrealistic. The effects of the boundary conditions may be tested by changing specified head conditions to specified flux and vice versa (see Problem 4.6). If the effect on boundary heads and fluxes is insignificant, the boundaries are not influencing the solution. If the effects near the boundaries are significant, enlargement of the modeled area and selec-

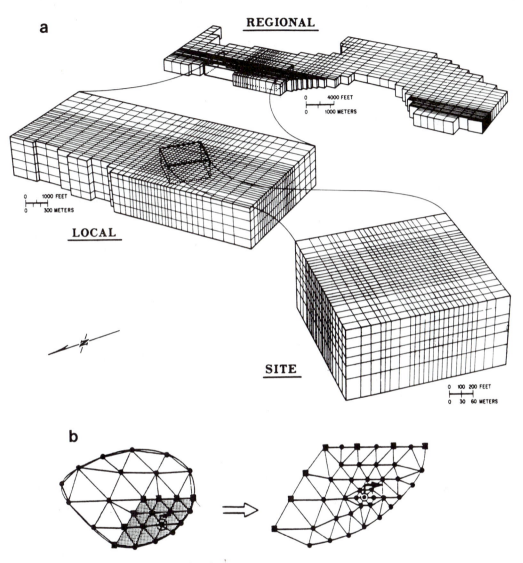

Fig. 4.5 Telescopic mesh refinement.
(a) Boundaries for a regional finite difference grid are defined from information about the regional flow system. The local and site grids have hydraulic boundaries defined from simulation results (Ward, Buss, Mercer and Hughes, Water Resources Research, 23(4), pp. 603–617, 1987, copyright by the American Geophysical Union).
(b) Finite element grids for regional and local scale models. The grids match along the nodes shown by squares. Boundary conditions along these nodes are determined from the solution of the regional scale problem (Townley and Wilson, 1980).

tion of new hydraulic boundaries may eliminate the error. If the changes do not eliminate the error, it may be necessary to model the regional flow system.

4.3 Simulating Boundaries

In finite element grids and mesh-centered finite difference grids, nodes always fall directly on the boundary (Fig. 3.10c–e). In block-centered finite difference grids, specified head boundaries are located directly at the node but flux boundaries are located at the outside edge of the block (Figs. 3.10b and 4.6). Values for heads and fluxes at boundaries must be determined from field data. It is usually easier to measure heads; direct measurement of fluxes is possible using seepage meters placed in rivers or lakes (Cherkauer and McBride, 1988; Lee, 1977) or by measurement of baseflow or springflow.

SPECIFIED HEAD

A specified head boundary is simulated by setting the head at the relevant boundary nodes equal to known head values. When the boundary is a river, the head along the boundary will vary spatially, whereas for lakes and reservoirs the boundary is described by constant head conditions (Fig. 4.7). In two-dimensional areal simulations, specified head boundary nodes represent fully penetrating surface water bodies or the vertically averaged head in the aquifer at hydraulic boundaries. In profile and full three-dimensional models, specified head nodes may represent the water table (Fig. 3.17a) or surface water bodies (Fig. 4.7).

In MODFLOW, specified head cells are identified by assigning values less than zero (usually -1) to entries in the IBOUND array. Active nodes have IBOUND values greater than zero (usually 1) and inactive nodes have values equal to zero. When an IBOUND value is less than zero, the code sets the unknown head for that cell equal to the specified head value, which is given in the starting head array (SHEAD). In PLASM, the user specifies a high value (usually 1E21) for the storage coefficient at specified head nodes. In this way a large volume of water is made available, preventing the assigned heads at specified head nodes from changing during the simulation.

It is important to recognize that a specified head boundary represents an inexhaustible supply of water. The groundwater system may pull water from the boundary or may discharge water into the boundary without changing the head at the specified head node. In some situations, this may be an unrealistic approximation of the response of the system. Of course, it is possible to change the head at the boundary as the simulation progresses provided a new value for the boundary head can be justified. MODFLOW's General Head Boundary Package provides a convenient way to change boundary head values during the simulation (Box 4.1). AQUIFEM-1 also allows the user to change boundary heads during the simulation.

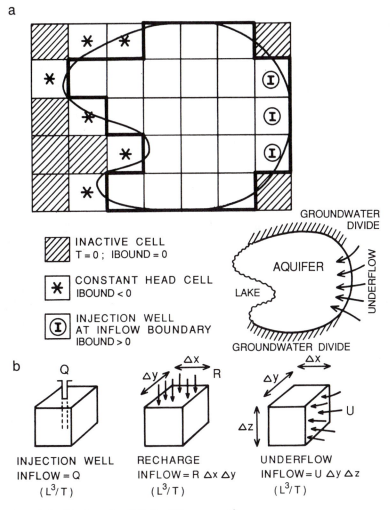

Fig. 4.6 Block-centered finite difference grid.
(a) Flux boundaries correspond to the edges of the boundary cells and constant-head boundaries pass through the nodes (adapted from McDonald and Harbaugh, 1988).
(b) Representation of fluxes. Volumes of water are placed into the block or extracted from the block using wells (Q), areal recharge, or leakage (R $\Delta x\,\Delta y$ or U $\Delta y\,\Delta z$).

SPECIFIED FLOW

Specified flow conditions are used to describe fluxes to surface water bodies, springflow, underflow (Fig. 4.8a), and seepage to or from bedrock underlying the modeled system (Fig. 4.8b). Specified flow conditions can also be used to simulate hydraulic boundaries defined from information on the regional flow

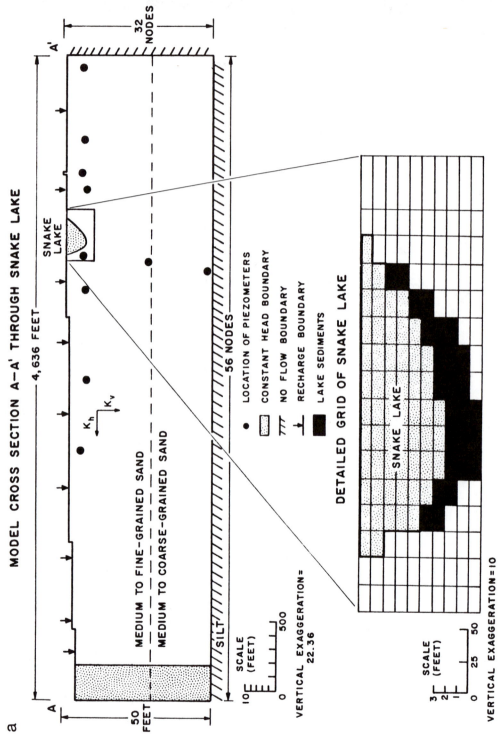

MODEL CROSS SECTION A–A' THROUGH SNAKE LAKE

4,636 FEET

32 NODES

A'

A

SNAKE LAKE

56 NODES

K_h

K_v

MEDIUM TO FINE-GRAINED SAND

MEDIUM TO COARSE-GRAINED SAND

SILT

50 FEET

SCALE (FEET)

10   0   500

VERTICAL EXAGGERATION= 22.36

a

LOCATION OF PIEZOMETERS
CONSTANT HEAD BOUNDARY
NO FLOW BOUNDARY
RECHARGE BOUNDARY
LAKE SEDIMENTS

DETAILED GRID OF SNAKE LAKE

SNAKE LAKE

SCALE (FEET)

3   2   1
0   25   50

VERTICAL EXAGGERATION=10

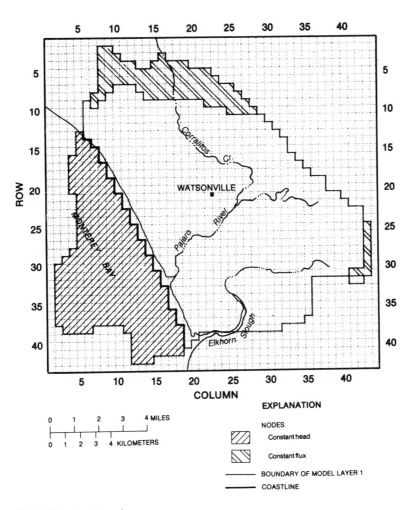

Fig. 4.7b (continued)

Fig. 4.7 Specified head boundaries.
(a) Specified head nodes used to represent a lake in a profile model of Snake Lake in northern Wisconsin (Anderson and Munter, Water Resources Research, 17(4), pp. 1139–1150, 1981, copyright by the American Geophysical Union).
(b) Specified head boundary nodes used to represent Monterey Bay in a three-layer model. The number of inactive cells outside the boundary varies depending on the configuration of the boundary between the aquifer and the bay. Layer 1 is the top layer; aquifers in layers 2 and 3 crop out in the walls of the Monterey submarine canyon (Johnson et al., 1988).

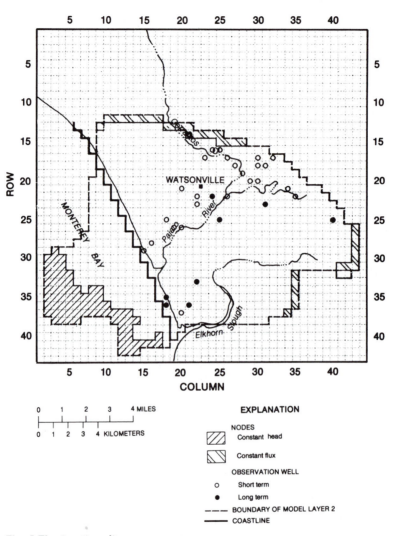

Fig. 4.7b   (continued)

system (Fig. 4.8a). Whenever possible, specified head conditions are selected over specified flow, however, because it is easier to measure head than to measure flow. Specified head conditions are also helpful in achieving calibration. In some situations, however, it may be advisable to use specified flow conditions. For example, flows into a system may be constant whereas heads along the boundary may be expected to change during the simulation.

Jorgensen (1989a) found that when simulating rivers in models of regional flow systems, the use of specified head conditions led to large errors in the calculated heads around the stream and consequently in flow rates between the stream and the aquifer. The errors were caused by the use of a coarse grid in the

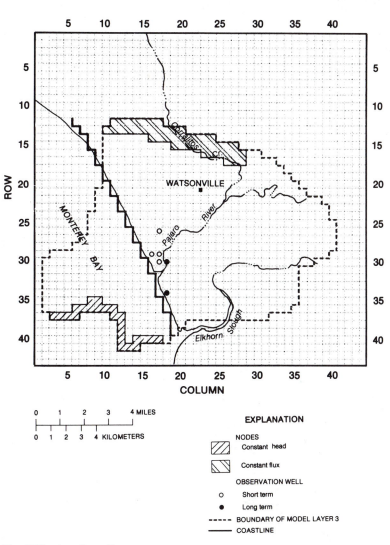

Fig. 4.7b *(continued)*

vicinity of the stream. When coarse nodal spacing was used around a specified head boundary representing a drain, Fipps et al. (1986) found large errors in heads and flows to the drain at a scale much smaller than a regional one. Both Jorgensen (1989a) and Fipps et al. (1986) had better success with coarse grids if a specified flow condition was used to represent the river or drain.

In finite difference models, specified flow boundaries are simulated by using injection or pumping wells to inject or extract water at the specified rate (Figs. 4.6 and 4.8). Inflows are treated as volumes of water "placed" into the cell. Conceptually, water may enter the top of the block as groundwater re-

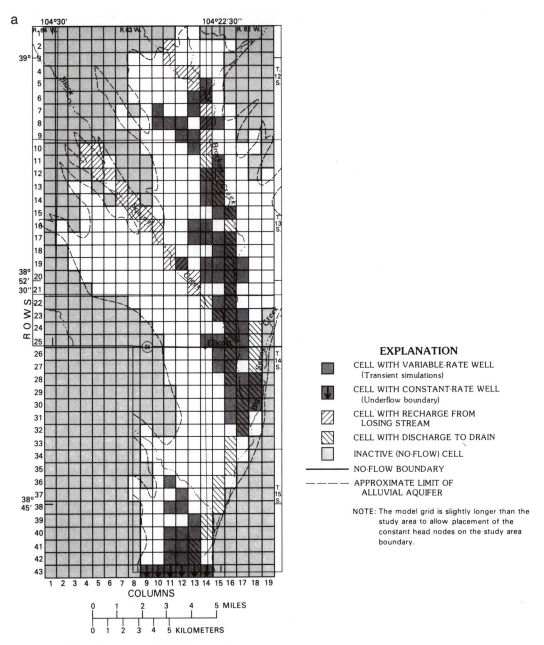

Fig. 4.8 Specified flux boundaries.

(a) Discharge wells are used to represent underflow from a groundwater basin (row 43) located in east central Colorado (Buckles and Watts, 1988).

(b) Injection wells in columns 28–30 are used to represent lateral seepage from adjacent bedrock into the unconfined upper layer of a two-layer model of the area around Anchorage, Alaska. Also note the use of drain cells to represent wetland areas adjacent to Cook Inlet and stream cells to represent leakage to partially penetrating streams. Constant-head nodes (row 1) and no-flow conditions are also used to define boundary conditions (Patrick et al., 1989).

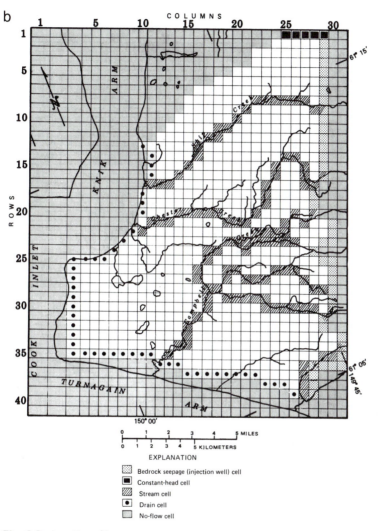

Fig. 4.8 (continued)

charge or the side of the block as underflow (Fig. 4.6b). The flux is assumed to be uniformly distributed over the face of the cell. In finite element models, the user assigns a flux to the portion of the boundary between two nodes (Fig. 4.9b). The code then reassigns fluxes to nodes as shown in Fig. 4.9c.

Both finite difference and finite element codes typically have a separate option that allows recharge to be input directly as a rate (L/T). For example, MODFLOW's recharge option is activated by using the Recharge Package (Section 5.2). PLASM and AQUIFEM-1 also allow fluxes across the water table to be input directly.

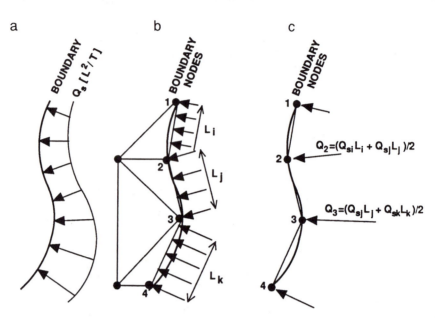

Fig. 4.9 Flux boundaries as represented in a finite element grid with triangular elements (Townley and Wilson, 1980).
(a) Continuous distribution of flow (Q_s) along a boundary.
(b) Discrete representation of flows along the sides of elements.
(c) Assignment of flows to nodes.

In a block-centered grid, flow boundaries coincide with the edge of the block as shown in Figs. 3.10b and 4.6a. PLASM requires the user to input values of transmissivity between nodal points, whereas MODFLOW requires values of transmissivity (or hydraulic conductivity) for the cell. By adjusting PLASM's input transmissivities and storage terms at boundary nodes, it is possible to move flux boundaries to the nodes (Box 4.2). MODFLOW cannot simulate mesh-centered boundaries without code modification.

No-Flow Boundaries

No-flow conditions occur when the flux across the boundary is zero. A no-flow boundary may represent impermeable bedrock (Fig. 4.10a), an impermeable fault zone (Fig. 4.10b), a groundwater divide (Fig. 4.2), or a streamline (Fig. 4.4a). A no-flow boundary also can be used to approximate the freshwater/saltwater interface in coastal aquifers (Fig. 4.10b). Groundwater in coastal aquifers discharges to the ocean through a zone of dispersion that forms the saltwater interface (Fig. 4.11). Rigorous treatment of the saltwater interface requires a model that allows for density effects as well as diffusion and dispersion of saltwater (Chapter 12). However, standard flow codes can simulate a saltwater

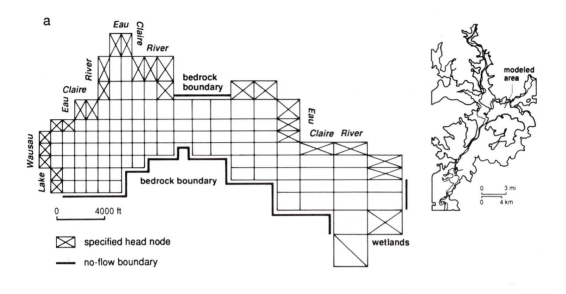

Fig. 4.10 No-flow boundaries used to represent impermeable conditions.
(a) A bedrock outcrop forms a no-flow boundary for a model of an aquifer in central Wisconsin. Also note the use of specified head boundaries for wetlands, Lake Wausau, and the Eau Claire River (Kendy and Bradbury, 1988).
(b) The Organ Mountain fault, a normal fault, forms a no-flow boundary for a New Mexico basin aquifer. A freshwater/saline-water interface forms another no-flow boundary along the other side of the basin (Risser, 1988).

interface approximately as a streamline boundary represented by no-flow conditions (Figs. 4.10b, 4.12 and 3.1a). Other examples of the use of a no-flow boundary for the saltwater interface are found in Hamilton and Larson (1988) and Guswa and Le Blanc (1985), who described a profile model and illustrated a technique for finding the location of the saltwater interface during the simulation.

In a block-centered finite difference grid, no-flow boundaries are simulated by assigning zeros to the transmissivities (or hydraulic conductivities) in the inactive cells just outside the boundary. In this way the boundary is set at the edge of the first active block (Figs. 4.6a, 4.8a). In finite element models, boundary fluxes are treated as illustrated in Fig. 4.9. At no-flow boundaries the flux is simply set equal to zero. Most finite difference and finite element models automatically assume no-flow boundaries around the edge of the model. The user must activate another type of boundary condition to cancel the no-flow boundaries. For example, the groundwater divides in Fig. 4.6a would be simulated automatically as no-flow boundaries without any special coding of input data.

b

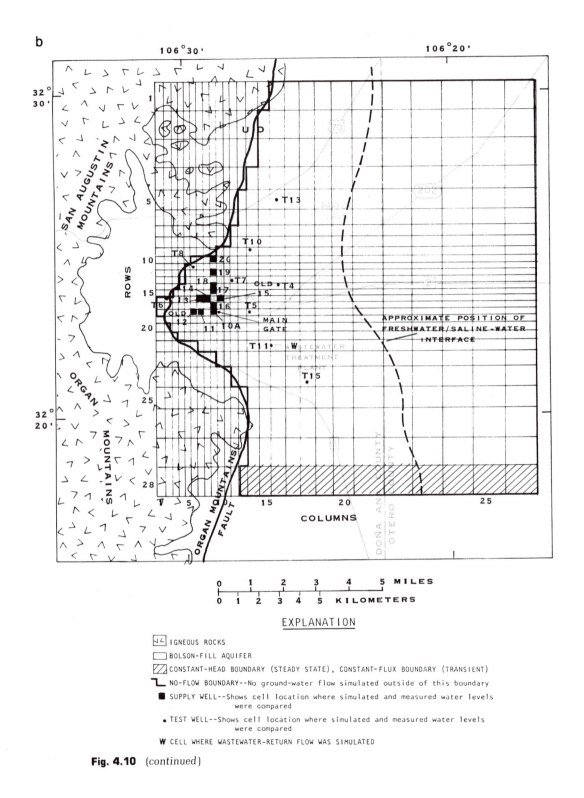

Fig. 4.10 (continued)

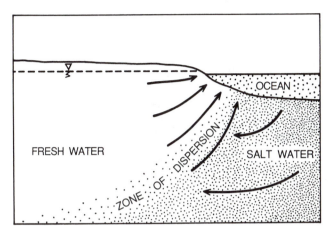

Fig. 4.11 The saltwater/freshwater interface. Upward flow through the zone of dispersion creates a no-flow boundary at the saltwater interface (Cooper et al., 1964).

HEAD-DEPENDENT FLOW

The flux across this type of boundary is dependent on the difference between a user-supplied specified head on one side of the boundary and the model-calculated head on the other side. The treatment of head-dependent flows in finite difference and finite element models is similar except that in finite element models fluxes occur to the node, whereas in finite difference models fluxes are calculated for the cell. Compare Figs. 4.6b and 4.9c.

Leakage to or from a river, lake, or reservoir (Figs. 4.2, 4.8, 4.13a) can be simulated using head-dependent conditions. The flux or leakage rate (L in Eqn. 2.1) is calculated as follows:

$$L = Q_L/A = K_z'/b'(h_{source} - h) \tag{4.1}$$

where Q_L is the volumetric flux and A is the area of the cell through which leakage occurs; h_{source} is the head in the source reservoir (e.g., a lake or river) and h is the head in the aquifer immediately below or adjacent to the source. K_z' is the vertical hydraulic conductivity of the interface (e.g., riverbed sediments) separating the aquifer from the source, b' is the thickness of the interface. The user specifies K_z', h_{source}, and b'. The current model-calculated value of h is used to calculate the leakage rate as the solution progresses. Evapotranspiration across the water table also can be represented as a head-dependent boundary, where the flux across the boundary is proportional to the depth of the water table below the land surface.

Another example of a head-dependent boundary is flow to a drain (Figs. 4.8a,b, 4.13b). Leakage to the drain is simulated whenever h in Eqn. 4.1 is

a

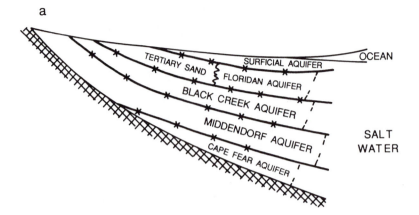

b

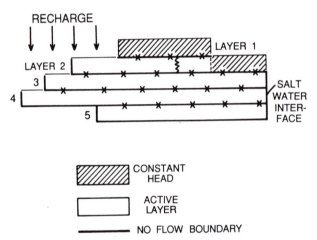

Fig. 4.12 Use of a no-flow boundary to represent the saltwater interface (Aucott, 1988).
(a) Dipping coastal plain sediments in South Carolina form aquifer systems with saltwater occupy-ing the eastern portions (right).
(b) A five-layer quasi three-dimensional model of the flow system uses a no-flow boundary set at the location of water with a chloride concentration of 10,000 mg/l.

Fig. 4.13 Head-dependent conditions.
(a) Head-dependent conditions used to allow groundwater exchange between Goose Lake, Drews and Cottonwood Reservoirs, and rivers in the interior of the Goose Lake Basin located in northern California and southern Oregon (Morgan, 1988).
(b) Drain nodes are used to represent groundwater flow into underground mine workings and rivers in southwestern Wisconsin (Toran and Bradbury, 1988). See Fig. 4.8b for an example of the use of drain nodes to represent head-dependent seepage to wetlands. See also Box 5.1.

a

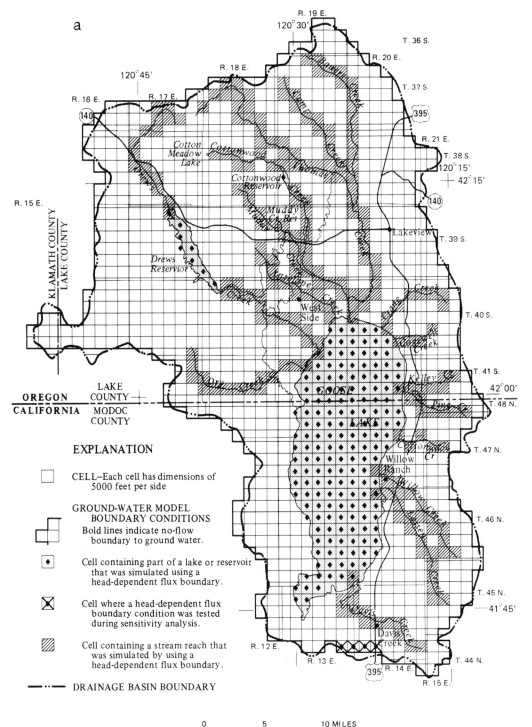

EXPLANATION

☐ CELL--Each cell has dimensions of
     5000 feet per side

☐ GROUND-WATER MODEL
     BOUNDARY CONDITIONS
     Bold lines indicate no-flow
     boundary to ground water.

◆ Cell containing part of a lake or reservoir
     that was simulated using a
     head-dependent flux boundary.

☒ Cell where a head-dependent flux
     boundary condition was tested
     during sensitivity analysis.

▨ Cell containing a stream reach that
     was simulated by using a
     head-dependent flux boundary.

—··— DRAINAGE BASIN BOUNDARY

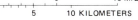

0   5   10 MILES
0   5   10 KILOMETERS

Fig. 4.13a

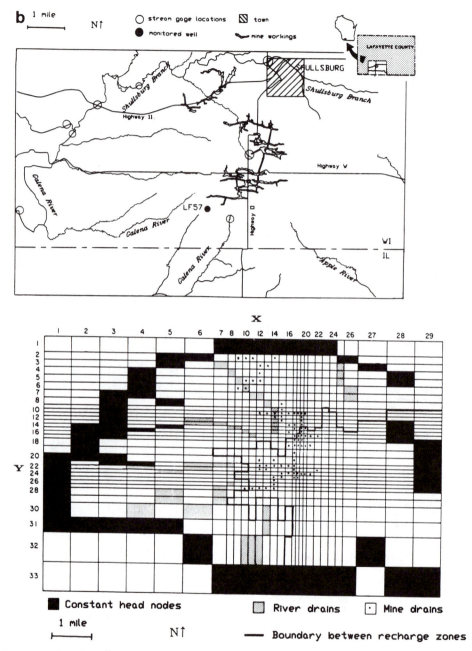

Fig. 4.13 (continued)

greater than h_{source} (the elevation of the drain). Leakage rate equals zero if h is less than h_{source}. The parameters K'_z and b' represent the properties of the interface between the drain and the aquifer. Specified head nodes may also be used to represent a drain if the head in the aquifer never falls below the drain.

Options for treating head-dependent conditions in MODFLOW, PLASM, and AQUIFEM-1 are presented in Box 4.1 and briefly summarized below. In

MODFLOW, leakage to and from rivers is handled with one of the river packages, RIV or STR1. Drains are simulated with the Drain Package; the Evapotranspiration Package is used for evapotranspiration across the water table. MODFLOW also has a General Head Boundary Package that allows the heads at the boundary to change in a transient simulation provided the user specifies a set of boundary head values at the beginning of the simulation. PLASM handles all head-dependent boundaries by input of leakance (K'_z/b'), source bed head (h_{source}), and a reference elevation. AQUIFEM-1 allows simulation of partially penetrating rivers, springs, and drains as head-dependent conditions and also allows for time-dependence of the associated boundary heads.

SPECIAL PROBLEMS

Two types of boundaries require special attention: (1) the water table and (2) seeps, springs, and seepage faces.

Water Table

Two-dimensional areal models that use the Dupuit assumptions calculate the head at the water table. In profile and full three-dimensional simulations, however, the water table is part of the boundary (Figs. 3.16a, 4.1, and 4.7a). Head (h) is equal to the sum of pressure head and elevation head (z). At the water table, pressure is equal to atmospheric pressure and pressure head is set to zero. Consequently, a mathematical statement of conditions along the water table is

$$h(x, y) = z(x, y) \qquad (4.2)$$

Equation 4.2 is not very helpful as a boundary condition because the location of the water table (z) is usually unknown. The water table is the most difficult of all boundaries to specify, because often it is the very feature we would like the model to calculate. This is particularly a problem in transient simulations that aim to predict the effects of pumping or changes in recharge on the location of water table.

There are two ways to avoid the problem of a water table boundary:

1. Model the problem using an unsaturated/saturated model. The unsaturated and saturated zones are simulated as a continuous flow field so that the water table is calculated as the surface of zero pressure head. The upper boundary condition is the flux across the land surface, or the infiltration rate, which is presumed known. If evapotranspiration is neglected, the infiltration rate is equal to the groundwater recharge rate, which is difficult to estimate (Section 5.2). Use of a model that includes the unsaturated zone also creates other complications discussed in Chapter 12, Section 12.3.

2. Use the Dupuit assumptions to model flow in the top layer of a three-dimensional or profile model. MODFLOW and AQUIFEM-N use this approach, making it is possible to specify only side and bottom bound-

ary conditions for the problem domain (Fig. 4.1b). The model calculates the location of the water table as part of the solution. Some error is introduced in MODFLOW simulations because of the necessity of specifying fixed values for VCONT at the beginning of the simulation (Box 3.1).

In a steady-state problem, it may be preferable to specify the heads at the water table, if known. The grid is designed so that the water table nodes are located at the elevation of the water table, i.e, $h = z$ (Box 6.1 and Fig. 6.5). The elevation of the water table may be estimated from a water table contour map, for example. Alternatively, the water table can be represented as a specified flux boundary by quantifying the spatial variation of recharge (or discharge) across the water table (Box 8.4). Specified flux boundary conditions on the water table in the system shown in Fig. 4.2, for example, would require that the recharge and discharge rate for each node along the water table be specified as model input. Unfortunately, it is difficult to identify groundwater recharge/discharge patterns and to measure rates. A review of techniques for estimating groundwater recharge concluded that "no single comprehensive estimation technique can yet be identified from the spectrum of methods available; all are reported to give suspect results" (Simmers, 1988). Methods for estimating recharge for use in modeling are discussed in Section 5.2.

Seeps, Springs, and Seepage Faces

A seep or spring forms when the water table intersects the land surface (e.g., line DC in the inset to Fig. 4.2). The saturated surface of contact formed when the water table coincides with the land surface is a seepage face. A seepage face also can form when the water table intersects a large-diameter well casing. The mathematical requirement along the seepage face is the same as for the water table, namely that $h(x, y) = z(x, y)$. The seepage face is neglected in most modeling applications because typically it is small compared to the scale of the problem. However, representation of the seepage face may be important in some regional groundwater problems (Potter and Gburek, 1987; Helgesen et al., 1982) and in some engineering applications (e.g., flow through a dam or flow to a well).

Springs or seeps are usually simulated by drain nodes. For example, MODFLOW's Drain Package can simulate flow from a spring by representing the point of emergence of the spring (the land surface) as the drain elevation. This option was used to represent seepage to wetlands around Cook Inlet in Fig. 4.8b. The drain no longer flows when the water table falls below the drain. Hence, drain nodes can be specified for the general area where seepage is likely to occur; the drain nodes will be activated only if the water table rises up to the level of the drain. In this way locations of seeps can be identified during the simulation as the drain nodes become activated. The head-dependent boundary option in PLASM and AQUIFEM-1 can be similarly used to represent seeps and springs. AQUIFEM-1 also has a "rising water table" option that allows the user to specify an upper limit on the amount of water table rise. Normally the upper limit is the elevation of the land surface. When the water level at a rising water table node reaches the specified limit, seepage occurs at the land surface. The head is fixed at that level for the duration of the time step and the node in effect becomes an internal specified head node. See Box 4.1 for more discussion of springflow.

The methods for dealing with seeps described above are not rigorous in that standard models like MODFLOW, PLASM, and AQUIFEM-1 use the Dupuit assumptions. Theoretical justification for incorporating a seepage face into a solution based on the Dupuit assumptions was presented by Potter and Gburek (1986). Finite element models like FREESURF (Neuman, 1976; Neuman and Witherspoon, 1970) and AQUIFEM-N (Townley, 1990) rigorously solve for the position of the seepage face. The elevation of the water table is calculated iteratively by allowing the top elements in the grid to deform as the water table and seepage face move (Fig. 4.14). The top nodes in the profile are fixed at the water table or seepage face by requiring that $h(x, y) = z(x, y)$ along the upper boundary of the system.

4.4 Internal Boundaries

All of the options for simulating boundary conditions also can be used in the interior of the problem domain to simulate sources or sinks of water or internal boundaries. For example, head-dependent conditions may be used to simulate rivers, lakes, and drains located within the interior of the grid (Fig. 4.13). The distinction between head-dependent *boundary conditions* and *internal* head-dependent conditions sometimes becomes blurred because the same model options are used to represent both. Internal head-dependent conditions and other kinds of sources and sinks are discussed in Chapter 5.

Surface water bodies in the interior of the grid also can be simulated using specified head conditions (Fig 4.15b). In two-dimensional areal simulations, the use of specified head nodes implies a fully penetrating surface water body.

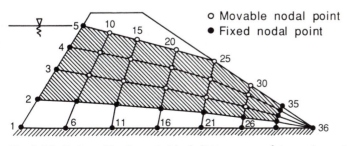

Fig. 4.14 Deformable elements (shaded) to accommodate moving water table nodes when simulating flow through a dam. Nodes 25 and 30 are seepage nodes (adapted from Neuman, 1976).

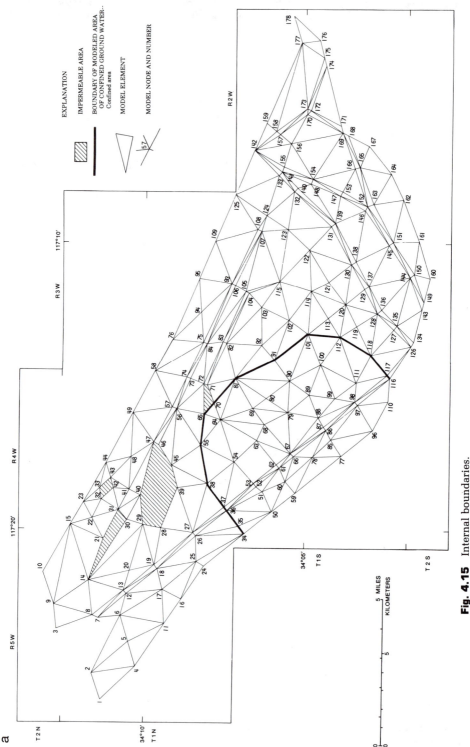

Fig. 4.15 Internal boundaries.

(a) Bedrock forming the Shandin Hills in the San Bernardino Valley, California, is represented by impermeable areas within a finite element grid (Hardt and Freckleton, 1987). See Fig. 3.22b for another example of bedrock hills that form impermeable internal boundaries.

(b) Specified head nodes are used to represent a fully penetrating river and a fully penetrating lake in the interior of a two-dimensional finite element grid (Townley and Wilson, 1980).

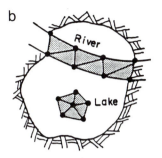

b

River

Lake

Fig. 4.15 (*continued*)

If a stream that crosses the problem domain is simulated as a fully penetrating internal boundary it should be recognized that it will in effect divide the problem domain into two independent regions that could be modeled separately (Fig. 4.15b). This may or may not be what the user intends. Impermeable rocks or faults within the problem domain form internal boundaries that can be simulated by setting transmissivity or hydraulic conductivity equal to zero at the appropriate nodes (Fig. 4.15a and Box 3.3).

Box 4.1
Head-Dependent Conditions

Head-dependent boundary conditions simulate flow across boundaries according to Eqn. 4.1. Strictly speaking, however, head-dependent fluxes are not always boundary conditions. For example, in the grid shown in Fig. 4.13a, the boundary is located along the perimeter of the basin. The stream and lake cells in the interior of the grid are more properly considered sink/source nodes. Hence, options for simulating head-dependent conditions can be used both for boundaries and for sources and sinks in the interior of the grid.

In this box, the options for head-dependent conditions used in MODFLOW, PLASM, and AQUIFEM-1 are discussed. The treatment in all three models is conceptually the same but the details of execution differ.

MODFLOW

MODFLOW has separate subroutines, or packages, to handle the following types of head-dependent conditions: rivers, drains, evapotranspiration, and boundary heads. Leakage through a confining bed may be simulated using the River Package or using VCONT arrays (Box 3.1).

River Package The River Package is used to simulate the flow of water between an aquifer and an overlying (or underlying) source reservoir which is usually a river or lake (Fig. 1). The code described by McDonald and Harbaugh (1988) contains the RIV package. It allows water to flow from the aquifer to the source reservoir, thereby removing water from the model by seepage to gaining stream reaches. Water can also flow out of the stream into the aquifer but the seepage out of the stream is independent of the stream discharge. Thus, a losing reach of stream could recharge the aquifer with more water than is being carried in the stream. No adjustment is made in stream stage. Even with these limitations, the RIV package

a

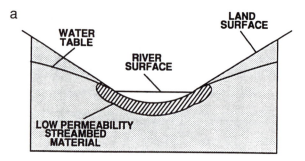

b

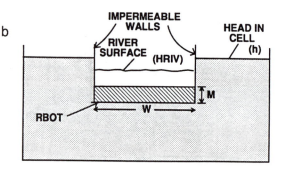

Fig. 1
(a) Stream-aquifer system.
(b) Representation of the stream-aquifer system in the RIV Package.

adequately represents most surface-ground-water systems.

The RIV package uses the streambed conductance (CRIV) to account for the length (L) and width (W) of the river channel in the cell, the thickness of the riverbed sediments (M), and their vertical hydraulic conductivity (K_r) where

$$CRIV = K_r LW/M \qquad (1)$$

The rate of leakage between the river and the aquifer (QRIV) is calculated from

$$QRIV = CRIV\ (HRIV - h) \qquad h > RBOT \qquad (2a)$$

where HRIV is the head in the source reservoir and h is the head in the aquifer directly below the source reservoir (Fig. 1b). When the water table falls below the bottom of the streambed (RBOT), leakage stabilizes and QRIV is calculated from

$$QRIV = CRIV\ (HRIV - RBOT)$$
$$h \le RBOT \qquad (2b)$$

When the RIV package is used to simulate a leaky confined aquifer, CRIV is the leakance parameter of the confining bed and HRIV is the head in the overlying (or underlying) source bed.

A new river package, the Streamflow-Routing Package (STR1), which allows leakage to and from the stream in the same manner as RIV, also considers the volume of streamflow in each river segment, and will increase streamflow in areas of gaining reaches and reduce streamflow by water lost through riverbed seepage in losing reaches. The reach will go dry if leakage or surface water diversions for a given reach exceed streamflow. In this case, leakage is set to zero and downstream reaches are prevented from leaking until additional water is added by tributaries or groundwater seepage. STR1 has an option to calculate stream stage in each river reach instead of using a preassigned value. The stage (d) is calculated from Manning's equation:

$$d = (Qn/CwS^{1/2})^{3/5} \qquad (3)$$

The user must input the discharge in each reach (Q), Manning's roughness coefficient (n), and the slope (S) and width (w) of the stream channel. C is a constant equal to 1.486 if Q is in ft^3/s and 1.0 if Q is in m^3/s. Equation 3 is valid for rectangular channels for which w is much greater than d. The use of the STR1 package includes a number of other assumptions and limitations outlined by Prudic (1989). Ozbilgin and Dickerman (1984) described a similar package that also calculates changes in water levels in ponds. A water budget for the pond considers rainfall and evaporation directly to the pond surface as well as in coming streamflow and outflow from the pond to streams. Based on the results of the water balance, changes in pond level are computed for each time step.

Drain Package The Drain Package is used to simulate both open and closed drains (Fig. 2). The Drain Package works in much the same way as the River Package, except that leakage from the drain to the aquifer is not allowed. That is, discharge to the drain is zero when heads in the cells adjacent to the drain are equal to or less than the assigned drain elevation (d). The River Package allows leakage to continue even when the head falls below the streambed sediments (Eqn. 2b).

When the head (h) in the aquifer is above the drain, water is removed from the model at a rate (QD) calculated from the assigned conductance of the drain (CD) and the difference between h and d:

$$QD = 0 \quad for\ h \le d \qquad (4a)$$

$$QD = CD(h - d) \quad for\ h > d \qquad (4b)$$

For an open drain (Fig. 2a), CD is analogous to CRIV in Eqn. 1 and is computed the same way. For a closed drain (Fig. 2b), CD is affected by the size and density of openings in the drain, the presence of chemical precipitates around the drain, and the hydraulic conductivity and thickness of backfill around the drain. In this case, CD is often estimated during calibration by requiring that simulated flows match measured flows to the drain.

Simulation of Spring Flow Springs and seeps are normally simulated with the Drain Package; the elevation of the spring or seep as it emerges at the land surface is the elevation of the drain. Diffuse flows, such as seepage to wetlands, can be simulated by specifying drain nodes in the general area where seepage is likely to occur. The drain nodes will be activated only when the head in the aquifer equals or exceeds the land surface elevation. Conductance terms are calculated from field-measured discharge and head or estimated during model calibration. The RIV package also can simulate a spring (or a drain) by setting RBOT equal to HRIV.

Evapotranspiration Package Evapotranspiration (ET) of groundwater may occur when the water table is close to the land surface or when phreatophytes draw water from below the water table. The Evapotranspiration Package requires the user to assign a maximum ET rate (R_{ETM}) to each cell from which ET may occur. The maximum rate is used when the water table in a cell equals an assigned head value, normally equal to the elevation of the land surface (h_s). No evapotranspiration occurs when the water table declines below an assigned "extinction" depth (d). In between these two extremes the ET rate is assumed to be linear. Dis-

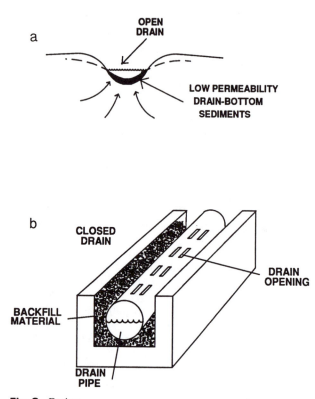

Fig. 2 Drains.
(a) Open drain.
(b) Closed drain.

charge by evapotranspiration (Q_{ET}) is calculated as follows:

$$Q_{ET} = Q_{ETM} \quad \text{for } h > h_s \qquad (5a)$$

where

$$Q_{ETM} = R_{ETM} \, \Delta x \, \Delta y$$

$$Q_{ET} = 0 \quad \text{for } h < (h_s - d) \qquad (5b)$$

$$Q_{ET} = Q_{ETM}[h - (h_s - d)]/d \qquad (5c)$$
$$\text{for } (h_s - d) \leq h \leq h_s$$

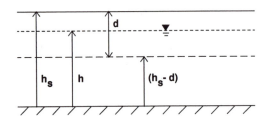

Fig. 3 Representation of evapotranspiration in MODFLOW.

where h is the elevation of the water table cal-culated by the model (Fig. 3). The extinction depth (d) is normally 6 to 8 feet below land surface but may be deeper if deep-rooted phreatophytes are present. Thomas et al. (1989) set d equal to 12 feet beneath a playa in Nevada and 30 feet in the area around the playa in which phreatophytes were growing. Danskin (1988) used 12 feet for the extinction depth in Owens Valley in Southern California.

General Head Boundary Package The General Head Boundary (GHB) Package is simi-lar in concept to the Drain and River Packages. Flow through the boundary (Q_b) is calculated as the product of the conductance of the bound-ary (C_b) and the difference between the head at or beyond the boundary (h_b) and in the aquifer (h):

$$Q_b = C_b(h_b - h) \qquad (6)$$

The conductance term (C_b) is analogous to streambed conductance (Eqn. 1), representing resistance to flow between the boundary and the aquifer (Fig. 4). When the GHB Package is used to simulate a specified head boundary, C_b is set to a large value (Box 8.1).

The GHB Package is more restrictive than the River and Drain Packages in that it assumes continuous linear discharge or leakage (Q_b),

whereas the River and Drain Packages allow for limits to discharge depending on the head in the aquifer relative to RBOT (Eqn. 2) or d (Eqn. 4).

PLASM

PLASM was originally written (Prickett and Lonnquist, 1971) so that each head-dependent condition required a slightly different version of the code. In the PC version of PLASM, all head-dependent conditions may be simulated using the following three variables: leakance of the source bed sediments (R), head in the source reservoir (RH), and a reference elevation (RD). Leakance is calculated from Eqn. 3.1. R, RH, and RD may not vary with time but the code could be modified to allow for time de-pendence. All head-dependent flows are as-sumed to follow a linear relation where the flux is calculated from

$$Q = R(RH - h)\,\Delta x\,\Delta y \quad \text{for } h > RD \quad (7a)$$

$$Q = R(RH - RD)\,\Delta x\,\Delta y \quad \text{for } h \le RD \quad (7b)$$

where h is the head in the aquifer as calculated by the model.

Leakage The head-dependent option in PLASM can be used to simulate leakage

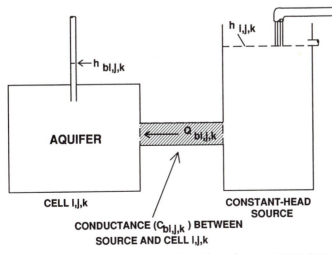

Fig. 4 Representation of a general head boundary in MODFLOW.

through a confining bed or leakage through river or lakebed sediments. When simulating leakage through a leaky confining bed, R is the leakance of the confining bed as defined in Eqn. 3.1, RH is the head in the overlying aquifer, and RD is the elevation of the bottom of the confining bed. When the head in the aquifer falls below RD, the aquifer converts to unconfined conditions and the model switches to the unconfined value of storage coefficient specified by the user. When simulating leakage to a river or lake, R is the leakance of the river or lakebed sediments, RH is the head in the river or lake, and RD is the elevation of the bottom of the river or lakebed sediments. Leakage reaches a maximum limiting value when the head in the aquifer is equal to RD. Leakage continues at this maximum rate when the head in the aquifer falls below RD (Eqn. 7b). Thus, the leakage option in PLASM produces the same effect as the RIV package in MODFLOW (Eqn. 2b). In MODFLOW, however, the user inputs CRIV as a leakance parameter, which depends on the length and width of the river reach. PLASM assumes that the length and width of the river reach are equal to the dimensions of the cell, Δx and Δy. If this is not the case, leakance (R in PLASM) can be adjusted as described in Section 5.3.

Drains Drains are simulated by setting RH equal to the elevation of the drain; RD is the elevation at which flow to the drain stops and is equal to the elevation of the drain (RH), so that RH = RD. R represents the conductance of the drain pipe and backfill and is similar to CD in MODFLOW's Drain Package. When the head in the aquifer is greater than the drain elevation, water flows into the drain and is removed from the model (Eqn. 7a). When the head in the aquifer is equal to or less than the drain elevation ($h <$ RD; RD = RH) no water flows into the drain (Eqn. 7b).

Springs Springs are simulated as drains where RH represents the seepage elevation, usually the ground surface; RD is set equal to RH. When the head around the spring node is at or above the ground surface, seepage occurs according to Eqn. 7a. If the head at the node is

lower than RH (= RD) no flow occurs (Eqn. 7b). The value of R controls the rate of flow and is estimated from field measurements of head and spring discharge or is estimated during model calibration. The value assigned to R should maintain the head at the land surface and reproduce the field measured spring discharge.

Evapotranspiration Evapotranspiration (ET) is assumed to cease when the water table drops below RD, here called the extinction elevation. The reference elevation, RH, is set equal to the land surface. Note that RH − RD equals d in Fig. 3. R is calculated by the user as follows:

$$R = Q_{MAX}/(RH - RD) \qquad (8a)$$

where Q_{max} is the maximum discharge from ET. Then, discharge by evapotranspiration (Q_{ET}) is calculated by the model from

$$Q_{ET} = Q_{MAX} - Q \qquad (8b)$$

where Q is determined from Eqn. 7a or 7b depending on the value of h relative to RD. Then $Q_{ET} = 0$ when $h <$ RD and $Q_{ET} = Q_{MAX}$ when $h =$ RH. Q_{MAX} is entered in an areal withdrawal array. The reader can verify that when RH $\geq h \geq$ RD, Eqns. 7a, 8a, and 8b can be combined to produce an equation identical in form to Eqn. 5c.

AQUIFEM-1

AQUIFEM-1 classifies boundaries as 1st type (specified head), 2nd type (specified flow), and 3rd type (head-dependent boundaries) and allows time-dependent boundary values to be specified in the input.

Leakage Leakage to or from aquifers overlying or underlying the aquifer being simulated is represented using the leaky aquifer option (Fig. 5). The user inputs the vertical hydraulic conductivity (K') and thickness of the leaky confining bed (B') and AQUIFEM-1 calculates a nodal leakage parameter (GKL) from

$$GKL = (K'/B') \, ARN \qquad (9a)$$

where ARN is calculated as the area associated with each node receiving or generating leakage (Fig. 6). Flux is then calculated from

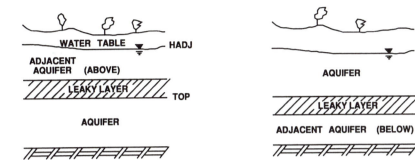

Fig. 5 Definition of the leakage option in AQUIFEM-1.

$$\text{FLXLKY} = \text{GKL (HADJ} - h) \qquad (9b)$$

where HADJ is the head in the source aquifer and h is the head in the aquifer being simulated. If the source aquifer is above the aquifer being simulated and the head in that aquifer falls below the bottom of the leaky confining bed (TOP), the leakage is calculated from

$$\text{FLXLKY} = \text{GKL (HADJ} - \text{TOP)} \qquad (9c)$$

The leaky aquifer option simulates flow to elements. This option also can be used to simulate leakage to and from lakes and wide rivers that are represented by clusters of elements (Fig. 7). When using this option, aquifer properties must be input by elements rather than by node (Box 3.3). For narrow rivers it is more ap-

propriate to simulate flow to a node rather than to the entire element.

Leakage to/from Narrow Rivers The user inputs the width (W'') and length (L) of the source reservoir, the thickness (B'') and vertical hydraulic conductivity (K'') of the source bed sediments, the elevation of the bottom of the source bed sediments (BEDLVL), and the head in the source reservoir (HBC3) (Fig. 8a). AQUIFEM-1 calculates a leakage parameter for adjacent segments along a reach where leakage occurs and then reassigns the flux to the common node. For example, consider the two segments shown in Fig. 8b. The leakage parameter (GKL3) for the two-segment reach is

$$\text{GKL3} = [(K''W''L/B'')_a + (K''W''L/B'')_b]/2 \quad (10a)$$

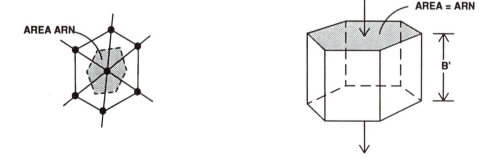

Fig. 6 Definition of leakage area (ARN) in AQUIFEM-1.

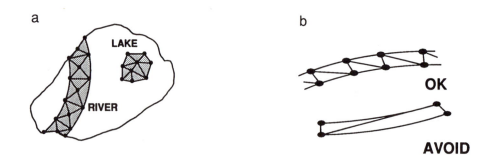

Fig. 7
(a) Leakage to/from lakes and wide rivers represented by clusters of elements.
(b) Avoid unfavorable aspect ratios when designing clusters.

The nodal flux is calculated from

$$\text{FLXLKY} = \text{GKL3(HBC3} - h) \qquad (10b)$$
$$\text{for } h > \text{BEDLVL}$$

$$\text{FLXLKY} = \text{GKL3(HBC3} - \text{BEDLVL}) \qquad (10c)$$
$$\text{for } h < \text{BEDLVL}$$

Drains, Springs, and Seeps Drains, springs, and seeps may be simulated using the option for simulating leakage to nodes as described above for narrow rivers except that HBC3 = BEDLVL. If seepage occurs over a large area, the leaky aquifer option should be used with HADJ = TOP.

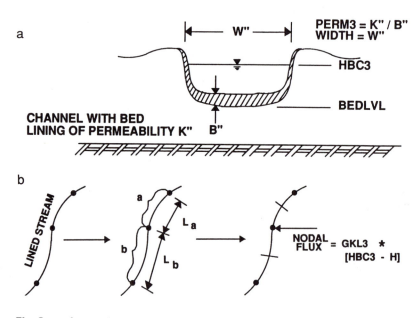

Fig. 8 Leakage to/from narrow rivers.
(a) Definition of terms.
(b) Definition of nodal leakage rate.

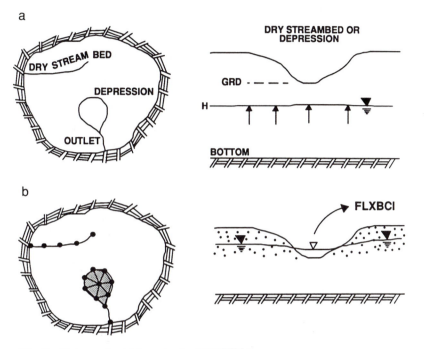

Fig. 9 Rising water table nodes in AQUIFEM-1.
(a) When $h <$ GRD, the rising water table option is not activated.
(b) When $h >$ GRD, discharge = FLXBC1.

Springs and seeps also may be represented by rising water table nodes (Fig. 9). The elevation of the ground surface (GRD) is specified. If the head in the aquifer rises above GRD, water seeps out of the model at rate equal to FCXBC1. The head at the node is fixed at GRD until the next time step when it may fall lower. Rising water table nodes also can be used to simulate flowing wells where GRD is equal to the elevation of the top of the casing.

Evapotranspiration Evapotranspiration is not in the standard version of AQUIFEM-1 but directions with Fortran code, are included in the manual by Townley and Wilson (1980). The evapotranspiration feature as executed in AQUIFEM-1 would be conceptually the same as in MODFLOW and PLASM.

Box 4.2
Mesh-Centered Boundaries

In finite element and mesh-centered finite difference grids, the boundary always falls on the node (Figure 3.10c–e). In block-centered finite difference grids, specified head boundaries fall on the node but flux boundaries fall at the edge of the cell (Figure 3.10b). All types of models calculate the head at the node.

For most regional problems, the exact placement of the boundary is unimportant. However, in some problems it may be crucial that the node fall directly on the boundary. For example, the objective of the modeling study may be to calculate the head at a groundwater divide. In a block-centered grid, the edge of the block would fall on the groundwater divide (Figure 3.10b) but the model would calculate the head at the node located a distance of one-half of the cell width away from the divide. If flux boundaries must coincide with boundary nodes, it is advisable to use a finite element code or a mesh-centered finite difference code. FLOW-PATH (Franz and Guiguer, 1990) and HST3D (Kipp, 1987) are special-purpose mesh-centered finite difference codes discussed in Chapter 12. In mesh-centered codes, flux boundaries are simulated by nodes placed outside the problem domain, variously called imaginary, fictitious, or ghost nodes. Details are given by Huyakorn and Pinder (1983) and Wang and Anderson (1982).

In block-centered finite difference grids, flux boundaries may be forced to coincide with the nodes if the transmissivity is input to represent the transmissivity *between the nodes*. Recall that the definition of areas of influence used in PLASM (Box 3.2) is different from the approach used in MODFLOW. PLASM requires the user to specify internodal transmissivities directly, whereas MODFLOW computes these values as the harmonic mean of the values specified for the cell (Eqn. 4, Box 3.1). PLASM's definition of transmissivities may cause an erroneous nonsymmetric distribution of heads for problems with radial symmetry (Potter and Gburek, 1987). However, the nonsymmetric definition of transmissivity input allows flux boundaries to be moved to the boundary nodes. This is done by adjusting nodal transmissivities, storage coefficients, and recharge terms at the boundary cells. Two examples are given below to illustrate the procedure. Because both are steady-state problems, it was not necessary to adjust the storage coefficient. Nevertheless, directions for adjusting the storage coefficient are given for completeness. Both problems were solved using PLASM. Five time steps were

used to reach steady state; the initial time step was 1000 days. The water balance error was set to 0.1%. Units of meters and days are used in Example 1. Example 2 uses inconsistent units of gallons-feet-days, which are the standard English units used in PLASM.

Example 1

The grid for a profile model (Chapter 6) of the unconfined aquifer considered in Box 2.1 is shown in Fig. 1. The profile is 1 meter thick so that $K_x = T_x$ and $K_y = T_y$. The aquifer has a uniform hydraulic conductivity of 10 m/day and a mix of specified head and no-flow boundary conditions. Adjustments to T_x, T_y, S, leakance (R), and discharge rate (Q) can be rationalized by studying the orientation of the boundary cells relative to the areas of influence shown in Fig. 1c. For example, consider the cells along the left-hand side boundary, with nodes designated by triangles in Fig. 1b. The area of influence for T_x for these cells is entirely within the problem domain, so no adjustment is necessary. The areas of influence for T_y and S, however, are half inside and half outside the boundary. Therefore, each value is divided by two. Note that it is necessary to make adjustments for the top corner nodes, which lie on both specified head and flux boundaries. Adjustments are necessary because these nodes are "connected" to other nodes in the grid by the area of influence for T_y, which is halved along the flux boundaries.

The head solution is shown in Fig. 2a. The solution for a block-centered grid is also shown for comparison (Fig. 2b). Note the disparity in the size of the problem domain. Head gradients in the block-centered solution are higher because the larger problem domain means that more water enters the system through the specified head nodes at the water table boundary.

Example 2

A two-dimensional areal model of a confined aquifer is shown in Fig. 3. A mix of constant-head and no-flow boundary conditions was used. The transmissivity was equal to 74,800 gal/day/ft (10,000 ft²/day) and a uniform re-

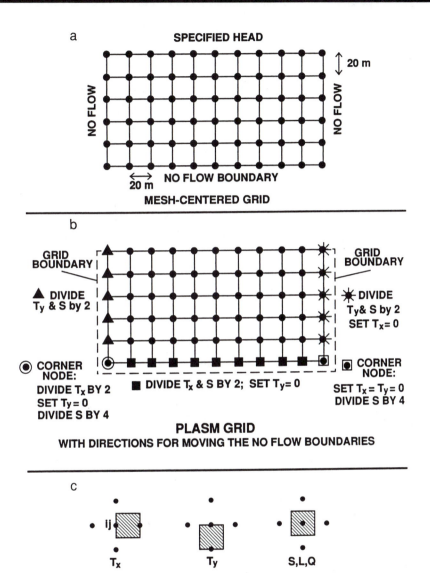

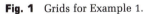

Fig. 1 Grids for Example 1.

(a) True mesh-centered grid with boundary conditions noted.

(b) Grid defined by PLASM. The three no-flow boundaries are in effect moved to coincide with the nodal points by adjusting values of transmissivity and storage coefficient as indicated.

(c) Areas of influence for node (i, j) as defined in PLASM for transmissivities (T_x, T_y), storage coefficient (S), leakage (L), and sink/source terms (Q).

a

HEAD (m)

100.00 101.00 102.00 103.00 104.00 105.00 106.00 107.00 108.00 109.00 110.00

101.60 101.96 102.60 103.35 104.17 105.00 105.83 106.65 107.40 108.04 108.40

102.47 102.65 103.08 103.65 104.31 105.00 105.69 106.35 106.92 107.35 107.53

102.98 103.10 103.41 103.87 104.41 105.00 105.59 106.13 106.59 106.90 107.02

103.25 103.35 103.61 104.00 104.48 105.00 105.52 106.00 106.39 106.65 106.75

103.34 103.43 103.67 104.04 104.50 105.00 105.50 105.96 106.33 106.57 106.66

FLUX BOUNDARIES MOVED TO COINCIDE WITH NODES

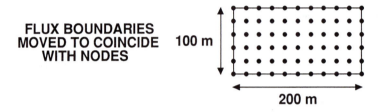

b

HEAD (m)

100.00 101.00 102.00 103.00 104.00 105.00 106.00 107.00 108.00 109.00 110.00

101.34 101.84 102.54 103.32 104.15 105.00 105.85 106.68 107.46 108.16 108.66

102.18 102.49 102.98 103.61 104.29 105.00 105.71 106.39 107.01 107.51 107.81

102.72 102.94 103.32 103.82 104.39 105.00 105.61 106.18 106.68 107.06 107.28

103.05 103.22 103.54 103.96 104.46 105.00 105.54 106.04 106.46 106.78 106.95

103.20 103.35 103.64 104.04 104.50 105.00 105.50 105.96 106.36 106.65 106.80

BLOCK-CENTERED GRID

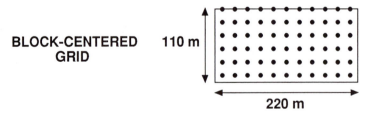

Fig. 2 Results for Example 1.
(a) Flux boundaries moved to coincide with nodes.
(b) Block-centered grid.

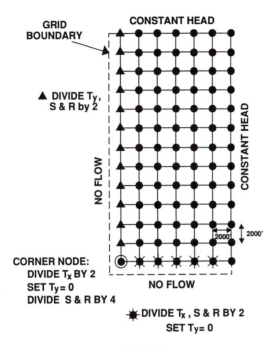

PLASM GRID

Fig. 3 Grid defined by PLASM for Example 2 with boundary conditions noted. The two no-flow boundaries are in effect moved to coincide with the nodes by adjusting values of transmissivity, storage coefficient, and recharge rate as indicated.

a

HEAD (ft)

0.00	0.00	0.00	0.00	0.00	0.00	0.00
4.87	4.77	4.45	3.91	3.07	1.85	0.00
8.71	8.51	7.91	6.87	5.30	3.09	0.00
11.71	11.43	10.58	9.11	6.95	3.98	0.00
14.04	13.69	12.63	10.82	8.19	4.63	0.00
15.83	15.42	14.20	12.11	9.12	5.11	0.00
17.19	16.75	15.39	13.10	9.82	5.48	0.00
18.23	17.74	16.29	13.83	10.34	5.75	0.00
18.99	18.48	16.95	14.38	10.72	5.95	0.00
19.54	19.01	17.43	14.77	11.00	6.09	0.00
19.90	19.37	17.74	15.03	11.18	6.19	0.00
20.12	19.57	17.93	15.17	11.29	6.24	0.00
20.18	19.64	17.99	15.22	11.32	6.26	0.00

**FLUX BOUNDARIES MOVED
TO COINCIDE WITH NODES**

24,000 ft

12,000 ft

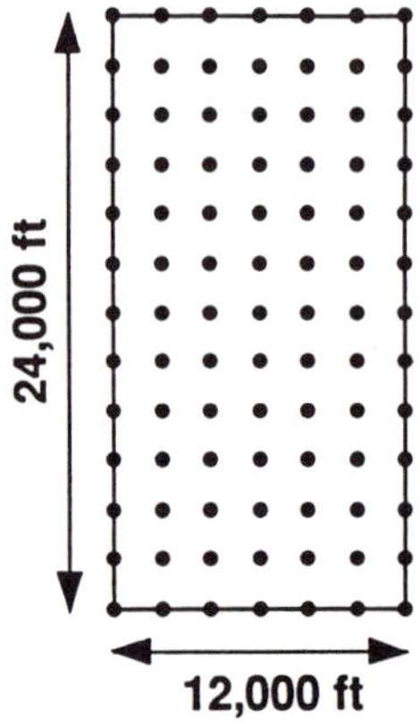

b

HEAD (ft)

0.00	0.00	0.00	0.00	0.00	0.00	0.00
5.29	5.10	4.71	4.10	3.20	1.91	0.00
9.54	9.18	8.42	7.24	5.55	3.21	0.00
12.93	12.40	11.33	9.66	7.31	4.15	0.00
15.60	14.94	13.60	11.53	8.66	4.86	0.00
17.70	16.93	15.37	12.98	9.68	5.39	0.00
19.34	18.48	16.75	14.10	10.47	5.80	0.00
20.60	19.67	17.80	14.95	11.07	6.11	0.00
21.56	20.58	18.60	15.60	11.53	6.35	0.00
22.27	21.25	19.19	16.07	11.86	6.52	0.00
22.77	21.72	19.60	16.41	12.10	6.64	0.00
23.09	22.02	19.87	16.62	12.25	6.72	0.00
23.24	22.16	19.99	16.72	12.32	6.76	0.00

BLOCK-CENTERED GRID

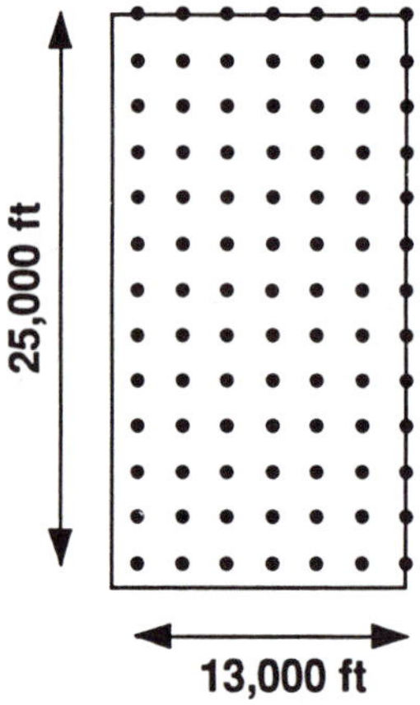

25,000 ft

13,000 ft

Fig. 4 Results for Example 2.
(a) Flux boundaries moved to coincide with nodes.
(b) Block-centered grid.

charge rate of −0.0230384 gal/day/ft² (−13.5 inches/year) was used. Note that in PLASM, recharge is negative by convention. In this example, as in Example 1, there are two nodes that lie along both a specified head boundary and a no-flow boundary. But in this case, only the upper left-hand node is connected to other nodes in the grid. Therefore, adjustments must be made for only this node. The head solution is shown in Fig. 4a. The head solution for a block-centered grid is shown in Fig. 4b for comparison. The larger problem domain produces higher heads, which are needed to transmit the higher discharge.

Box 4.3
Selecting Boundary Conditions: Case Studies

It is sometimes impossible to choose the physical boundaries of the aquifer when designing a model for a hydrogeologic site investigation. Then hydraulic boundaries are selected based on knowledge of the flow system. Typically, information is available on the general configuration of the regional flow system but site information is insufficient to deduce long-term flow patterns. One approach to this dilemma is to use what Ward et al. (1987) call *telescopic mesh refinement* (TMR). The approach consists of designing a nested set of grids (Fig. 4.5) so that the site grid, which has the finest nodal spacing, is embedded in a regional grid with coarser nodal spacing. The solution of the regional model is then used to define boundary conditions for the site model.

Ward et al. (1987) applied the technique to a hazardous waste site in the Miami River valley near Dayton, Ohio. For this application they used a nested series of three models designed at the regional, local, and site scales (Fig. 4.5a). A steady-state, two-dimensional areal model was used to simulate the aquifer at the regional scale (Fig. 1). Boundaries consisted of the bedrock wall of the valley to the west, specified head boundaries at the northern and southern ends, and a combination of specified head and no-flow boundaries along the eastern side of the model. The river was simulated by a series of leakage nodes. The aquifer was assumed to be homogeneous and isotropic with $K = 350$ ft/day. Recharge was assumed to occur uniformly at a rate of 6 inches/year.

The steady-state solution of the regional model was used to set boundary conditions for the steady-state, three-dimensional model at the local scale. This model had five layers and two zones of hydraulic conductivity (Fig. 2a). As before, the western boundary of the model coincided with the impermeable bedrock wall of the valley. The other side boundaries were specified heads taken from the solution of the regional model (Fig. 2b). An auxiliary computer program calculated boundary heads for the lo-

cal model by linear interpolation of regional heads in eight adjacent grid blocks. Because the regional model did not provide information on the vertical variation of head, the heads along the side boundaries were assumed to be constant with depth. The bottom boundary was set at bedrock and assumed to be impermeable. Recharge to the water table occurred at a rate of 6 inches/year.

The grid at the site scale was oriented differently from the regional and local scale grids (Fig. 2b). This was permissible because there is no horizontal anisotropy of hydraulic conductivity. Boundary conditions for the site model were determined from solution of the local model and consisted of specified head conditions along the bottom and side boundaries. The site model had six layers but simulated only the upper 80 feet of the aquifer, whereas the local model simulated the upper 180 feet of the aquifer. Hence, the lower boundary of the site model was designed to allow movement of water in and out of the model. Recharge was applied to the top layer at a rate of 6 inches/year. The hydraulic properties of the aquifer were the same as in the local model. The site model was used to simulate remediation by a pump and treat system with 15 pumping wells and 7 injection wells. Because the pumped water was reinjected, the effects of pumping did not extend to the boundaries of the model. The site model involved solving a solute transport model to predict the extent of capture of the contaminants.

In the application reported by Ward et al. (1987), the physical boundaries of the system, except for the bedrock wall of the valley, were not used even at the regional scale. Buxton and Reilly (1986) used TMR with a regional model constructed for a much larger region, thereby allowing the model to extend to the physical limits of the aquifer. They used a nested set of two transient three-dimensional models to predict the effects of the loss of groundwater recharge owing to the installation of sanitary

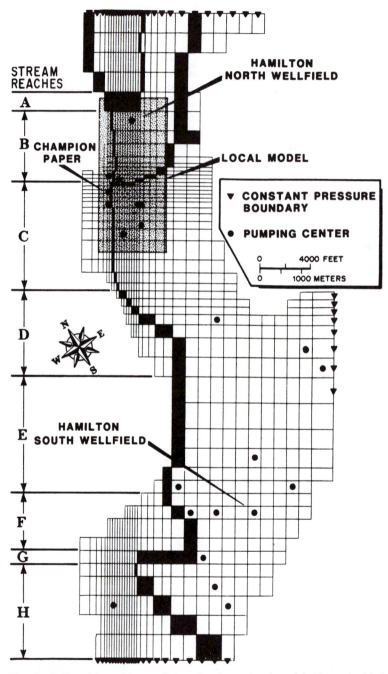

Fig. 1 Grid and boundary conditions for the regional model. The embedded grid for the local model is shaded (Ward et al., 1987).

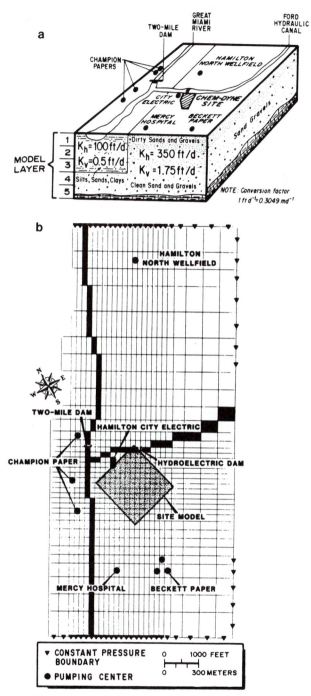

Fig. 2

(a) Schematic block diagram for the local model showing the five layers and two hydraulic conductivity zones (Ward et al., 1987).

(b) Grid and boundary conditions for the local model. The embedded grid for the site model is shaded (Ward et al., 1987). Note the different orientation of the site grid.

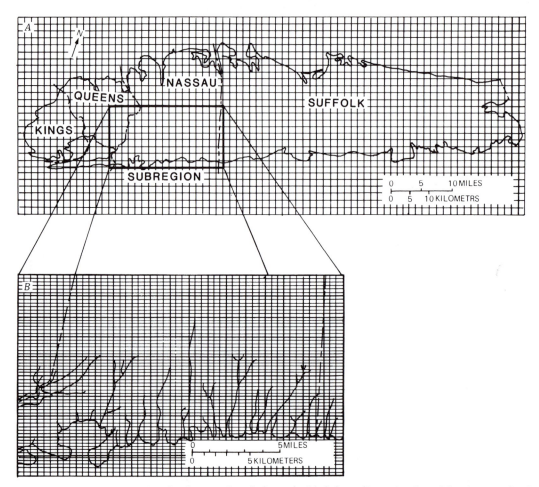

Fig. 3 Regional and subregional grids for embedded three-dimensional models of Long Island, New York (Buxton and Reilly, 1986).

sewers with ocean outfalls in a portion of Nassau County, Long Island (Fig. 3). The boundaries of the regional model were Long Island Sound and the Atlantic Ocean. The north, east, and west boundaries of the subregional model were determined from the solution of the regional model. The Atlantic Ocean formed the southern boundary. In contrast to the simulation of Ward et al. (1987), stresses to the aquifer during the predictive simulation extended to the boundaries, making it necessary to run both the regional and subregional models under transient conditions. The procedure of assigning boundary conditions for the subregional model was done for each time step because both the rate and direction of groundwater flow across the boundaries changed with time.

Problems

4.1 The town of Hubbertville is planning to expand its water supply by constructing a well in an unconfined aquifer consisting of sand and gravel (Fig. P4.1). The well was designed to pump constantly at a rate of 20,000 m³/day. Well construction was halted by the State Fish and Game Service, managers of

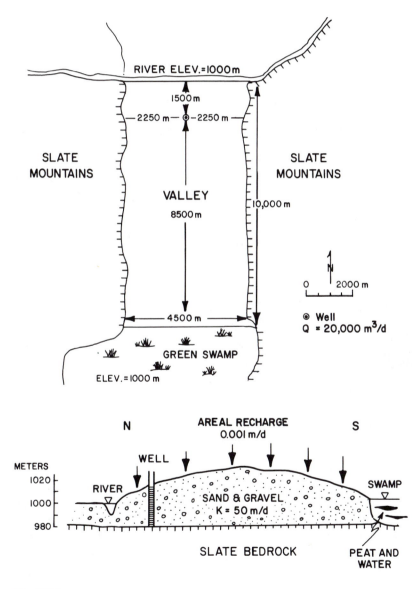

RIVER ELEV.=1000 m

1500 m

2250 m — ⊙ — 2250 m

SLATE MOUNTAINS

SLATE MOUNTAINS

VALLEY

8500 m

10,000 m

N

0   2000 m

⊙ **Well**
Q = 20,000 m³/d

4500 m

GREEN SWAMP

ELEV.=1000 m

N   AREAL RECHARGE   S
0.001 m/d

WELL

METERS
1020

RIVER   SWAMP

1000   SAND & GRAVEL
K = 50 m/d

980

SLATE BEDROCK   PEAT AND WATER

Fig. P4.1

the Green Swamp Conservation area. The state claimed the pumping would "significantly reduce" the groundwater discharge to the swamp and damage waterfowl habitat. The town claimed that the river and the groundwater divide located somewhere near the center of the valley would prevent any change in flow to the swamp.

(a) Construct a two-dimensional plan view steady-state model of the area between the river and swamp prior to pumping using the information provided in Fig. P4.1. Justify your assignment of boundary conditions and produce a contour map of head distribution and a N–S water table profile. Label the groundwater divide simulated between the river and the swamp. Save your input files.

(b) Using the steady-state heads derived in (a), locate the groundwater divide in the central portion of the valley and run the model using first a no-flow boundary and then a specified head boundary at this location. Compare results with those in part (a). Note the effect of this internal boundary on the results.

(c) Run the model under steady-state conditions to show the effects of the pumping well under each of the three possible boundary conditions:

   (i) physical boundaries shown in Fig. P4.1,
   (ii) no-flow boundary at the internal groundwater divide,
   (iii) specified head boundary at the internal groundwater divide.

Discuss the effect of fixing the location of the hydrologic divide on the resulting head distribution. Compare north–south water table profiles drawn through the node representing the pumping well for the simulations. Calculate and compare the total steady-state groundwater discharge to the swamp under the pre-pumping scenario described in (a) with the results under the pumping scenarios described in (c). In light of the modeling results consider what might be meant by "significantly reduced" as used by the state. Devise a more quantitative way to describe or judge the impact of the proposed well on the swamp system.

4.2 (a) Replace both the river and swamp boundaries of the model you designed in Problem 4.1a with specified flux boundaries. You can use the water balance results from Problem 4.1a to calculate the fluxes. Run this model with initial head guesses of 1000 m and then a second time with guesses of 2000 m. Compare the results with those in 4.1a and explain the differences.

(b) Take one of the models you designed in part (a) and replace one constant flux node on either the river or swamp boundary with a specified head node equal to 1000 m. Run the model to steady state. Compare the results with those in part (a) and with 4.1a. Explain the differences.

4.3 Replace the specified head nodes representing the river in Problem 4.1a with a head-dependent flux boundary represented as a river with a stage of

1000 m and a width of 500 m. Set the hydraulic conductivity of the river bottom sediments equal to 5 m/day. The thickness of the sediments is 1 m and the elevation of the bottom of the sediments is 995 m.

(a) Run the model and compare results with those of Problem 4.1a.

(b) What is the effect of using a river width of 5 m instead of 500 m?

4.4 Models of the setting in Fig. P4.1 typically assume the base of the aquifer is essentially impermeable. How could you incorporate a uniform upward rate of seepage from bedrock into the water table aquifer in a two-dimensional plan view model? How would you simulate this seepage at the node representing the pumping well?

4.5 The problems modeled in 4.1 and 4.2 used the Dupuit assumptions to represent an unconfined aquifer.

(a) Discuss how these simplifying assumptions effect the position of the water table.

(b) Read the paper by Potter and Gburek (1987). How was the degree of error in their model formulation determined?

4.6 It is often tempting to define a problem domain within an areally extensive aquifer by using a hydraulic no-flow boundary condition to represent a flow line. Assignment of no-flow boundaries in this way defines a flow tube. One way to determine if this boundary assignment will adversely affect the modeling results is to replace the zero-flux boundaries with specified heads and determine if flow occurs to or from these nodes. This method can be illustrated by using the modeling results from 4.1a. Let us assume that the Slate Mountains do not exist. Instead the steady-state model represents a 4500-m-wide flow tube within a larger areally extensive aquifer.

(a) Replace the no-flow boundaries that delineate a flow tube in 4.1a with specified heads taken from the steady-state flow field. Run this model to reproduce the steady-state heads of 4.1a. Calculate the east–west fluxes from the flow tube boundary and the discharge to the Green Swamp. Be sure your modeled area is 4500 m wide.

(b) Simulate the steady-state flow field with the pumping well using the boundaries from 4.6a. Examine the flow to and from the specified head boundaries that delineate the flow tube and compare them to the pre-pumping rates calculated in Problem 4.6a. What is the discharge to the Green Swamp? How would you redefine your model boundaries after this analysis?

SOURCES AND SINKS

"Among the many problems in heat-conduction analogous to those in
ground-water hydraulics are those concerning sources and sinks, sources being
analogous to recharging wells and sinks to ordinary discharging wells."
—C.V. Theis, 1935

"...speak to the rock and it will yield its water..."
—*The Old Testament, Book of Numbers, v.20, 7–8*

Water may enter or leave a model in one of two ways—through the boundaries,
as determined by the boundary conditions, or through sources and sinks within
the interior of the grid. In Chapter 4, we examined boundary conditions and in
this chapter we focus on interior sources and sinks. The sink/source terms
discussed in this chapter are represented by R^* in Eqn. 2.2.

The same model options are used to represent boundary sources and sinks
as to represent internal sources and sinks. For example, finite difference
models simulate specified flux boundaries by means of pumping or injection
wells placed along the boundary (Fig. 4.6). Yet, the reader should remember
that internal sources and sinks are not boundary conditions. It should be clear
that a sink in the interior of the grid is not a boundary condition, but confusion
sometime arises over other kinds of interior sources and sinks. Groundwater
recharge to the water table, for example, is a boundary condition in some
profile and full three-dimensional models but it is a source term in areal two-
dimensional models and in profile and three-dimensional models that use the
Dupuit assumptions to simulate flow in the upper layer of the model (Section
4.3). Specified head nodes are used to represent specified head boundary con-

ditions, but specified head nodes also may be placed within the problem domain to represent lakes, rivers, drains, or other types of internal sources and sinks of water (Fig. 4.15b).

5.1 Injection and Pumping Wells

An injection or pumping well is a point source or sink and is represented in a model by a node. The user typically specifies an injection or pumping rate in units of volume of water per time for each node so designated. In a full three-dimensional model, the node can come close to representing a point in space. In a quasi three-dimensional or two-dimensional areal model, however, the node represents the thickness of the aquifer. Hence, it is implicitly assumed that the well penetrates the full thickness of the aquifer. Simulating pumping or injection in a profile model is complicated because radial flow to the well is not possible in a conventional profile model. Radial flow to a well can be simulated, however, using an axisymmetric profile model. Details are given in Chapter 6.

FINITE DIFFERENCE MODELS

In finite difference models, the node represents the finite difference cell. Hence, a point source or sink of water is injected or extracted over the volume of aquifer represented by the cell that contains the point source or sink. The diameter of a well is typically much smaller than the dimensions of the cell. To represent the effects of a point sink more accurately, small cells around pumping nodes are preferred. But field problems generally require large grids and can seldom accommodate cells as small as the actual well diameter. The head gradient across most of the cells in the grid will be small so that the average head in the cell approximately equals the head at the node. The head gradient near a well node, however, is likely to be relatively large. A finite difference model does not simulate this gradient accurately because the model extracts or injects water to the entire cell rather than to the nodal point. The head calculated by the model is not a good approximation of the head in the well, but heads at nodes away from the point source or sink are correct.

The model-calculated head can be thought to represent the head at some distance (r_e) from the well node. An estimate of the head in the well can be obtained from formulas based on the steady-state Thiem equation, which can be applied to quasi steady-state conditions when the rate of removal of water from storage near the pumping well is zero. In transient simulations we can assume that after a short period of time, release of water from storage is negligible in the immediate vicinity of the well and the Thiem equation will apply there. First consider two-dimensional areal or quasi three-dimensional simulations for which wells fully penetrate the aquifer. The head in the well may be calculated from the following form of the Thiem equation:

$$h_w = h_{i,j} - \frac{Q_{WT}}{2\pi T} \ln\left(\frac{r_e}{r_w}\right) \tag{5.1}$$

where Q_{WT} is the total pumping or injection rate from the well; h_w is the head in the well; $h_{i,j}$ is the head computed by the finite difference model for the well node; r_e is the radial distance, measured from the node, at which the head is equal to $h_{i,j}$; and T is the transmissivity. The radius r_e is called the *effective well block radius*.

Equation 5.1 can be used to calculate the head in the well (h_w) during a transient simulation, given the model-calculated head ($h_{i,j}$) and an estimate of r_e. It can be demonstrated that $r_e = 0.208a$ when there is a regular grid in the vicinity of the well node, i.e., $\Delta x = \Delta y = a$ (Prickett, 1967; Trescott et al., 1976). To verify this relation refer to Fig. 5.1, which shows a portion of a two-dimensional areal grid in the vicinity of a well node. Assuming that drawdown is symmetric in the vicinity of the well node, the volumetric flow rate through each side of the cell is $Q_{WT}/4$. We apply Darcy's law to calculate flow through the right-hand side face of the cell, so that

$$\frac{Q_{WT}}{4} = aT \frac{h_{i+1,j} - h_{i,j}}{a} \tag{5.2}$$

Application of the Thiem equation between $r = \Delta x = a$ and $r = r_e$ yields

$$\frac{Q_{WT}}{4} = \frac{\pi T}{2} \frac{h_{i+1,j} - h_{i,j}}{\ln(a/r_e)} \tag{5.3}$$

By equating Eqns. 5.2 and 5.3 we obtain

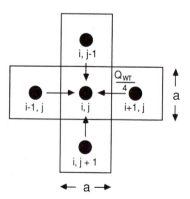

Fig. 5.1 Two-dimensional areal grid in the vicinity of a pumping or injection node (i, j). Pumping or injection rate is Q_{WT}.

$$\frac{a}{r_e} = e^{\pi/2} = 4.81 \qquad \text{or} \qquad r_e = 0.208a \tag{5.4}$$

Beljin (1987) presented an example that shows the importance of correcting model-calculated heads. Pritchett and Garg (1980) presented formulas that allow calculation of r_e for other grid geometries. For a grid that is expanded using a constant expansion factor of N, where $N = \Delta x_{i+1,j}/\Delta x_{i,j}$,

$$r_e = \frac{\Delta x_{ij}}{\sqrt{\pi}} C \tag{5.5}$$

Values of C are given in Table 5.1 for selected values of N. For an irregular grid with Δx and Δy equal to constants but $\Delta x \neq \Delta y$, and α defined to be equal to the maximum value of the aspect ratio ($\Delta x/\Delta y$ or $\Delta y/\Delta x$)

$$r_e = \sqrt{\frac{\Delta x\, \Delta y}{\pi}} E \tag{5.6}$$

Values of E for selected values of α are shown in Table 5.1.

By modifying the Thiem equation for unconfined conditions, a formula analogous to Eqn. 5.1 for the unconfined case can be written as follows:

$$h_w = \sqrt{h_{i,j}^2 - \frac{Q_{WT}}{\pi K} \ln\left(\frac{r_e}{r_w}\right)} \tag{5.7}$$

In two-dimensional areal simulations, effects of partial penetration of pumping or injection wells are usually ignored because the effects of partial penetration are limited to a radius equal to $1.5(K_x/K_z)^{1/2}$ times the saturated thickness of the aquifer. Furthermore, errors introduced by approximating the well by a finite difference block and other errors associated with discretization generally overwhelm errors caused by ignoring partial penetration.

In three-dimensional models it is possible to simulate the effects of partial penetration because pumping or injection nodes can be placed in any layer. The user specifies the pumping or injection rate and location of the well screen. In MODFLOW, data for pumping or injection wells are input via the Well Package. Finite difference models can simulate the effects of pumping from a well that penetrates more than one aquifer provided that the user supplies the pumping rate for *each* layer. That is, the pumping or injection rate for the multilayer well must be apportioned among the individual layers penetrated by the well such that the total pumping or injection rate for the well (Q_{WT}) is equal to the sum of the pumping rates from the individual layers ($\Sigma Q_{i,j,k}$). The pumping or injection rates for each layer ($Q_{i,j,k}$) can be approximately calculated from

$$Q_{i,j,k} = T_{i,j,k}(Q_{WT}/\Sigma T_{i,j,k}) \tag{5.8}$$

where $T_{i,j,k}$ is the transmissivity of a layer and $\Sigma T_{i,j,k}$ is the sum of the transmissivities of all layers penetrated by the well (McDonald and Harbaugh, 1988).

Table 5.1
Computation of Effective Well Radius (r_e) for Irregular Grids

Axisymmetric case, effect of grid nonuniformity		Cartesian case, effect of zone aspect ratio (α)		
N	C	α	Grid size (zones)	E
0.0	0.606531	1.1	21 × 23	1.002
0.1	0.606024	1.2	21 × 25	1.008
0.2	0.604518	1.3	21 × 27	1.017
0.3	0.602035	1.4	21 × 29	1.028
0.4	0.598614	1.5	21 × 31	1.041
0.5	0.594302	1.75	21 × 36	1.077
0.6	0.589156	2.0	15 × 29	1.118
0.7	0.583235	2.5	15 × 36	1.204
0.8	0.576603	3.0	15 × 43	1.290
0.9	0.569323	3.5	15 × 50	1.375
1.0	0.561457	4.0	15 × 57	1.456
1.1	0.553068	4.5	15 × 64	1.535
1.2	0.544214	5.0	15 × 71	1.610
1.3	0.534953	6.0	8 × 43	1.754
1.4	0.525338	7.0	8 × 50	1.887
1.5	0.515420	8.0	8 × 57	2.012
1.6	0.505246	9.0	8 × 64	2.129
1.7	0.494862	10.0	8 × 71	2.240
1.8	0.484308			
1.9	0.473623			
2.0	0.462843			
2.2	0.441125			
2.4	0.419385			
2.6	0.397814			
2.8	0.376572			
3.0	0.355786			
3.5	0.306432			
4.0	0.261541			
5.0	0.186447			
6.0	0.129980			
7.0	0.089093			
8.0	0.060262			
9.0	0.040327			
10.0	0.026751			

$$\alpha = \left(\frac{\Delta x}{\Delta y}, \frac{\Delta y}{\Delta x}\right)_{max} \qquad \Delta x = \text{a constant}$$
$$\Delta y = \text{a constant}$$

$$r_e = \sqrt{\frac{\Delta x\, \Delta y}{\pi}}\, E$$

$$\Delta x = \Delta y \qquad r_e = \frac{\Delta x_{ij}}{\sqrt{\pi}}\, C$$

$$N = \frac{\Delta x_{i+1,j}}{\Delta x_{i,j}}$$

(Pritchett and Garg. *Water Resources Research*, **10**(2), pp. 295–302, 1980, copyright American Geophysical Union)

It should be recognized that Eqn. 5.8 is an approximation because it does not allow for the fact that $Q_{i,j,k}$ is a function of the head ($h_{i,j,k}$), which is an unknown calculated during the solution. Conventional finite difference equations do not recognize that a well that penetrates more than one aquifer or stratigraphic layer forms a pathway for water movement between layers. Consequently, the head in a multiaquifer well is a composite average of the heads in all the layers it penetrates (Papadopulos, 1966). Bennett et al. (1982) presented a modified form of the finite difference governing equation to be used when multiaquifer well effects are important. Kontis and Mandle (1988) presented an equation to compute the effective well block radius for each layer in regional three-dimensional models that have more than one well per nodal block. They also gave an equation for computing the effective water level in the wells for each finite difference block where wells are present. An addition to MODFLOW's Well Package, to simulate the effects of multiaquifer wells, is in preparation (McDonald, 1984).

FINITE ELEMENT MODELS

In a finite element model, the pumping or injection rate is assigned to the node itself if the well is located at the node. If the source or sink is not located at a node, the flow is divided among the nodes of the element that contains the well. Details are given by Istok (1989). In either case, water is extracted from nodes rather than the area around the nodes. For this reason finite element models can simulate point sinks and sources more accurately than finite difference models, particularly when the well is located directly on the node. Hence, finite element models calculate the head in the pumping or injection well directly so that there is no need for the correction formulas used in finite difference modeling. Figure 5.2 shows a comparison of drawdowns computed by two finite difference codes with a solution from a finite element code. The solutions from all three models agree except at the pumping node.

5.2 Flux across the Water Table

Flux across the water table is treated as an internal source or sink if the model uses Dupuit assumptions to calculate the water table. Otherwise, flux across the water table is a boundary condition. In either case, the user must input an array of flux values.

ESTIMATING WATER TABLE FLUXES

Recharge refers to the volume of infiltrated water that crosses the water table and becomes part of the groundwater flow system. Discharge refers to groundwater that moves upward across the water table and discharges directly to the

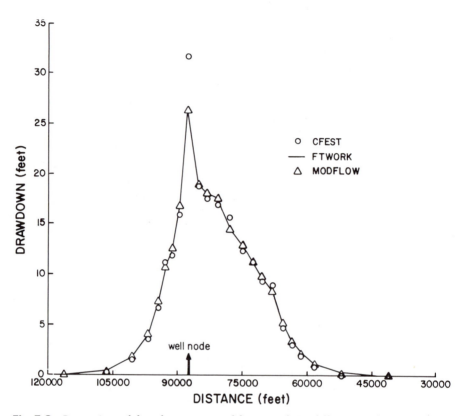

Fig. 5.2 Comparison of drawdown computed from two finite difference codes (MODFLOW and FTWORK) with a finite element solution from CFEST (Sims et al., 1989). The head calculated by the finite difference codes represents the average head for the cell. The head calculated by the finite element code is a better approximation of the head in the well itself.

surface or to the unsaturated zone. No one has yet devised a universally applicable method for estimating groundwater recharge. Numerous methods have been proposed (Simmers, 1988) but most have met with limited success. For lack of a way to quantify the spatial distribution of recharge and discharge, modelers have traditionally assumed a spatially uniform recharge rate across the water table equal to some percentage of average annual precipitation. The recharge rate is often adjusted during calibration.

    Recent field studies aimed at studying the recharge process, however, demonstrate that there are significant spatial and temporal variations in groundwater recharge rates (Stoertz and Bradbury, 1989; Stephens and Knowlton, 1986; Sophocleus and Perry, 1985; Steenhuis et al., 1985). Moreover, a typical groundwater basin contains areas where the net flux across the water table is upward (e.g., Fig. 4.2). In fact, discharge areas can be expected to occupy 5 to

30% of the surface area of a groundwater basin (Freeze and Cherry, 1979, p. 197). Recent modeling studies incorporate spatial variation in recharge by defining recharge zones (Fig. 5.3 and Box 3.4, Fig. 3). Typically, there is little hydrogeologic information to use in defining recharge zones and in assigning recharge rates to each zone. Instead recharge zonation is usually justified on the basis of a successful calibration.

Another complicating factor in estimating fluxes across the water table is the need to account for scale effects. The net groundwater recharge for an entire groundwater basin will differ from the net groundwater recharge defined for the area represented by one nodal block within a grid of the entire basin. Under some conditions, nodal recharge rates may approach or exceed the average annual precipitation rate, whereas the average recharge for the entire basin is always less than precipitation. The nodal spacing also will affect the net recharge or discharge rate appropriate for the cell. When a coarse grid is used to simulate a basin, an individual cell may include areas experiencing recharge as well as areas experiencing discharge; recharge added to the cell may discharge within the same cell. This is the case for the cells in Fig. 5.3 that are assigned a zero recharge rate. This intracell flow is modeled only as a net flux for the cell (Feinstein and Anderson, 1987; Stoertz and Bradbury, 1989; Jorgensen, 1989a,b).

Water Balance Modeling

Stoertz and Bradbury (1989) described a computerized water budget method that calculates the spatial distribution of recharge and discharge rates given a water table contour map and estimates of hydraulic conductivity and saturated thickness. The method may be applied in two or three dimensions using any groundwater flow model that includes a water balance calculation. All water table nodes are given specified head values estimated from the contour map. The water balance calculation gives the fluxes between the water table cells and adjacent cells in the model. The difference between inflow and outflow for any water table cell is equal to the recharge or discharge that must occur vertically across the water table. In areal two-dimensional simulations, all nodes in the grid are specified head nodes. The finite difference or finite element equations are not solved because all heads are known. The water balance portion of the code is used alone to calculate the fluxes from which recharge is deduced. In profile and three-dimensional simulations, the finite difference or finite element equations are solved because it is necessary to compute the heads at nodes below the water table before calculating fluxes for the water table nodes.

The method can be used with any flow model that calculates fluxes between specified head nodes as part of the water balance. (MODFLOW does not calculate these fluxes and modifications to the program, given in Table 5.2, are necessary to use the method.) The model-calculated nodal fluxes can be contoured to produce a recharge/discharge map (Fig. 5.4).

The recharge/discharge pattern obtained in this way could be used to esti-

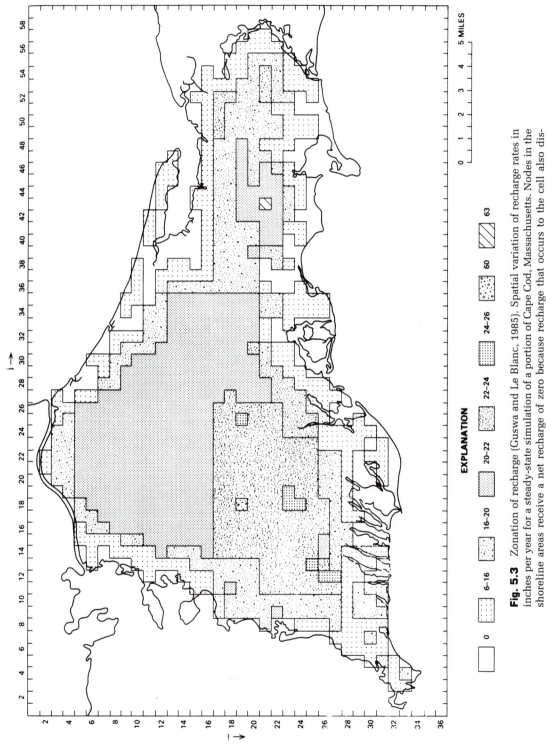

Fig. 5.3 Zonation of recharge (Guswa and Le Blanc, 1985). Spatial variation of recharge rates in inches per year for a steady-state simulation of a portion of Cape Cod, Massachusetts. Nodes in the shoreline areas receive a net recharge of zero because recharge that occurs to the cell also discharges within the cell.

EXPLANATION

| 0 | 6–16 | 16–20 | 20–22 | 22–24 | 24–26 | 60 | 63 |

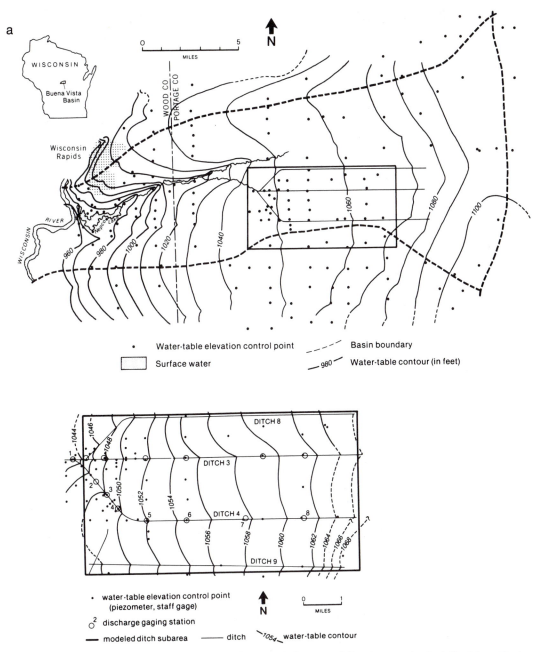

Water-table elevation control point   ------ Basin boundary

Surface water   980 Water-table contour (in feet)

• water-table elevation control point
(piezometer, staff gage)

○² discharge gaging station

— modeled ditch subarea   —— ditch   ⌐1054⌐ water-table contour

Fig. 5.4 Mapping of recharge areas by water balance modeling for a sandy glaciofluvial aquifer in central Wisconsin (Stoertz and Bradbury, 1989).
(a) Water table maps of a groundwater basin and a subarea used as input to the water balance.
(b) Recharge/discharge maps generated by water balance modeling for the groundwater basin and a subarea. Steady-state heads and hydraulic conductivities are specified and the model calculates water balances for the cell which are interpreted to be either recharge or discharge. Note the effect of scale dependence by comparing the recharge/discharge distribution at the basin scale to the distribution for the subregion. Dimensions of the grids are indicated by column and row numbers.

b

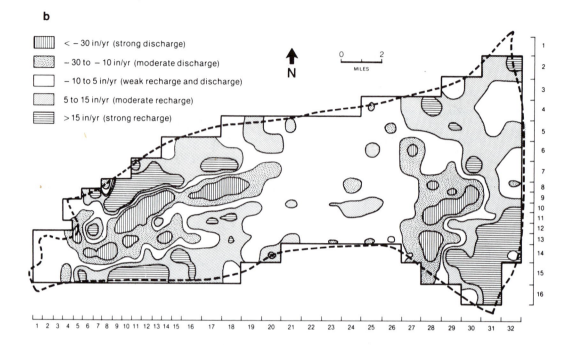

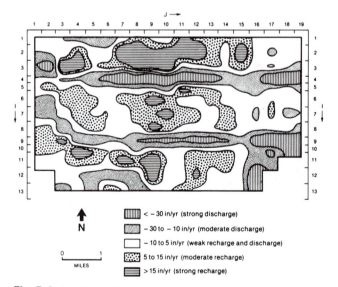

Fig. 5.4 *(continued)*

Table 5.2
Code Modifications to MODFLOW to Calculate Fluxes between
Specified Head Nodes

Main Program

Remove the following 19 lines:

```
DO 300 KPER=1,NPER
DO 200 KSTP=1,NSTP
DO 100 KITER=1,MXITER
IF(IUNIT(9).GT.0) CALL SIP1AP(...5 LINES TOTAL...)
IF(IUNIT(11).GT.0) CALL SOR1AP(...4 LINES TOTAL...)
IF (ICNVG.EQ.1) GO TO 110
100 CONTINUE
KITER=MXITER
100 CONTINUE
IF(ICNVG.EQ.0)STOP
200 CONTINUE
300 CONTINUE
```

Subroutine SBCF1F

In all of the following lines, change .LE to EQ.

```
IF(IBOUND(J-1,I,K).LE.0)GO TO 30
IF(IBOUND((J+1,I,K).LE.0)GO TO 60
IF(IBOUND(J,I-1,K).LE.0)GO TO TO 90
IF(IBOUND(J,I+1,K).LE.0)GO TO 120
IF(IBOUND(J,I,K-1).LE.0)GO TO 150
IF(IBOUND(J,I,K+1).LE.0)GO TO 180
```

From Stoertz and Bradbury, 1989.

mate the spatial distribution of fluxes. These fluxes could then be used as specified flux boundary conditions (see Problem 6.1b) in a predictive simulation that calculates the head at the water table. However, care must be taken to use this procedure only if the recharge/discharge pattern is not expected to change as a result of stresses applied to the model during the predictive simulation. Furthermore, it should be recognized that fluxes computed in this manner are highly dependent on assumed values for hydraulic conductivity and interpolated heads taken from the water table contour map. Errors in hydraulic conductivity and head estimates may cause highly inaccurate flux estimates. The calculated fluxes also will be dependent on scale effects discussed in the section above.

Modeling the Unsaturated Zone

Another possibility for simulating recharge is to use an unsaturated zone flow model to route infiltrated water to the water table. Unsaturated zone models are theoretically more complex and require more field parameters than saturated zone models (Section 12.3). Furthermore, infiltration introduced at the upper boundary of the unsaturated model will eventually arrive at the water table as groundwater recharge unless evapotranspiration is simulated. Hence, unsaturated zone models still require knowledge of groundwater recharge rates to specify the upper boundary condition. For these reasons, unsaturated zone models typically are not used unless there is some other reason to justify the additional complexity involved in this type of modeling.

Some investigators have used unsaturated zone models to simulate the timing of recharge arrival at the water table (Krishnamurthi et al., 1977; Watson, 1986; Stoertz et al., 1991). Output from an unsaturated model may also be routed as recharge to a saturated zone model (Pikul et al., 1974). The disadvantage of this approach is that the unsaturated and saturated models are uncoupled. That is, the length of the unsaturated zone is not allowed to change in response to the transient movement of the water table.

Alternatively, a model of the entire unsaturated/saturated zone could be used to simulate the response of the water table to recharge (Freeze, 1969, 1971). The advantage of this type of model is that the flow field between the two zones is continuous; the water table is calculated as the surface of zero pressure head. The disadvantage of using this type of model or any model that involves the unsaturated zone is the introduction of considerable additional complexity (Section 12.3).

FINITE DIFFERENCE MODELS

In finite difference models fluxes across the water table represent a volume of water applied to the top area of the cell per unit of time (Fig. 5.5). The model may have a separate array set aside for recharge, as in MODFLOW's Recharge Package, or recharge may be simulated by injection wells. In three-dimensional simulations recharge is most conveniently handled when the water table is always expected to be in the top layer of the model. Then the top layer is designated as unconfined and an array of recharge/discharge rates is specified for that layer (Fig. 5.5). When the water table cuts across layers, it is necessary to use a different approach. One alternative is to place a recharge node (injection well) at every water table node. Problems may arise, however, because finite difference models do not check to ensure that $h = z$ at the water table (Eqn. 4.2). For example, suppose that the user specifies where the water table will be located throughout the simulation as in Fig. 5.6. The model will continue to apply recharge to the designated water table nodes even if this causes the head to exceed the elevation of the top of the layer. MODFLOW's Recharge Package deals, in part, with this difficulty. When the water table is falling, the Recharge Package checks to ensure that recharge is applied to the top of the saturated column of any (i, j) location, i.e., the top active node (Fig. 5.6). Consequently, the user does not have to predetermine the layer to which re-

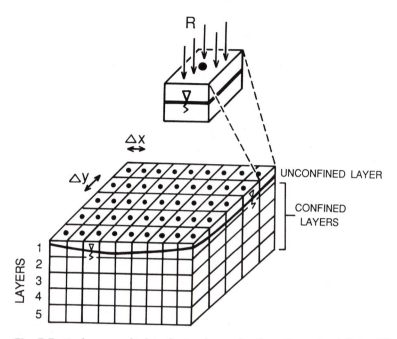

Fig. 5.5 Recharge applied to the top layer of a three-dimensional finite difference model. Recharge is specified as a rate (L/T). The code computes the volumetric rate of water added to the model by multiplying nodal recharge rates by the area of the top of the cell ($\Delta x\ \Delta y$).

charge should be added; the appropriate layer is automatically selected by the model. However, the current version of the Recharge Package is not able to simulate a rising water table since the code cannot move the water table into a dry cell. Consequently, when the water table is rising MODFLOW continues to apply recharge to the layer receiving recharge at the beginning of the simulation and may calculate heads higher than the top of the layer, thereby violating the boundary condition that $h(x, y) = z(x, y)$. A new block-centered flow package, BCF2, that will allow the water table to rise into dry layers is in preparation (McDonald, personal communication, 1991).

FINITE ELEMENT MODELS

Most finite element models allow the user to assign a distributed source like recharge to an element, whereas point sources and sinks may be assigned to a node. Assembly of the matrix equation within the code, however, requires that distributed sources and sinks be reassigned to nodes. The code multiplies user-specified element fluxes by the area of the element; the water is then apportioned to each node in the element (Fig. 4.9). Details of the procedure are given by Istok (1989).

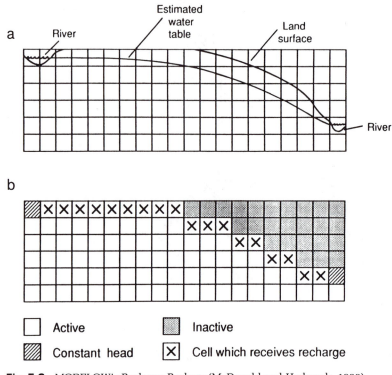

Fig. 5.6 MODFLOW's Recharge Package (McDonald and Harbaugh, 1988).
(a) Cross section showing field situation with finite difference grid superimposed.
(b) MODFLOW's Recharge Package allows recharge to be applied to the top active cell in each column of nodes. The package can follow a falling water table by shifting recharge downward to lower layers but cannot follow a rising water table because the code cannot saturate nodes that are designated as inactive.

Some finite element codes like FREESURF (Neuman, 1976) and AQUIFEM-N (Townley, 1990) accommodate a moving water table by deforming the top elements of the grid to ensure that $h(x, y) = z(x, y)$ at the water table nodes (Fig. 4.14). Of course, the user must take care to construct the finite element grid with the water table movement in mind so that element deformation does not increase the aspect ratio beyond its recommended limit (Section 3.3).

5.3 Leakage

Leakage refers to the movement of water through a layer of material that has a vertical hydraulic conductivity lower than that of the aquifer. Leakage may enter or leave the aquifer depending on the relative difference in heads between the aquifer and the source reservoir on the other side of the leaky layer (Fig. 5.7). Furthermore, the direction and volume of leakage may change during the simulation as the head in the aquifer changes. The source reservoir may be an unconfined aquifer, a river, or a lake.

Leakage is a type of head-dependent boundary condition (Box 4.1). In two-dimensional areal simulations, leakage through confining beds is simulated with a leakage term (Section 3.2). Use of leakage terms also is a convenient way to simulate partially penetrating rivers and lakes in two-dimensional areal models. In a simulation employing the leakage option, the source reservoir is not explicitly represented in the grid. Leakage is added to or extracted from the node that is directly beneath the source reservoir (Fig. 5.7a). If the head in the aquifer is greater than the head in the source reservoir, water is removed from the model. Otherwise, water is added via leakage. In either case, the volume of leakage (Q_L^*) is calculated from

$$Q_L^* = -K_z' \left(\frac{h_{\text{source}} - h}{b'} \right) wl \tag{5.9}$$

where K_z' is the vertical hydraulic conductivity of the leaky layer and b' is its thickness. The ratio K_z'/b' is the *leakance*. The parameters w and l are the width and length, respectively, of the source reservoir; h_{source} is the head in the source reservoir and h is the head in the aquifer, which is calculated by the model. The user supplies values of K_z', b', h_{source}, w, and l.

The volumetric leakage rate (Q_L^*) must be converted to an areal leakage rate applied to the top of the cell. Herein arises a problem. The source reservoir (lake or river, for example) may be narrower than the blocks or elements of the model (Fig. 5.7b). The areal leakage rate (L) must be adjusted so that the volume emanating from the source reservoir through the area wl is the same as the volume applied to the area of the cell or element $(\Delta x \, \Delta y)$. That is, $Q_L^* = L \, \Delta x \, \Delta y$, or

$$L = Q_L^*/\Delta x \, \Delta y \tag{5.10}$$

MODFLOW requires the user to input leakance as well as w and l and then performs the necessary adjustments within the code for each river segment (Fig. 5.7b). AQUIFEM-1 also has options for treating narrow rivers (Box 4.1). PLASM assumes that w and l have the same dimensions as the cell. If this is not the case, the user must compute adjusted leakance values $(K_z'/b')_m$ *as follows*:

$$\left(\frac{K_z'}{b'} \right)_m = \frac{K_z'}{b'} \frac{lw}{\Delta x \, \Delta y} \tag{5.11}$$

and input these values to PLASM.

Leakage to drains (Box 5.1) is handled in a similar way except that drains act only as sinks of water. That is, there is no leakage of water from the drains to the aquifer when the head in the aquifer is less than the head in the drain. Drains can be broadly interpreted to represent dewatering from the operation of

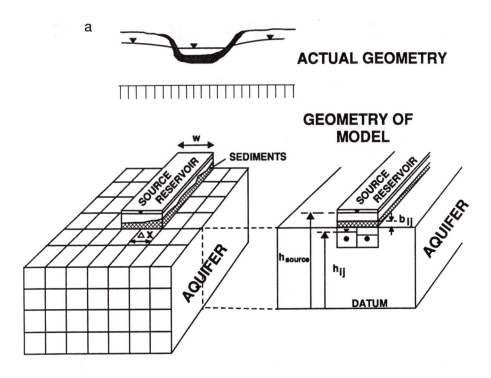

ACTUAL GEOMETRY

GEOMETRY OF MODEL

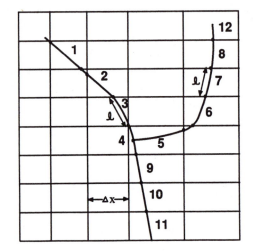

Fig. 5.7 Representation of leakage.

(a) The source reservoir located over the aquifer. In this example the width of the source reservoir (w) is equal to 2 Δx. The head in the source reservoir is h_{source} and the head in the aquifer below the source reservoir is $h_{i,j,k}$. The sediments between the source reservoir and the aquifer are shown by hatch marks.

(b) Discretization of a river acting as a source reservoir (McDonald and Harbaugh, 1988). This example shows a situation where $w \ll \Delta x$, which is more typical than the situation shown in (a). Numbering of the reaches of the source reservoir for input to MODFLOW's River Package is also shown. The length (l) and width (w) of each segment together with the leakance of the source bed sediments and the head in the source reservoir are input.

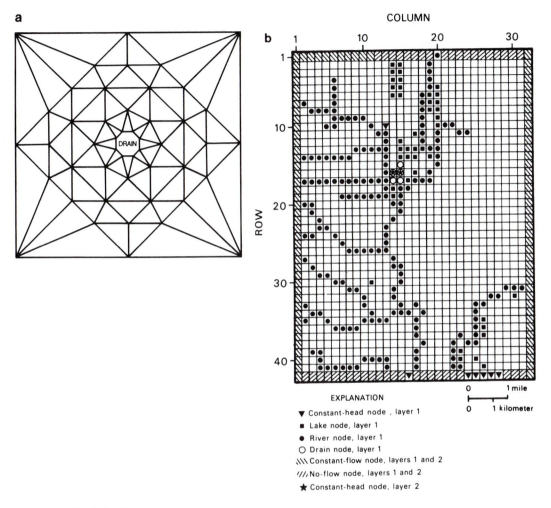

Fig. 5.8 Drains.
(a) The finite element grid for a 0.2 by 0.2 m area around an agricultural drain (Fipps, Skaggs, and Nieber, Water Resources Research, 22(11), pp. 1613–1621, 1986, copyright by the American Geophysical Union).
(b) Use of drain nodes near the center of a finite difference grid to simulate dewatering of a quarry near Columbus, Ohio (Eberts and Bair, 1990). Three drain nodes in layer 1 and two constant head nodes in layer 2 were used to represent the quarry. Also note the use of river and lake nodes to simulate internal sources and sinks of water.

underground mines (Fig. 4.13b), and quarries (Fig. 5.8b), agricultural drainage tiles (Fig. 5.8a), and drains around landfills. Springs and seeps also can be treated as drains (Section 4.3 and Box 4.1).

Input of leakage terms in MODFLOW is handled through the River Package, the Drain Package, or the General Head Boundary Package (Box 4.1). PLASM and AQUIFEM-1 also have options to accommodate leakage to and from source reservoirs and drains (Box 4.1).

Box 5.1
Case Study Using Drains

Yager (1987) used drain nodes in a simulation of groundwater flow beneath a nuclear fuel reprocessing facility to simulate boundary conditions as well as head-dependent conditions within the interior of the model. The purpose of the model was to predict flow paths of contaminants (tritium) that had leaked from storage areas and reprocessing facilities. The site is located in a glaciated section of the Appalachian Plateau in southwestern New York, in which surficial deposits of silty sand and gravel overlie glacial till, which overlies shale bedrock. All but the southwestern portion of the site borders incised stream channels, creating a plateau. Water in the surficial sand and gravel layer is unconfined. The saturated thickness decreases from over 5 m in the central portion of the plateau to less than 1 m at seepage faces at the plateau margins. Spring locations and water table contours are shown in Fig. 1.

A two-dimensional finite difference flow model with 442 active cells, each 30.5 × 30.5 m, was designed (Fig. 2) with most of the boundary formed by the edge of the plateau and simulated by drain nodes to represent springs and seepage faces. Specified flow conditions were used along the rest of the boundary. Inter-

nal to the model, drain nodes were used to simulate the french drain, as well as small streams, a wetland area, and a waste lagoon (lagoon #1). Drains were simulated using MODFLOW's Drain Package (Box 4.1). All drain elevations were assumed to be 0.3 m above the base of the surficial aquifer to ensure sufficient saturated material for flow. Drain conductances were estimated by trial and error during model calibration by matching measured and simulated heads and measured and simulated discharges. Drain conductance was calculated by multiplying drain conductivity (estimated to be 0.03 m/day) by the average cross-sectional area of the drain (Fig. 3) and dividing by the thickness of the interface (taken to be 0.3 m).

Discharge rates to the drains were calculated to be 2.1 cm/yr to the french drain, 2.2 cm/yr to the waste lagoon, 14.5 cm/yr to the stream leading to the wetland area, and 9.0 cm/yr to seepage along the edge of the plateau. Groundwater flowing from potential sources of tritium was predicted to discharge into the stream channel above gaging station NP-3 (Figs. 1 and 2), the french drain, and through seepage faces along the southeastern border of the plateau.

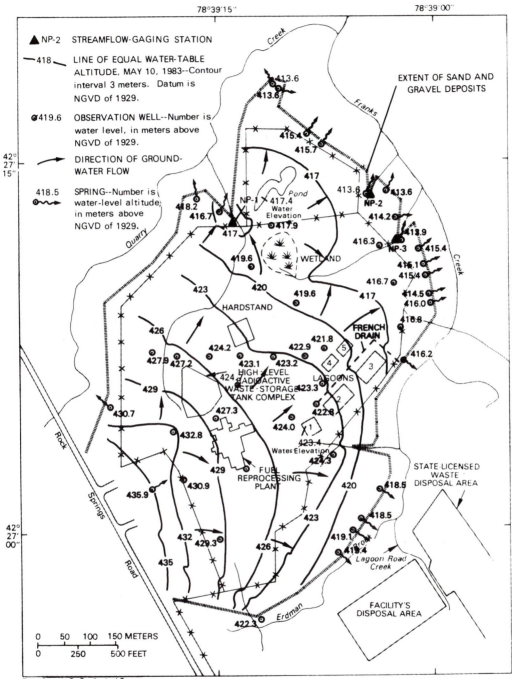

Fig. 1 Water table contours and inferred direction of groundwater flow (Yager, 1987). The locations of the french drain, springs, and local streams near the nuclear fuel reprocessing facility in southwestern New York State are also shown, as is the extent of the surficial sand and gravel deposits.

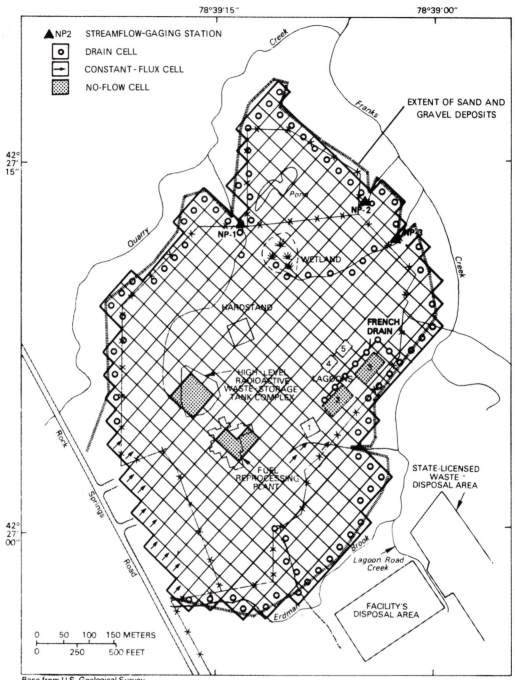

Fig. 2 The finite difference grid, assignment of boundaries, and location of drain nodes (Yager, 1987).

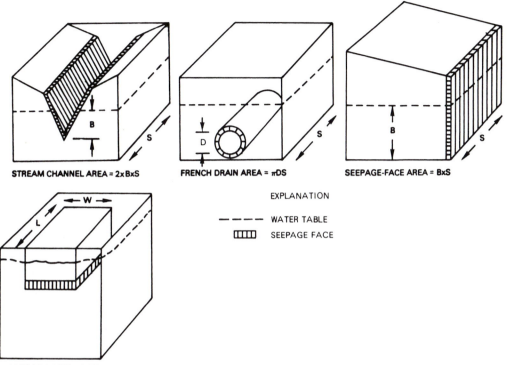

STREAM CHANNEL AREA = 2xBxS   **FRENCH DRAIN AREA = πDS**   **SEEPAGE-FACE AREA = BxS**

EXPLANATION

– – – – –   WATER TABLE

▥▥▥   SEEPAGE FACE

LAGOON 1 AREA = LxW

Fig. 3   Representation of streams, french drain, seepage faces, and waste lagoon #1 as drains with the appropriate cross-sectional area shown (Yager, 1987).

Problems

5.1   This problem is designed to illustrate how a well is represented in a numerical model and the effect of boundaries and space discretization on model results. Figure P5.1 shows a confined aquifer surrounded by no-flow boundaries.

(a) Design a regular grid with uniformly spaced nodes 1000 m apart. Determine the drawdown in the pumping well node and at a distance 1000 m from the pumping well after 10 days of continuous pumping at 530 m³/day. (Use 10 time steps each equal to 1 day and assume static initial conditions.) Plot a drawdown profile from the well node to the east boundary.

(b) Change the nodal spacing to 200 m. Using the same number of time steps, run this model and determine the drawdown at the well and

NO-FLOW BOUNDARY

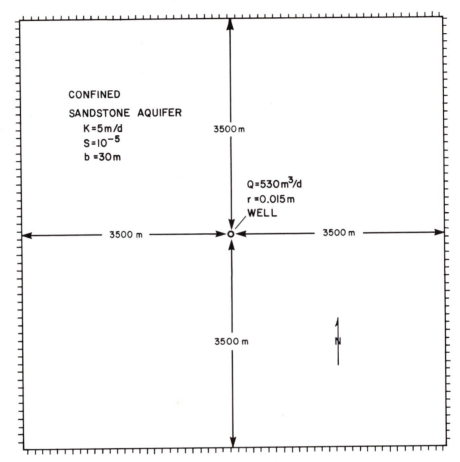

CONFINED
SANDSTONE AQUIFER
  K =5m/d
  S =10^{-5}
  b =30m

3500 m

Q=530m^3/d
r =0.015 m
WELL

3500 m

3500 m

3500 m

3500 m

N

Fig. P5.1

1000 m to the east after 10 days of continuous pumping. Plot the drawdown profile for this model on the graph prepared for Problem 5.1a.
- **(c)** Create a variable grid with nodal spacing equal to 10 m in the vicinity
  of the well. Using the time steps and conditions in 5.1a and 5.1b,
  determine the drawdown at the well node and 1000 m east of that node.
  Again, plot a drawdown profile with those of 5.1a and 5.1b. Explain
  why profiles are not identical.
- **(d)** If you used a finite difference model to solve this problem, calculate the
  actual drawdown in the well for 5.1a, 5.1b, and 5.1c using Eqn. 5.1.
  Why are these values different from the value calculated by the model

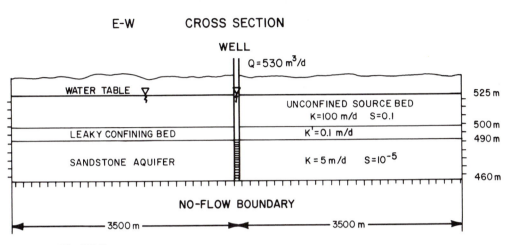

Fig. P5.2

for the cell representing the well? Solve the problem using a finite element model and compare results with those from the finite difference model.

(e) Why would use of the Theis equation to predict drawdown at the well and 1000 m to the east after 10 days of continuous pumping be inappropriate in this case?

5.2 Add two overlying layers to the aquifer modeled in 5.1, one a leaky confining unit and the second an unconfined aquifer, as shown in Fig. P5.2. The pumping rate and time steps remain the same. Use the variable grid design from Problem 5.1c.

(a) Simulate this problem using a two-dimensional areal model with leakage, a quasi three-dimensional model, or a full three-dimensional model. If you use a full three-dimensional model assume that the storage coefficient of the confining bed is 10^{-6}. Determine the drawdown in the confined aquifer after 10 days of pumping and compare your results with those in Problem 5.1c. How would the results from the two models you did not select compare with your results?

(b) Suppose the well is completed in both the unconfined and confined aquifers. Explain how you would model this dual aquifer well if it pumped a total of 530 m³/day.

5.3 Simulation of the effects of pumping a well screened through more than one aquifer has been reported by a number of authors. Read USGS Water Supply Paper 2205 (Hearne, 1985) and outline the steps taken to assign pumping rates to individual layers.

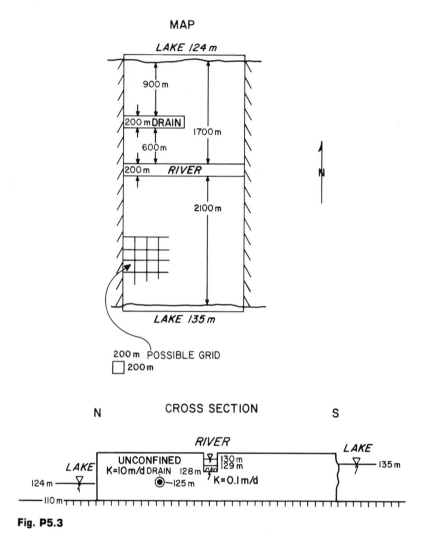

Fig. P5.3

5.4 This problem illustrates the use of head-dependent river and drain source and sink terms. An unconfined aquifer between two lakes is shown in Fig. P5.3.

(a) Design a steady-state model of this problem and determine the head distribution. You will need to determine the drain envelope and drain pipe equivalent hydraulic conductivity required to obtain a modeled water table elevation at the drain nodes of 125 m.

(b) Determine the rate of groundwater flow to or from the river and to the drain under steady-state conditions.

(c) Rerun Problem 5.4a assuming a river width of 50 m instead of 200 m.

How does this change affect the total groundwater flow to or from the river?

(d) Change the river nodes to drain nodes with an elevation of 130 m. Run the model again and describe the effect of this change on the head distribution. Under what field situations would such a representation of a river be valid?

5.5 Suppose that a mine is to be constructed in the aquifer shown in Fig. P5.1. Assume that the location of the mine shaft will coincide with the location of the well shown in Fig. P5.1.

Predict the quantity of water which must be pumped for the first 5 days after a 10 by 10 m wide shaft penetrates the top of the aquifer at an elevation of 1950 m. Use the variable grid of Problem 5.1c, an initial static head of 2000 m, and five time steps with $\Delta t = 1$ day. Assume that the opening of the shaft causes the head in the mine to drop instantaneously to the base of the shaft. Use the model to calculate daily pumping rates required to keep the shaft dry and to predict the head distribution at the end of the first 5 days of pumping. What additional methods could be used to simulate this dewatering problem?

5.6 Read the report by Danskin (1988) and then briefly describe and comment on his use of internal source and sink terms. Specifically, how were ephemeral streams and evapotranspiration simulated?

6

PROFILE MODELS

"To simulate ground-water flow in the x-z plane, the longitudinal axis of the profile model must be oriented along a ground-water flow path, because in general the part of the aquifer to be modeled must function as a two-dimensional flow system."
—*S.G. Robson, 1978*

Cross-sectional or profile models are useful when vertical flow is important but a full three-dimensional model is unwarranted. Profile models are also helpful in testing the validity of the conceptual model of the system prior to designing a full three-dimensional model. Profile models have been used frequently in an interpretive sense to study patterns in regional flow systems (e.g., Freeze and Witherspoon, 1967; McBride and Pfannkuch, 1975; Winter, 1976, 1983). A profile model assumes that all flow occurs parallel to and in the plane of the profile. In other words, there is no component of flow at an angle to the profile. For this reason, standard profile models are not useful in simulating point sinks or sources of water (e.g., wells) imbedded within the profile. Axisymmetric profiles (Section 6.3) may be used for this purpose, however.

6.1   Orienting the Profile

The main consideration in orienting the profile is to align the model along a flow line (Fig. 6.1a). Frequently, this is not the orientation the user wants. All the points of interest such as observation wells may not fall along one flow line (Fig. 6.1b). If the profile is not oriented along a flow line, there will be errors in the results because the model cannot simulate components of flow at an angle to the cross section. Such errors tend to get lumped with errors caused by other

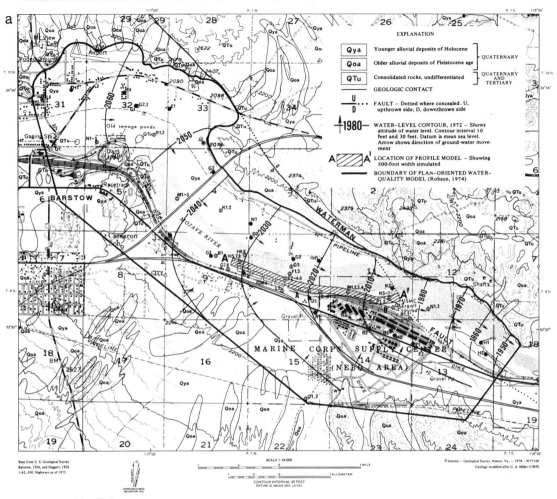

Fig. 6.1

(a) Orientation of a profile along a flow line (**A–A′**) in an alluvial aquifer to simulate transport of contaminants from sewage treatment ponds at Barstow, California (Robson, 1978). The width of the profile was 500 feet to accommodate the width of the treatment ponds. The ratio of horizontal to vertical hydraulic conductivity for the profile was estimated to be 200 based on the longitudinal and vertical extent of the contaminant plume and the head distribution within the profile.

(b) Example of a profile aligned to coincide with observation wells for a model of the Snake Lake area in northern Wisconsin. The profile does not always align parallel to the flow, e.g., when crossing Highway 51. The errors caused by this alignment were not believed to be significant for the purposes of the simulation. The grid and boundary conditions of the profile model are shown in Fig. 4.7a. (Anderson and Munter, Water Resources Research, 17(4), pp.1139–1150, 1981, copyright by the American Geophysical Union.)

b

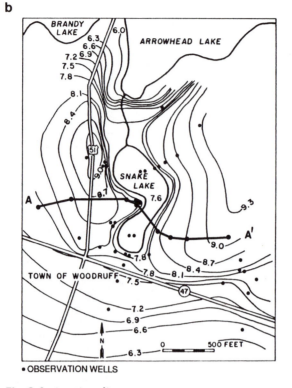

• OBSERVATION WELLS

Fig. 6.1 (continued)

types of discrepancies between the modeled system and the field system. Such discrepancies produce errors during model calibration. Attempts to fit simulated heads to observed heads during calibration may lead to erroneous selection of the calibration parameters, usually the vertical anisotropy ratio of hydraulic conductivity. Calibration errors caused by incorrect alignment of the profile model may or may not be important, depending on the purpose of the simulation, but the potential for error should be recognized if it is necessary to orient all or part of the profile model at an angle to flow.

Both two- and three-dimensional codes can be used to simulate flow in cross section. Most two-dimensional codes are designed to simulate aquifers in areal view. The governing equation for this type of simulation is given in Eqn. 2.1. The governing equation for a profile simulation is

$$\frac{\partial}{\partial x}\left(K_x \frac{\partial h}{\partial x}\right) + \frac{\partial}{\partial z}\left(K_z \frac{\partial h}{\partial z}\right) = S_s \frac{\partial h}{\partial t} - R^* \tag{6.1}$$

Comparison of Eqns. 2.1 and 6.1 indicates that they are the same if $y = z$ and if values for K, S_s, and R^* are equal to T, S, and R. The model assumes that the

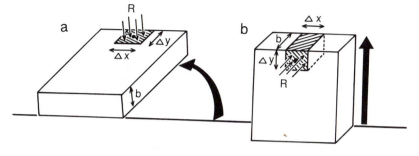

Fig. 6.2 Conceptualization of a profile model.
(a) Standard orientation of an areal two-dimensional model. Recharge is applied to the area $\Delta x\, \Delta y$.
(b) Viewpoint needed for a profile model. In a field setting, recharge occurs to the area $\Delta x b$. The volume of water represented by $R\, \Delta x b\, \Delta t$ must be applied to the area $\Delta x\, \Delta y$ in the model.

grid for the profile simulation is oriented as shown in Fig. 6.2a and solves Eqn. 2.1, but the user views the profile as shown in Fig. 6.2b and wants to solve Eqn. 6.1. In order to make the transition from the user's perspective to the requirements of the model, it is necessary to let b equal the thickness of the profile and to calculate transmissivities for Eqn. 2.1 by multiplying K by the thickness of the profile. It is also necessary to use the confined option in the model in order to ensure that the model solves Eqn. 2.1 rather than Eqn. 2.3. Adjustments to recharge rates (R) and storage coefficients (S) are also needed (Section 6.2).

Generally the profile is taken to be one unit thick but other thicknesses may be used. A profile of variable thickness also could be simulated. A thick profile may allow more observation wells to be included in the plane of the profile or may be useful when it is necessary to consider flow through a given section of aquifer, as in Fig. 6.1a, where the profile is taken to be 500 feet thick.

When a three-dimensional model is used to simulate a profile, the profile model can be cut out of the three-dimensional grid in one of two ways (Fig. 6.3). The model can have one layer (layer orientation), or a vertical slice (slice orientation) can be cut out of the layers. If layer orientation is selected, the model operates the same as a two-dimensional areal model, using one confined layer and solving Eqn. 2.1. The adjustments to transmissivity discussed above as well as the adjustments to recharge and storage coefficients discussed in Section 6.2 are necessary.

In slice orientation, the model has several layers as shown in Fig. 6.3. The governing equation is Eqn. 2.2. The thickness of the profile is now Δy, which is then set equal to one for a profile one unit thick. Because the vertical slice includes several model layers, data assembly for profile models in slice orientation is more cumbersome. However, slice orientation is a more natural one for profile modeling because the model solves the appropriate governing equation, Eqn. 2.2. No adjustment to hydraulic conductivity, storage coefficient, or recharge is needed in slice orientation.

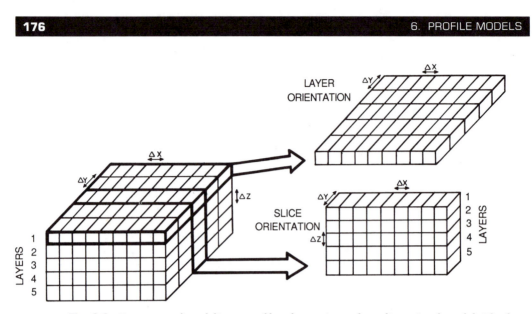

Fig. 6.3 Two ways of modeling a profile when using a three-dimensional model. The layer orientation is the same as in an areal two-dimensional model (Fig. 6.2a). The slice orientation uses a three-dimensional grid where Δy is equal to the thickness of the profile.

6.2 Parameter Adjustment for Profile Modeling

The values of storage coefficient and recharge must be adjusted before input to a profile model (Leake, 1977) except when using a three-dimensional model in slice orientation. In layer orientation, the model applies recharge to the shaded portion of the cell as shown in Fig. 6.2a, i.e., to the area $\Delta x \, \Delta y$. However, it is clear from the profile shown in Fig. 6.2b that field values of recharge would be measured for water entering the system through the area $\Delta x b$. We must instruct the code to take the volume of water equal to $R \, \Delta x b \, \Delta t$ and apply it to the area $\Delta x \, \Delta y$. In other words, the recharge rate required by the profile model (R_m) is

$$R_m = R \, \Delta x b / \Delta x \, \Delta y = Rb / \Delta y \qquad (6.2)$$

The user must calculate values of R_m from field-estimated recharge (R) and input these adjusted values to the model as recharge.

　　In transient simulations storage coefficients for the water table nodes must be adjusted in a similar fashion. The storage coefficient at the water table is a measure of the volume of water (ΔV) released from storage by drainage as the water table falls and is equal to the specific yield (S_y)

$$S_y = \Delta V / A \, \Delta h \qquad (6.3)$$

where A is area of aquifer and Δh is decline in head. The model requires the area to be equal to the shaded area $\Delta x \, \Delta y$ in Fig. 6.2a, but in the user's viewpoint (Fig. 6.2b) the area should be $\Delta x b$. Therefore, specific yield (S_y) must be adjusted to (S_y)$_m$ for input to the model

$$(S_y)_m = S_y b / \Delta y \qquad (6.4)$$

Equation 6.4 applies only to storage at the water-table nodes. Below the water table, water is derived from storage by compression of the aquifer and expansion of water so that the storage coefficient is a function of specific storage (Table 3.4). In Fig. 5.5, for example, all model layers below the top layer are designated as confined layers. If compressible storage is important, storage coefficients for the interior nodes may be calculated by multiplying specific storage by the thickness of the profile. This is the storage coefficient for confined conditions, which is small relative to specific yield and commonly is set equal to zero or close to zero in profile modeling.

6.3 Axisymmetric Profiles

Only infinitely long line sources and sinks, such as a canal oriented perpendicular to the profile, can be simulated with a standard profile model. It is not possible to simulate point sources and sinks because they create radial flow. A standard profile model cannot account for components of flow outside the cross section. Some special-purpose codes (Reilly, 1984; Stark and McDonald, 1980; Neuman, 1976) allow for radial coordinates and therefore can accommodate point sinks and sources within a profile where the source or sink is located at $r = 0$. This profile configuration is known as axisymmetric (Fig. 6.4a).

Standard models can be tricked into simulating axisymmetric profiles, however, by adjusting the thickness of the profile as shown in Fig. 6.4b. From the user's perspective this amounts to adjustment of transmissivity as described in Section 6.1. That is, one defines a profile of variable thickness (Fig. 6.4b) where T is a function of the radial distance from the well. Partial penetration can be simulated by inserting the perforated interval of the well into the appropriate model row (layer orientation) or layer (slice orientation). The pumping or injection rate of the well depends on the size of the pie-shaped piece simulated. If a 20° angle is simulated, as shown in Fig. 6.4b, the pumping rate simulated by the model is 20/360 or 1/18 of the actual pumping rate. Typically the edge of the model is set at the radius of influence of the well and may be simulated using a no-flow or constant-head boundary condition.

6.4 The Moving Boundary Problem

The water table typically forms the upper boundary of a profile model. In slice orientation, it is possible to use the Dupuit assumptions to calculate the head at the water table. If instead, specified head values are given along the water table,

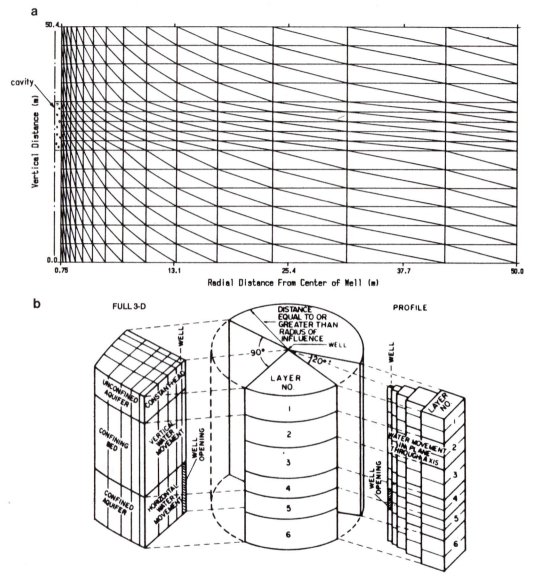

Fig. 6.4 Axisymmetric profiles.
(a) A finite element mesh for an axisymmetric profile model used to simulate transient ground-water flow into a cavity located in saturated till (Keller, Van der Kamp, and Cherry, Water Resources Research, 25(11), pp. 2299–2317, 1989, copyright by the American Geophysical Union).
(b) A finite difference grid for an axisymmetric profile used to simulate a partially penetrating pumping well is shown on the right side of the cylindrical section of aquifer (Land, 1977). The thickness of the profile (and the transmissivity assigned to a cell) increases with distance from the well. Adjustments to storage coefficients are also needed to reflect the change in profile thickness. The pumping rate is adjusted according to the angle (here equal to 20°) used to designate the aquifer wedge. The grid for a full three-dimensional model, which assumes radial flow and uses symmetry to model only one quarter of the aquifer, is shown on the left-hand side of the figure.

the grid is designed so that the condition $h(x, y) = z(x, y)$ is observed at the water table nodes (Fig. 6.5a). It is assumed that the water table will not move during the simulation. The location and rates of recharge and discharge can be calculated from the head solution (Fig. 6.5b).

If a flux boundary is assumed, recharge and discharge rates are specified and the location of the water table is unknown at the beginning of the simulation. Ideally the location of the water table is part of the solution and the water table nodes are moved during the simulation to maintain $h(x, y) = z(x, y)$. Such a moving boundary is handled only with difficulty by standard two-dimensional models like PLASM and AQUIFEM-1. When using a two-dimensional model or a three-dimensional model in layer orientation, the user must fix the location of the water table nodes at the beginning of the simulation. The fluctuation of the water table during the simulation ideally will be restricted to movement within the water table cells. For example, suppose that for the grid in Fig. 6.6a, a water table node is designated to occupy the fifth row. Suppose that Δz is 10 feet and that the elevation of the top of the cell (TOP) is 60 ft and the bottom (BOT) is 50 ft. It is implicitly assumed that the water table will fluctuate between 50 and 60 feet. This means that the user must estimate the final position of the water table within a range determined by the value of Δz. The simulation is run and if the head at the water table is not between TOP and BOT for the designated water table cell, the grid should be redesigned and the simulation run again (Fig. 6.6b).

For a three-dimensional model in slice orientation, the situation is somewhat different. In MODFLOW, the Recharge Package can be used to ensure that recharge is applied to the top active nodes (Fig. 5.6b). Use of the Recharge Package ensures that the code will move the water table down as needed. However, the current version of the MODFLOW cannot move the water table up into an overlying dry layer. Therefore, the user should set the initial position of the water table higher than the elevation of the expected final position (Fig. 6.7a).

Some finite element models (e.g., Townley, 1990; Mitten et al., 1988; Durbin and Berenbrock, 1985; Neuman, 1976) handle the moving water table boundary using elements that deform as the water table moves (Figs. 6.7b and 4.14). This is the best way to handle the moving boundary problem but care must be taken to ensure that the deformed elements always have an aspect ratio close to unity (Section 3.3).

Codes that simulate both the saturated and unsaturated zones (Section 12.3) offer another way to solve the moving boundary problem. In a saturated-unsaturated model, the position of the water table is an internal boundary in a problem domain that extends from the land surface to some distance below the water table (Fig. 6.8). The solution is formulated in terms of pressure head and the water table is the surface of zero pressure head. Freeze (1969, 1971) and Neuman and Witherspoon (1971) discuss the moving boundary problem in more detail.

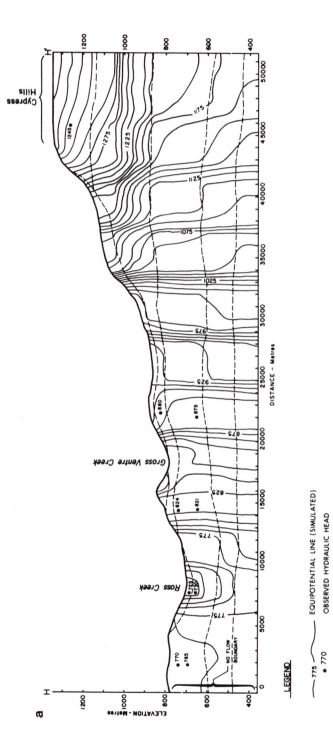

Fig. 6.5 Profile model for a steady-state regional flow system of the Ross Creek Basin in southeast-ern Alberta (Ophori and Toth, 1989). The water table boundary is simulated by specified head nodes.
(a) Equipotential lines generated by the finite element profile model.
(b) Flow lines were generated by using the flow model to calculate streamfunctions (Section 12.2 and Box 12.1). Recharge and discharge areas and the presence of local, intermediate, and regional flow systems are shown by the streamlines.

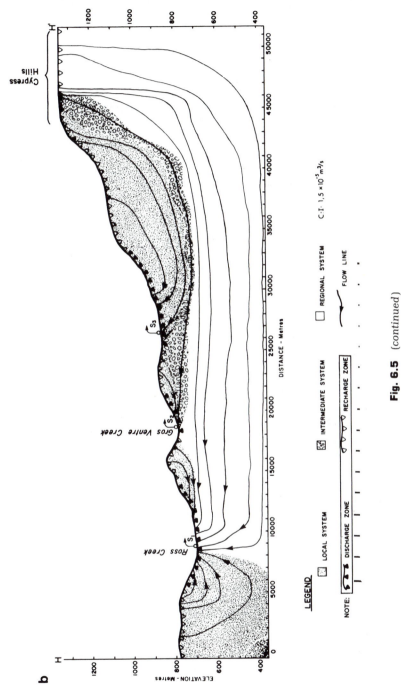

Fig. 6.5 (continued)

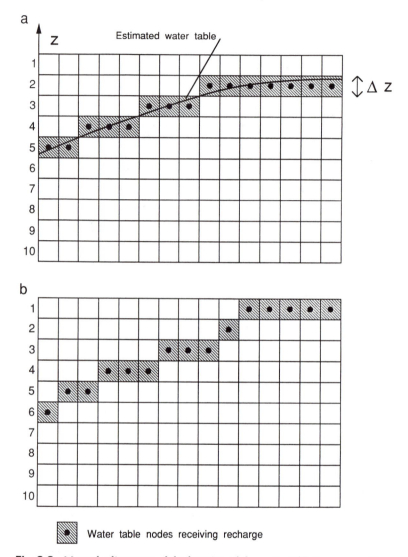

Fig. 6.6 Manual adjustment of the location of the water table in a conventional finite difference profile simulation in which the water table boundary is simulated as a flux boundary. The head at the water table must fall within the range defined by Δz, i.e., between the elevations of the top and bottom of the cell.

(a) Initial grid design based on an estimate of the elevation of the water table.

(b) Redesign of the grid in response to results from an initial run of the model. Water table boundary nodes are relocated so that the head at each water table node as predicted by the model falls within the range TOP $\geq h \geq$ BOT, where TOP and BOT are the top and bottom elevations of the cell.

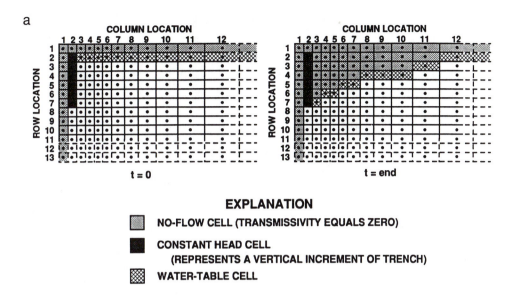

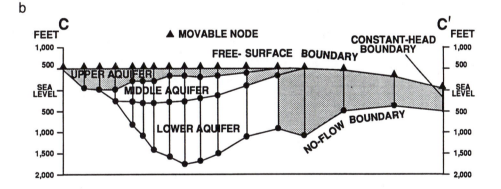

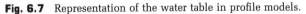

Fig. 6.7 Representation of the water table in profile models.

(a) Simulation of a falling water table caused by drainage to a trench shown at the left-hand side of the profile. The water table cells at the beginning and end of the simulation are shown (Leake, 1977).

(b) Movable nodes along the water table (free surface boundary) in a finite element model that allows for deformable elements (shaded) (Mitten et al., 1988).

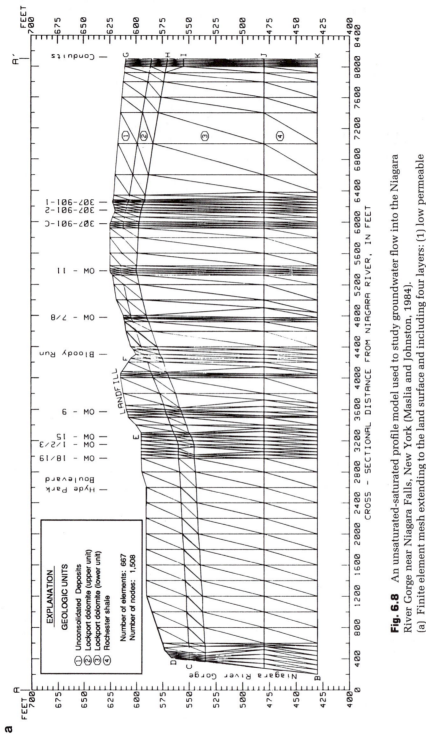

Fig. 6.8 An unsaturated-saturated profile model used to study groundwater flow into the Niagara River Gorge near Niagara Falls, New York (Maslia and Johnston, 1984).

(a) Finite element mesh extending to the land surface and including four layers: (1) low permeable till and lacustrine deposits; (2) permeable fractured upper unit of a dolomite aquifer; (3) a lower unit of the dolomite aquifer; (4) a shale.

(b) Location of the water table (heavy line) within the unsaturated-saturated problem domain and the head distribution in the profile.

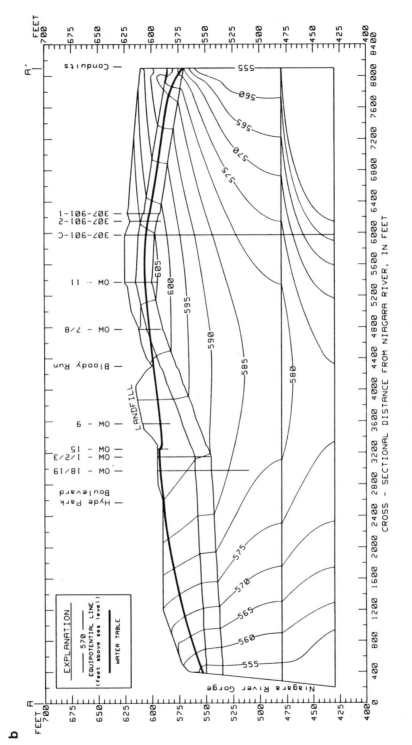

Fig. 6.8 (continued)

Box 6.1
Case Study of a Profile Model

Bailey (1988) used a profile model to study the flow system in a cross section through Bear Creek Valley, Tennessee and to evaluate the potential for migration of contaminants into Bear Creek and ultimately out of the Oak Ridge Reservation (Fig. 1). Because of few deep well data, however, the model could not be calibrated; consequently, it was not possible to evaluate the potential for off-site migration of contaminants. Nevertheless, this modeling exercise demonstrates the use of a preliminary model in an interpretive sense and shows how a model serves as a framework for organizing available field data and for identifying deficiencies in the existing data base. Connell and Bailey (1989) later used an automated inverse model to perform the calibration (Box 8.4).

The data base consisted of information from over 400 wells and test borings with water level data available for over 200 wells. However, 140 of the 200 wells were less than 100 feet deep and only one deep well was located in the cross section (Fig. 1c). The ridges to the north and south of the valley are groundwater divides so that groundwater flows toward the center of the valley in response to recharge from precipitation. Hydraulic conductivities were estimated from the 50th percentile values of the available hydraulic conductivity measurements for each of the seven bedrock units (Table 1). Flow in the overlying regolith was not simulated directly.

MODFLOW was used to simulate steady-state flow in the cross section. A variable grid was used (Fig. 1c), for which the smallest cells were 50 by 50 ft and the largest were 75 by 150 ft. The upper boundary of the model was simulated as a specified head boundary and was set in the regolith to represent the average October water table. The decision to locate the entire thickness of regolith within the specified head boundary nodes effectively removed the regolith from the model because the properties of the regolith were not used in the simulation. The side boundaries represented the groundwater divides paralleling the ridges and were

Table 1

Initial and Final Estimates for Hydraulic Conductivities Used in the Profile Model of Bear Creek Valley

Geologic unit	Bedrock hydraulic conductivity (ft/day)	
	Initial	Final
Knox Group (Copper Ridge Dolomite)	0.2	0.1
Maynardville Limestone	1.0	0.1
Nolichucky Shale	0.2	0.04
Maryville Limestone	0.07	0.01
Rogersville Shale/Rutledge Limestone	0.04	0.008
Pumpkin Valley Shale	0.05	0.01
Rome Formation	0.2	0.1

From Bailey, 1988.

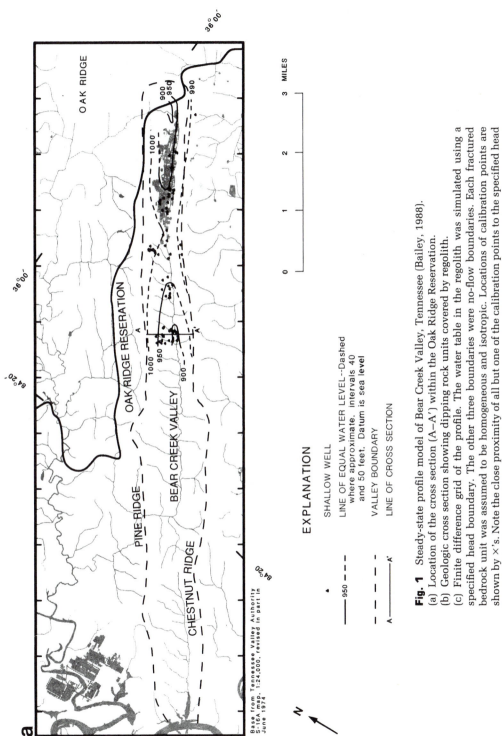

EXPLANATION

▲   SHALLOW WELL

——— 950 ———   LINE OF EQUAL WATER LEVEL--Dashed
where approximate. intervals 40
and 50 feet. Datum is sea level

– – –   VALLEY BOUNDARY

A ——— A'   LINE OF CROSS SECTION

Fig. 1 Steady-state profile model of Bear Creek Valley, Tennessee (Bailey, 1988).
(a) Location of the cross section (A–A') within the Oak Ridge Reservation.
(b) Geologic cross section showing dipping rock units covered by regolith.
(c) Finite difference grid of the profile. The water table in the regolith was simulated using a
specified head boundary. The other three boundaries were no-flow boundaries. Each fractured
bedrock unit was assumed to be homogeneous and isotropic. Locations of calibration points are
shown by ×'s. Note the close proximity of all but one of the calibration points to the specified head
boundary in the regolith.

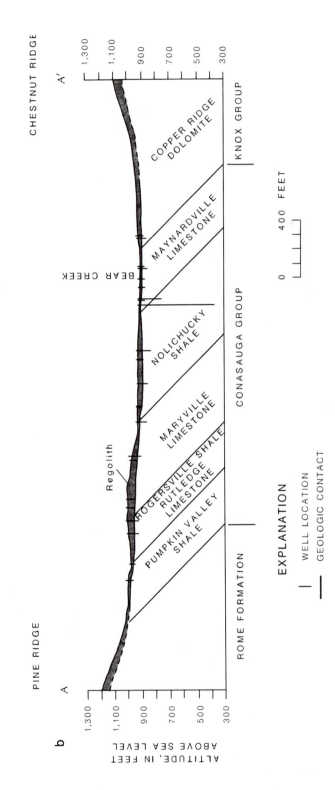

Fig. 1 (continued)

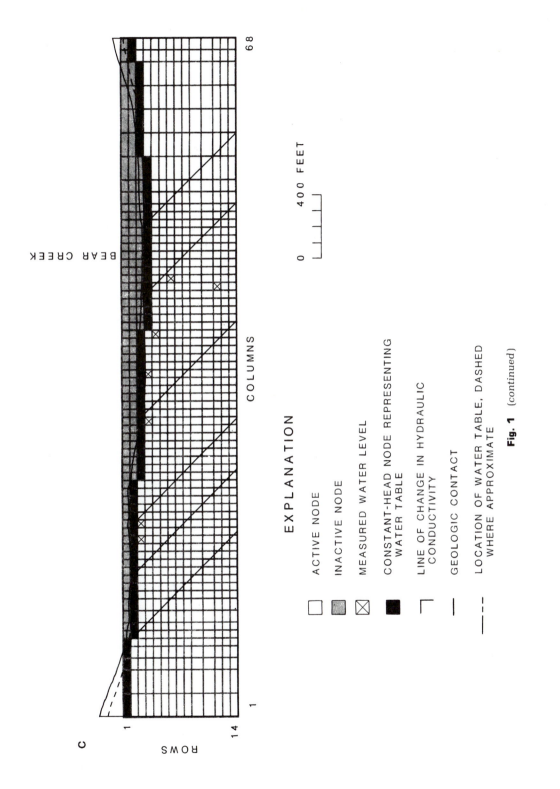

Fig. 1 (continued)

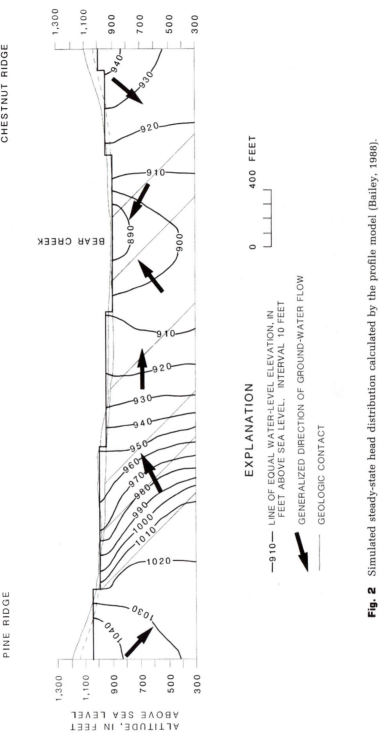

Fig. 2 Simulated steady-state head distribution calculated by the profile model (Bailey, 1988). Note the deep circulation of water within the profile.

simulated as no-flow conditions. The bottom boundary represented the depth at which impermeable rock was encountered and was simulated as a no-flow boundary.

During calibration, initially estimated hydraulic conductivities were reduced (Table 1) to achieve a match between measured and simulated heads. These changes were justified because lower hydraulic conductivities are typically observed in the flow direction normal to the strike of the bedrock units, which is the orientation of the profile model. No vertical anisotropy was assumed. The model was not sensitive to changes in parameters near the water table boundary where most of the calibration points are found (Fig. 1c). The model cannot be considered calibrated because of the close proximity of the calibration points to the specified head boundary. Nevertheless, the flow pattern produced by the model (Fig. 2) can be used to assess additional data needs and guide future field work. Results show circulation of groundwater to a depth of at least 500 feet. Hence, deeper piezometer nests and wells would be useful in future calibration attempts. In a later study, Connell and Bailey (1989) (Box 8.4), changed the upper boundary condition to a specified flux boundary so that the shallow water-level measurements become meaningful calibration points.

Applications of profile models were also reported by Robertson and Cherry (1989), Robertson et al. (1989), Hillman and Verschuren (1988), Forster and Smith (1988), Siegel (1988), Gupta and Shrestha (1986), Osborne and Sykes (1986), Pennequin (1983), Stephens (1983), Andrews and Anderson (1978), and Wilson and Hamilton (1978).

Problems

6.1 Design a profile model for the unconfined aquifer shown in Fig. P6.1. This west-to-east cross section terminates at groundwater divides on the sides and is underlain by bedrock. The water table is held at an elevation of 425 m at the center of the profile and then slopes with a gradient of 0.5 m/50 m to within 25 m of the side boundaries, where it is 432 m in elevation.

(a) Set boundary conditions, using specified head nodes at the water table,

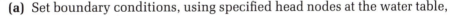

Fig. P6.1

and assign the appropriate parameters to determine the steady-state head distribution. Run the model and contour the resulting heads.

(b) Use Darcy's law to calculate the steady-state recharge flux across the water table and fluxes along the water table. Now compute net flux for each water table cell. Replace the specified head water table nodes that represent the area of recharge with the net recharge fluxes just calculated. Keep all the specified head nodes in the discharge area. Run the model and compare the steady-state heads with those of Problem 6.1a.

(c) Change the hydrologic conditions shown in Fig. P6.1 by assuming that the recharge nodes in 6.1b represent the rates and area of recharge to the pictured flow system, but that all recharge water discharges to a 100-m-wide river with a stage of 425 m located in the middle of the profile. Run the model and contour the resulting heads. Compare this flow field with the results from 6.1b.

(d) Don't lose sight of what recharge and river nodes represent in a profile model. That is, recharge is assumed to occur at the given rates infinitely far in a direction perpendicular to the cross section. Likewise, the elevation of the stream stage is assumed to be constant in the direction perpendicular to the profile. How would you conceptualize assigning recharge and river stage in a three-dimensional flow field of this system? How will the results of this simulation differ from the profile representation?

6.2 Describe how boundary conditions and parameters were assigned in the simulation reported by Ophori and Toth (1989).

6.3 Fig. P6.2 shows a profile of a confined aquifer. Boundary conditions consist of constant flux across the left boundary and a specified head of 100 m along the right side. The upper and lower boundaries are no-flow conditions.

(a) Use a profile model to calculate the steady-state head distribution assuming the aquifer is isotropic and homogeneous. Plot a head profile using the data from a depth of 70 m below the top of the aquifer.

(b) Insert the area of low hydraulic conductivity shown in Fig. P6.2 and

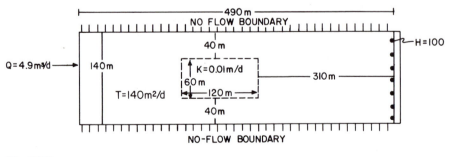

Fig. P6.2

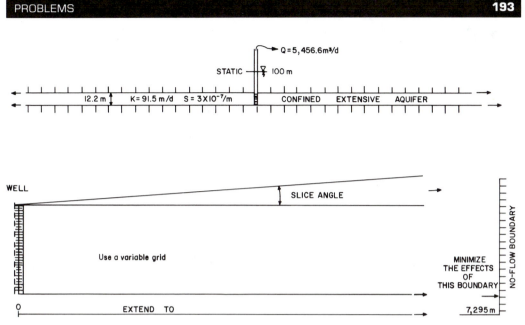

Fig. P6.3

rerun the model. Contour the steady-state head data and plot a head profile similar to the profile plotted in Problem 6.3a. Describe how the zone of low hydraulic conductivity affects the head distribution.

6.4 Axisymmetric profile models can be used to simulate the effects of pumping wells.

(a) The aquifer represented in Fig. P6.3 is confined and areally extensive, covering tens of kilometers. Follow the methods described in Sections 6.2 and 6.3 to construct a model of this confined aquifer. You will need to calculate the variable thickness of the profile based on an assumed angle for the pie-shaped wedge of the aquifer. The transmissivity and specific storage must be adjusted accordingly to calculate the transmissivity and storage coefficient for the cell. Use seven time steps to calculate the heads and drawdowns after 1 day of pumping.

(b) Plot a drawdown profile using the results of Problem 6.4a. Compare numerical results to the drawdown calculated using the Theis equation.

6.5 See Fig. 6.8 and then read the paper by Maslia and Johnston (1984). Outline how boundaries were defined and give the rationale for that assignment.

7

SPECIAL NEEDS FOR TRANSIENT SIMULATIONS

"And time ... must have a stop."
—*Henry IV, Pt. 1*

Transient simulations are needed to analyze time-dependent problems. A transient simulation typically begins with steady-state initial conditions and ends before or when a new steady state is reached. Transient simulations produce a set of heads for each time step, whereas steady-state simulations generate only one set of heads. Hence, transient simulations are more complicated simply in terms of data management. A computer-aided contouring package is an invaluable aid in analyzing the multiple sets of head data produced by a transient solution. Postprocessors that include contouring routines are available for many groundwater flow codes.

Transient head data can be displayed using a series of contour maps or hydrographs to show the transient variation of water level at selected nodes (Fig. 7.1).

Transient problems are inherently more complicated for reasons other than data management:

1. The storage characteristics of the aquifers must be specified. The storage characteristics of the confining layers must be specified when transient release of water from storage is important.
2. Initial conditions giving the head distribution in the aquifer at the beginning of the simulation must be specified in addition to boundary conditions.
3. Hydrologic stresses (e.g., pumping) may propagate out to designated

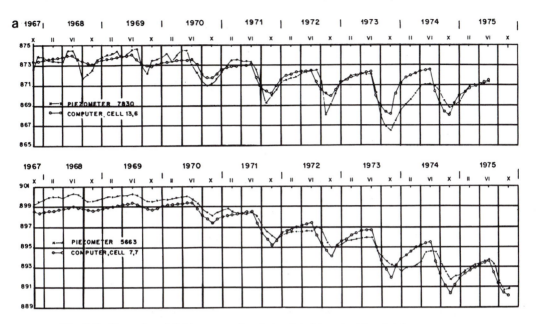

Fig. 7.1 Examples of field-measured and simulated transient head data.
(a) Simulated and observed heads for the Uluova Plain in southeastern Turkey (Karanjac et al., 1977). The observed drawdown averaged over a 7-year period and in 24 wells was 6.3 m. The simulated average drawdown at 24 nodes was 6.0 m.
(b) Simulated and observed drawdown in selected wells in a confined sandstone aquifer in southeastern Wisconsin (U.S. Geological Survey, 1976).
(c) Simulated and observed heads (numbers next to black dots) in a profile model of the Snake Lake area in northern Wisconsin. The contour map labeled $t = 0$ represents dynamic steady-state initial conditions. Transient simulations at 20, 23, and 36 days show the response of the system to changes in recharge rate (Anderson and Munter, Water Resources Research, 17(4), pp.1139–1150, copyright by the American Geophysical Union, 1981).

hydraulic boundaries of the model and cause the boundary conditions to become inappropriate.

4. The time dimension as well as the space dimension must be discretized.

Each of these special needs is discussed below.

7.1 Storage Parameters

During a transient simulation, water is released from or taken into storage within the porous material. Heads change with time as a result of this transfer of water. When the transfer to and from storage stops, the system reaches steady state and heads stabilize. In performing a transient simulation, it is necessary to specify the parameter that describes the capacity of an aquifer to transfer water

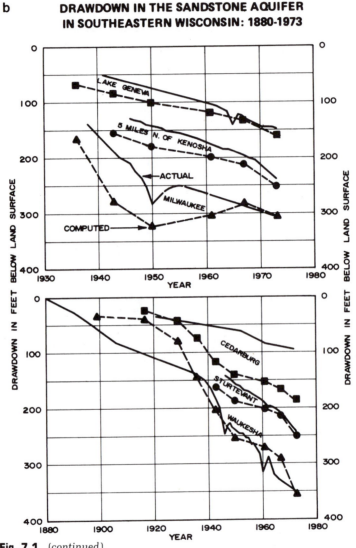

COMPARISON OF ACTUAL AND COMPUTED DRAWDOWN IN THE SANDSTONE AQUIFER IN SOUTHEASTERN WISCONSIN: 1880-1973

Fig. 7.1 (continued)

to and from storage. This property is known as storativity and is described by one of the following storage parameters: specific storage (S_s), storage coefficient (S), or specific yield (S_y).

Specific storage (S_s), which is used in the three-dimensional governing equation (Eqn. 2.2), is equal to the volume of water released from storage within a unit volume of porous material per unit decline in head. Storage coefficient (S) is used in two-dimensional areal simulations; it is a vertically averaged

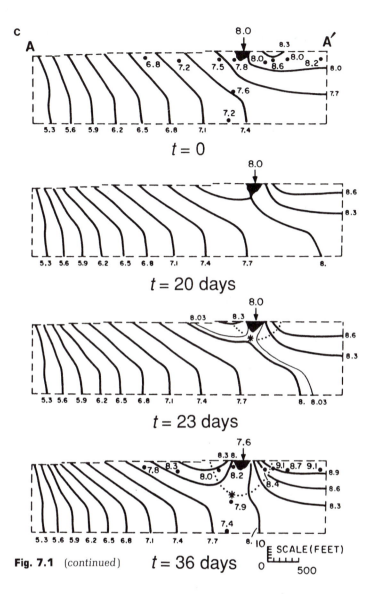

Fig. 7.1 (continued)

parameter equal to the volume of water released per unit area of aquifer per unit decline in head. Just as $T = bK$, $S = bS_s$, where b is aquifer thickness and S_s and S measure the volume of water released from aquifer compression and by expansion of water. The relevant storage parameter for unconfined aquifers is *specific yield* (S_y), which is a measure of the volume of water (ΔV) per volume of porous material ($A \, \Delta h$) released by gravity drainage in response to decline of the water table.

Values of storativity are not easily obtained. S and S_y can be measured during pumping tests, but these estimates, particularly of S_y, are subject to error (Neuman, 1979). Typical ranges in storage parameters are given in Tables 3.4 and 3.5.

Models require input of arrays of storativity values—one value for each node, cell, or element. In practice, storativity is often assumed to be uniform within an aquifer or confining bed. Fortunately, the range in any given storage parameter is small relative to the range possible for hydraulic conductivity (Table 3.3). For some problems, however, results may be sensitive to the storage parameter (Fig. 7.2).

Confining layers are not represented directly in two-dimensional areal and quasi three-dimensional simulations. Storage properties of the confining layers in these simulations are not normally considered but flow across the layer is represented by a leakage term. The steady-state leakage term is given in Eqn. 2.1. A transient leakage term $(L_{i,j,k})_T$ was derived by Bredehoeft and Pinder (1970) and used by Trescott et al. (1976). It is equal to the steady-state leakage $(L_{i,j,k})$ plus an additional term.

$$(L_{i,j,k})_T = L_{i,j,k} + [(h_{i,j,k})_0 - h_{i,j,k}] \frac{(K_z')_{i,j}}{\left(\frac{\pi}{3} t_D\right)^{1/2} b_{i,j}'} \left[1 + 2 \sum_{n=1}^{\infty} \exp\left(\frac{-n^2}{t_D}\right)\right]$$

$$(7.1)$$

$$t_D = \frac{t(K_z')_{i,j}}{b_{i,j}'^2 (S_s)_{i,j}}$$

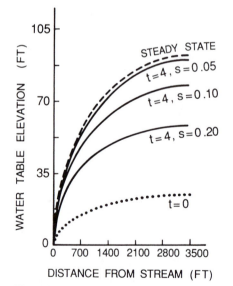

Fig. 7.2 Water table profiles showing the effect of storage coefficient (S) (here equal to specific yield) on a one-dimensional simulation of an unconfined aquifer receiving a constant rate of recharge equal to 2 ft/month. Groundwater discharges to a stream at distance equal to zero; t is time in months and the head at $t = 0$ is the initial condition. At steady state, the solution is independent of S (Zucker, Remson, Ebert and Aguado, Water Resources Research, 9(3), pp. 586–592, 1973, copyright by the American Geophysical Union).

where $(h_{i,j,k})_0$ is the head in the aquifer at the start of pumping; S_s is specific storage of the confining layer, K_z' is the vertical hydraulic conductivity of the confining layer, b' is its thickness, and t_D is dimensionless time.

Transient leakage from confining beds is not included in the current versions of PLASM or AQUIFEM-1, or when MODFLOW is used in two dimensions or in quasi three-dimensional mode. If transient release of water from storage in the confining layer is judged to be important, these models could be modified to include Eqn. 7.1. Alternatively, full three-dimensional models can be used to simulate the confining beds as one or more separate layers, each with its own storage coefficient (e.g., Watts, 1989). Leake (1990) described a model that accounts for permanent compaction of compressible interbeds as water is released from storage. The model was applied to simulate groundwater flow in the Central Valley of California.

7.2 Initial Conditions

Initial conditions refer to the head distribution everywhere in the system at the beginning of the simulation and thus are boundary conditions in time. It is standard practice to select as the initial condition a steady-state head solution generated by a calibrated model. The reason for using this type of head distribution is explained by Franke et al. (1987) as follows:

> Use of model-generated head values ensures that the initial head data and the model hydrologic inputs and parameters are consistent. If the field-measured head values were used as initial conditions, the model response in the early time steps would reflect not only the model stress under study but also the adjustment of model head values to offset the lack of correspondence between model hydrologic inputs and parameters and the initial head values.

Two types of steady-state solutions can be used as initial conditions: static steady state and dynamic average steady state. Under *static steady-state* conditions, head is constant throughout the problem domain and there is no flow of water in the system (Fig. 7.3a). The static steady-state solution is used for so-called drawdown simulations (Section 4.3) where the intent of the simulation is to calculate drawdown in response to pumping. In this type of simulation, relative heads as measured by drawdown are of interest, rather than absolute values of head. The principle of superposition is used to justify calculation of drawdown from an arbitrary horizontal datum that represents the initial head distribution. Results from a simulation of drawdown in the Chicago area in which static steady state was used as an initial condition are shown in Fig. 7.4. Static steady-state conditions do not incorporate the regional flow caused by regional head gradients and for this reason may not be appropriate in some simulations. Most codes, including MODFLOW and AQUIFEM-1, will calculate drawdowns relative to any user-specified initial head distribution.

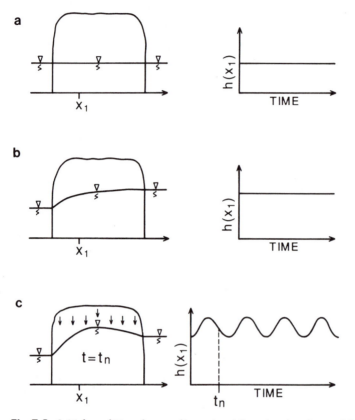

Fig. 7.3 Initial conditions for one-dimensional flow showing the spatial variation of head $h(x)$ on the left and the temporal variation of head at a specific point in space, $h(x_1)$, on the right.
(a) Static steady state; $h(x)$ is constant in space and time.
(b) Dynamic average steady state; $h(x)$ is constant in time but not in space.
(c) Dynamic cyclic conditions; $h(x)$ varies in both space and time.

Typically, the user-specified initial head distribution is calculated in a steady-state simulation run prior to the transient simulation. This initial head distribution represents dynamic average steady-state conditions. Under *dynamic average steady-state* conditions, head varies spatially and flow into the system equals flow out of the system (Fig. 7.3b). The dynamic average steady state is the initial condition used most frequently. For example, the solution of Problem 4.1a is used as the initial condition for Problem 7.3a.

In addition to the two types of steady-state initial conditions discussed above, there is the possibility of using a transient solution to generate *dynamic cyclic initial conditions*. Dynamic cyclic conditions consist of a set of heads that represent cyclic water level fluctuations. For example, the cycle might

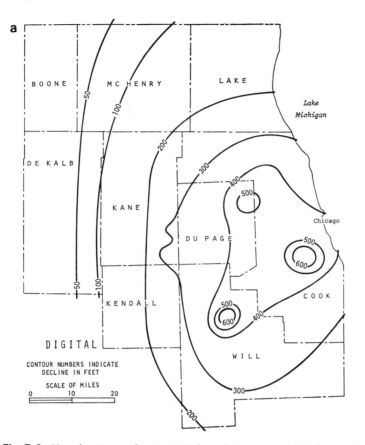

Fig. 7.4 Use of static steady-state initial conditions to simulate the transient effects of pumping around Chicago, Illinois (Prickett and Lonnquist, 1971).
(a) Contours showing the decline in water level from the assumed static steady-state initial condition for the period 1864–1958.
(b) The finite difference grid used in the simulation. The contour map shown in (a) is for the area that includes the pumping centers numbered from 1 to 7. No-flow conditions are used for all boundaries.

represent monthly head fluctuations for a series of average years. Because all years are the same, each cycle consists of the identical set of head values (Fig. 7.3c). That is, the heads in January of each cycle are the same, as are the heads in February, and so on through December. The starting condition is "dynamic" because the heads change monthly. The starting condition is a kind of steady state, however, because the cycle as a whole is at steady state. In short, average monthly heads for a particular month do not change with time although heads *do* change from month to month within an average year. Dynamic cyclic starting conditions may be generated by running the model with a set of cyclic inputs (monthly average recharge rates, for example) until the resulting heads

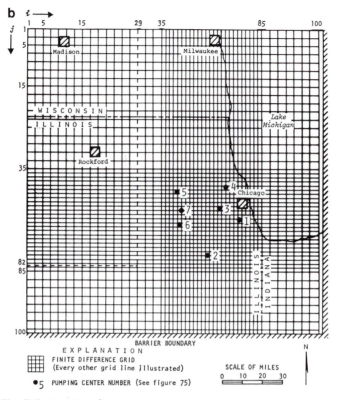

Fig. 7.4 *(continued)*

produce the desired cyclic pattern. Generating the dynamic cyclic starting condition requires a transient simulation and generally involves calibration to an observed hydrograph (Fig. 8.6). A predictive simulation would initiate some stress in the average yearly cycle (e.g., Problem 7.3c).

Another alternative in selecting a starting head distribution for a transient simulation is to use an arbitrarily defined head distribution and then run the transient model until it matches field-measured heads. Then, these calibrated heads are used as starting conditions in predictive simulations (Watts, 1989). The rationale behind this selection of initial conditions is that the influence of the initial conditions diminishes as the simulation progresses, so errors associated with selecting possibly erroneous initial conditions will be small provided sufficient simulated time has elapsed.

7.3 Boundary Conditions

In transient simulations it is important to monitor the way in which transient effects propagate out to the boundaries. Boundary conditions commonly are selected to achieve a dynamic average steady-state calibration, which is then used as the initial condition for a transient simulation. It is necessary to consider whether the stresses imposed on the system during the transient simulation will propagate out to the boundaries of the modeled system. A key concern is whether the stresses in the field would cause heads or flows to change in the section of the aquifer that is designated as a boundary in the modeled system. For example, stresses imposed on the system might cause a groundwater divide to shift position (Problems 4.1 and 7.3). The solution may be quite different if the boundary is fixed at the old position of the groundwater divide and represented by a no-flow boundary. If the hydrologic stresses propagate to the boundaries (Problem 4.6) and cause a simulated effect that is unrealistic, it will be necessary to change the boundary conditions by expanding the grid and moving the boundaries farther from the center of the grid.

In many modeling applications, it is impossible to locate the boundaries of the model at the physical boundaries of the system. In this case, hydraulic conditions are set at the boundaries of the problem domain (Section 4.2). If the transient simulation is run to steady state, the transient response will propagate to the boundary. The effects on the boundary should always be evaluated by checking the change in flow rates across specified head boundaries and the change in heads at specified flow boundaries. For specified head boundaries, the flow across the boundary should be the same for the initial condition and for the final time step of the transient solution. For specified flux boundaries, the heads along the boundary should be the same for the initial condition and the final step of the transient solution. The effects of the boundary conditions on the solution can also be evaluated by changing specified head conditions to specified flow and vice versa (Problem 4.6). If the resulting changes in the head solution are insignificant, the boundary does not affect the solution (de Marsily, 1986).

No-flow boundaries may be set far from the center of the grid to simulate the transient effects of pumping (Fig. 7.4b). In this case, the desired results are changes in head (drawdown) as a result of pumping, rather than the absolute values of head. Static steady-state initial conditions, where the heads are set equal to some arbitrary reference head, e.g., zero, are selected. Alternatively, other types of boundary conditions could be selected to produce dynamic average steady-state initial conditions. Again, the boundaries are set far from the center of the grid where pumping occurs. These drawdown simulations are valid until the effects of pumping reach the boundaries. Two case studies involving the selection of boundaries in transient simulations are discussed in Box 4.3.

Of course, it may be that interaction of the aquifer with the boundaries is realistic and appropriate. For example, the purpose of the model may be to simulate the effect of induced recharge from a large lake that is simulated as a constant-head boundary. When the cone of depression caused by pumping intersects the boundary, the model will simulate the effects of drawing water from the lake and into the aquifer.

7.4 Discretizing Time

CHOOSING THE TIME STEP

Selection of the time step (Δt) and construction of the grid are critical steps in model design because the values of the space and time discretization strongly influence the numerical results. Ideally, it is desirable to use small nodal spacing and small time steps so that the numerical representation better approximates the partial differential equation. The sensitivity of the solution to both space and time steps should be tested (Section 8.2).

Time steps may also be influenced by the requirements of a particular code. For example, the unsaturated flow equation (Section 12.3) is subject to numerical instabilities that may cause the model to calculate unrealistic oscillating values of pressure head. Similar difficulties may be encountered when solving the solute transport equation (Section 12.5). Numerical oscillations can usually be prevented by decreasing the time step. Groundwater flow codes are less prone to numerical instability (Section 8.2). Nevertheless, it is good modeling practice to make several trial runs of the model using different Δt's. Then the largest possible Δt that does not significantly change the solution can be used in production runs (Fig. 7.5).

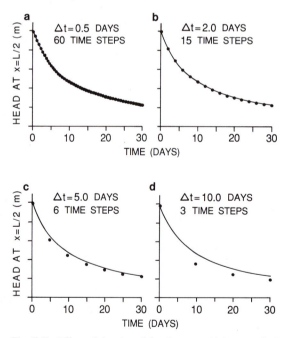

Fig. 7.5 Effect of the size of the time step (Δt) on a solution for the decay of a groundwater mound. Solid line is the analytical solution; circles represent the numerical solutions (adapted from Townley and Wilson, 1980).

Just as it is desirable to use small nodal spacing, ideally one would like to use small time steps to obtain an accurate solution. However, it is usually impractical to use extremely small time steps. A good order-of-magnitude estimate for the initial time step is the maximum time step allowed for an explicit formulation of the governing equation (de Marsily, 1986). This time step is sometimes called the critical time step (Δt_c). For a two-dimensional areal formulation of the governing equation using a regular grid ($\Delta x = \Delta y = a$) and assuming a homogeneous and isotropic aquifer,

$$\Delta t_c = Sa^2/4T \tag{7.2}$$

In more general applications Δt_c can be approximated by selecting a representative cell or element dimension, a, and representative cell or element properties, S and T. Then Δt_c is calculated from Eqn. 7.2.

The solution is sensitive to rapidly fluctuating water levels caused by introducing a stress, making it important to use time steps on the order of Δt_c to capture the early response of the system even if one is interested only in the solution at later times. For example, Fig. 7.5 shows that the solution at 10 and 20 days is inaccurate when a time step of 10 days is used. Note that even when a large time step is used, the solution becomes more accurate as steady state is approached. In fact, the procedure for obtaining a steady-state solution in some codes (e.g., PLASM) is to set the time step to some large value, so that the model takes one large time step to reach steady state. In MODFLOW and AQUIFEM-1 storage coefficients are set equal to zero to achieve a steady-state solution in one time step.

Instead of using small time steps when a stress is introduced, one could ignore results from the first few time steps. Prickett and Lonnquist (1971) compare a PLASM solution with the Theis analytical solution using time steps ranging from 0.5 to 10 days. In all cases the numerical solution agreed with the Theis solution after about six time steps. As a rule of thumb, the solution should proceed through five time steps, during which there are no significant changes in values of sources and sinks or boundary conditions, before the solution is considered accurate (de Marsily, 1986, p. 400).

Most codes allow the time step to increase as the simulation progresses. Increasing the time step is recommended when simulating a stress, such as pumping, that is applied to the aquifer at the beginning of the simulation. It is recommended, however, that Δt be reduced any time during the simulation that a new stress is imposed on the system.

Time steps typically are increased as a geometric progression of ratio 1.2 to 1.5. According to de Marsily (1986, p. 399) $\sqrt{2}$ is often a good choice. MODFLOW has a user-specified multiplier (TSMULT) to increase Δt as a geometric progression. The code computes the initial time step, DELT(1), using the following formula:

$$\text{DELT (1)} = \text{PERLEN (1-TSMULT) / (1-TSMULT**NSTP)} \tag{7.3}$$

where PERLEN is the length of a stress period (see discussion below) and NSTP is the number of time steps in a stress period. PLASM uses a similar formula.

AQUIFEM-1 allows for increase of the time step in a multiplicative or additive way with a user-specified factor F:

$$\Delta t_n = \Delta t_{n-1} * F \tag{7.4a}$$

$$\Delta t_n = \Delta t_{n-1} + F \tag{7.4b}$$

The total elapsed time after n time steps can be calculated using the following formulas:

$$t_n = t_0 + m \, \Delta t_0 \left(\frac{F^{n/m} - 1}{F - 1} \right) \tag{7.5a}$$

$$t_n = t_0 + n \, \Delta t_0 + \frac{n}{2} \left(\frac{n}{m} - 1 \right) F \tag{7.5b}$$

where 7.5a is for use with the multiplicative formula of 7.4a and 7.5b is for use with the additive formula of 7.4b. In Eqns. 7.5, t_0 is the starting time, Δt_0 is the initial time step, and m is the number of consecutive time steps for which Δt was held constant (typically $m = 1$). For the solution techniques in AQUIFEM-1, the stability and accuracy of the solution require that Δt_c (Eqn. 7.2) be less than $a^2 S/2T$ for good accuracy. The ratio $\Delta t T/a^2 S$ should be calculated for a number of elements, especially in areas of special interest. This ratio ideally should be less than 0.5 but never exceed 20. It should be used as a guide in selecting the initial time step.

STRESS PERIODS

Most flow codes allow the user the option of discretizing the simulation period into blocks of time of variable lengths. These smaller blocks of time are known as stress periods in MODFLOW and as withdrawal rate schedules in PLASM. AQUIFEM-1 refers to sequential blocks of time simply as time one, time two, and so on. We will refer to these blocks of time as stress periods.

The use of stress periods offers the option of changing some of the parameters or stresses while the simulation is in progress (Box 7.1). In PLASM changes are limited to the discharge or recharge rates. In MODFLOW one has the option of changing parameters associated with head-dependent boundary conditions in the River, Drain, Evapotranspiration, and General Head Boundary Packages (Box 4.1), as well as the recharge rates in the Recharge Package (Section 5.2) and pumping rates in the Well Package. The time step multiplier (TSMULT) may be reset at the beginning of each stress period and a new initial Δt is computed using Eqn. 7.3.

In AQUIFEM-1, all types of boundary conditions as well as sources and sinks can be reset at the beginning of a new stress period. The time stepping sequence as defined in Eqn. 7.4 is not restarted, however. It is advisable to use small uniform time steps if new stresses are imposed during the simulation. Alternatively, one can use a variable time step up until the new stress is introduced. The output can be saved and the simulation restarted with the new stress and a small time step.

Box 7.1
Case Study of a Transient Simulation

Lines (1976) used a two-dimensional finite difference model to predict changes in an unconfined aquifer in southeastern Wyoming during the period 1974–1979. The purpose of the simulation was to evaluate the effect of pumping from the aquifer by irrigation wells. The simulation involved a transient calibration to 1974 water levels and a predictive simulation to calculate water levels through 1979.

The aquifer is composed of sandstone and conglomerate of the Arikaree and Ogallala formations and the sandy siltstone of the White River formation in the western portion of the study area. Water levels in about 150 wells (Fig. 1) were measured in September 1973 to determine the potentiometric surface at the beginning of the calibration period. Groundwater flows from west to east across the study area, discharging to wells and the alluvium of streams and rivers. A regular grid with 59 rows and 35 columns and uniform nodal spacing of 0.5 mi was used to represent the 340 square mile study area, with boundary conditions as shown in Fig. 2.

Hydraulic conductivity was estimated from two aquifer tests and 12 specific capacity tests in the study area and from six tests outside the area. Values of hydraulic conductivity ranged from 0.4 to 0.7 ft/day. Specific yield was estimated by calculating the net volume of water that drained from the aquifer during the period January–February, 1974. The difference between the volume of water that recharged the aquifer and that discharged to streams provided an estimate of the net volume of water drained from the aquifer. Changes in water levels in wells were used to estimate the volume of porous material drained. The ratio of net volume of water drained to volume of porous material drained gives the specific yield, which in this case was 0.12.

Surface water discharge records collected at continuous recording gaging stations and from synoptic stream seepage studies (Fig. 1) were used to estimate leakage to and from streams and the North Laramie Canal. Underflow at the boundaries of the study area was also estimated. Monthly pumpage from the 22 irrigation wells shown in Fig. 1 was estimated from efficiency test data and from electrical power consumption. Recharge was assumed to be 6.5% of monthly precipitation. Recharge from irrigation water was assumed to be negligible. Monthly recharge and discharge totals for the 1974 water year are shown in Fig. 3.

For the transient calibration, each month in the 1974 water year was treated as a pumping or stress period and was divided into 14 time steps. October water levels were selected as initial conditions. During the trial-and-error calibration, adjustments were made to values of leakage factors and hydraulic conductivity until differences between simulated and measured heads and leakage rates to streams were acceptable. Then the model was used to predict the effects of continued pumping of the aquifer at 1974 rates. Stress periods 6 months in length were used to simulate the period 1975–1979; each pumping period was divided into 15 time steps. The first time step was about 5 hours and each successive time step was increased by a factor of 1.5. The monthly discharge rates calculated for the 1974 water year were averaged over 6-month periods for use in the predictive simulation. Recharge was set equal to 6.5% of the long-term average precipitation. The model predicted changes in the potentiometric surface after 5 years of withdrawal and changes in leakage from the aquifer to streams (Fig. 4).

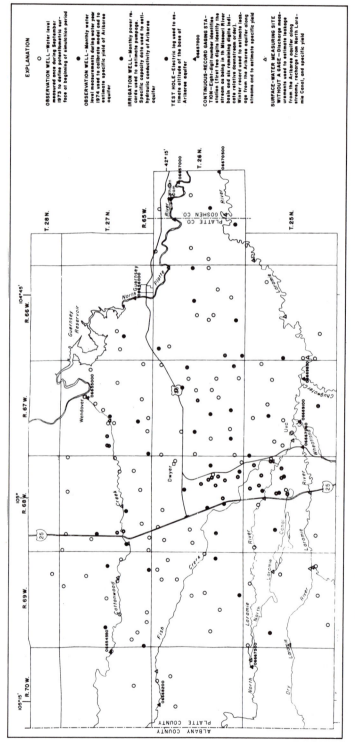

Fig. 1 Data collection network for the characterization and simulation of the Arikaree aquifer, southeastern Wyoming (Lines, 1976).

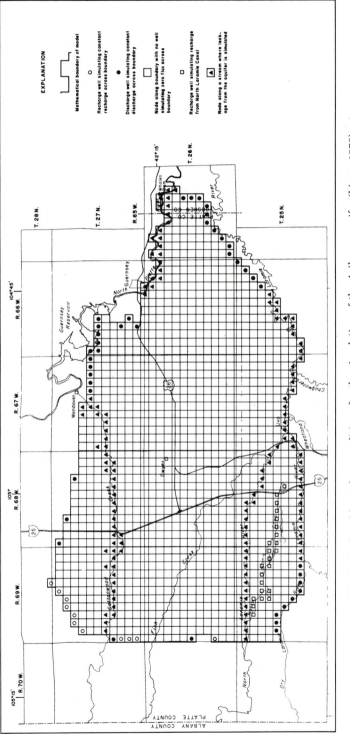

Fig. 2 Grid and boundary conditions for the simulation of the Arikaree aquifer (Lines, 1976).

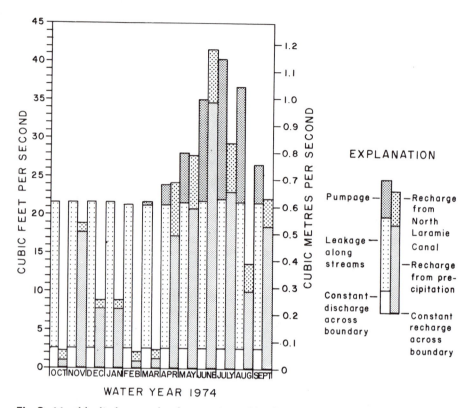

Fig. 3 Monthly discharge and recharge estimated for the Arikaree aquifer during water year 1974.

Problems

7.1 For many types of regional modeling problems, the magnitude and spatial distribution of the storage coefficient are not well defined. Model the situation described in Problem 5.1c using storage coefficient values of 10^{-6}, 10^{-5}, 10^{-4}, and 10^{-3} in separate runs of the model. Tabulate the drawdown at the well node and at a second node about 1000 m from the pumping well. Comment on how you would set an acceptable range of storage coefficients for use in a regional model.

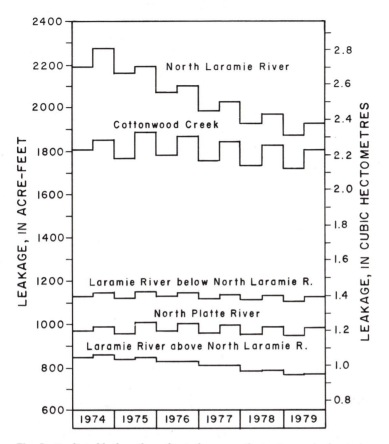

Fig. 4 Predicted leakage from the Arikaree aquifer to streams in the study area during water years 1974 through 1979 assuming no new development after water year 1974 (Lines, 1976).

7.2 Problem 5.1c can also be used to illustrate the effect of time discretization on simulated heads. Run the model with $S = 10^{-5}$ to predict the head and drawdown after 1 day of continuous pumping using 1, 5, and 10 time steps of uniform length. Compare and contrast your results. Then run the model with 10 variable-length time steps. Use smaller time steps at early times and fewer toward the end of the 1-day pumping period. What effect does using variable time steps have on the results?

7.3 Initial conditions for transient models may consist of a set of static steady-state heads as in Problem 5.1 or a dynamic average steady-state head distribution as in Problem 4.1a. Some problems will require a third representation of initial conditions, namely the use of dynamic cyclic conditions. In Problem 4.1 we attempted to evaluate the effect of pumping from a well penetrating a sand and gravel aquifer receiving areal recharge. The town has now proposed to complete the well and pump it for 1 year to meet emergency water supply demands. They argue that operation of the well will act as a long-term test so that the effect on the swamp can be assessed. You are asked to predict the effect prior to the 1-year test.

(a) Convert the steady-state pumping model in Problem 4.1c into a transient model using $S = 0.1$ and predict the head distribution and reduction in flow to the swamp caused by 1 year of continuous pumping. Heads should be calculated at the end of each month starting with January. Prepare a profile of the head distribution for each month. The profile should be oriented north-south and pass through the pumping node.

(b) Your model and predictions from Problem 7.3a have been reviewed and it is suggested that your initial conditions were inappropriate because they were based on January water levels. Thus, in effect you assumed an initial steady-state water table and a constant rate of recharge over the entire year. Inspection of well hydrographs suggests that the water table rises 2 m each year in early spring and then declines over the rest of the year to the initial January low water table. Monthly recharge rates estimated from hydrograph and water balance calculations are presented below. Produce the head distribution (without pumping) at the end of each month for a 1-year period starting with January. Construct a hydrograph of a node located on or near the groundwater divide for this 1 year of record.

Estimated Values of Recharge (m/day)

January	0.001
February	0.003
March	0.004
April	0.003
May	0.001
June	0.0005
July	0.00005
August	0.00005
September	0.00005
October	0.00005
November	0.00005
December	0.0005

(c) Using the dynamic cyclic initial conditions derived in 7.3b, determine the monthly head and drawdown distribution resulting from pumping the well for 1 year. Compare and contrast the monthly and 1-year results with those from Problems 4.1c and 7.3a. Look again at the hydrograph produced in Problem 7.3b and the table of recharge values provided above. Are your model results intuitive? Why or why not?

7.4 Review the report by Hamilton and Larson (1988).

(a) Outline the methods they used to select storage terms, boundary conditions, and initial conditions.

(b) Briefly describe the time discretization used for the transient model and the effect of pumping on internal sources and sinks.

8

MODEL EXECUTION AND THE CALIBRATION PROCESS

"There is something fascinating about Science. One gets such wholesale returns of conjecture out of such a trifling investment of fact."
—*Mark Twain, Life on the Mississippi*

"...judgement is selection of a fact. There are, in a sense, no facts in nature; or if you like, there are an infinite number of potential facts in nature, out of which the judgement selects a few which become truly facts by that act of selection."
—*Immanuel Kant, Critique of Judgement*

Recall from Chapter 1 that there are 12 steps in modeling. In previous chapters, we discussed steps 1, 2, and 4: establishing the purpose; developing the conceptual model; and designing the model including grid design, assignment of parameters to the nodes, cells, or elements of the model, assignment of boundary and initial conditions, and selection of time steps. In this chapter, code selection (step 3) and model calibration, sensitivity analysis, verification, and prediction (steps 5–9) are discussed. We also address problems that are typically encountered during the initial execution of the model. The end result of most modeling studies is usually a predictive simulation but the modeler must be prepared to defend the procedures and judgments used in all steps of the modeling process, particularly model calibration.

8.1 Code Selection

The following questions should be asked when selecting a computer code: (1) Has the accuracy of the code been checked (verified) against one or more analytical solutions? (2) Does the code include a water balance computation? (3) Has the code been used in other field studies; i.e., does it have a proven track record?

Codes for groundwater flow are *verified* by comparing the numerical results with one or more analytical solutions. Examples used in code verification are typically included in the user's manual. The purpose of code verification is to demonstrate that the numerical solution is relatively free of round-off and truncation errors, which, if uncontrolled, can lead to an unstable solution. The comparison of numerical results with an analytical solution will also depend on the choice of error criterion, grid spacing, and time step. PLASM, MODFLOW, and AQUIFEM-1 have been verified and produce numerically stable solutions. In general, the small round-off and truncation errors associated with numerically stable codes are not of concern in solving groundwater problems. Truncation errors are of concern in certain nonlinear problems, particularly in unsaturated flow modeling (Section 12.3) and when the solution of the flow problem is used in transport analysis.

A *water balance* calculation should be part of every modeling exercise. The water balance involves computation of flows across boundaries, to and from sources and sinks and storage (Table 8.1). The water balance gives information about discharge rates to surface water bodies or recharge rates across the water table. Some models provide a node-by-node printout of boundary fluxes and may compute fluxes between layers (Fig. 8.1). If the code does not contain a water balance computation, one should be added to the code or another code

Table 8.1

Estimated Volumes of Groundwater Recharge and Discharge for the Kirkwood-Conklin Aquifer System, New York

Sources	Volume[a]	Percent	Discharges	Volume	Percent
Recharge from precipitation on the floodplain	1.0	14	Production wells	1.2	17
Infiltration from upland areas	1.1	16	Net groundwater discharge to river	5.0	72
Underflow into study area	4.9	70	Underflow out of study area	0.8	11
Total	7.0	100		7.0	100

[a] Volumes are in millions of gallons per day.
Modified from Yager, 1986.

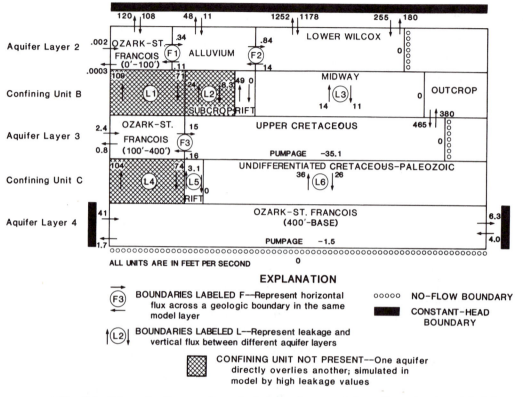

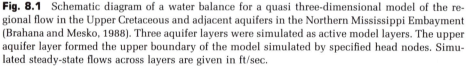

Fig. 8.1 Schematic diagram of a water balance for a quasi three-dimensional model of the regional flow in the Upper Cretaceous and adjacent aquifers in the Northern Mississippi Embayment (Brahana and Mesko, 1988). Three aquifer layers were simulated as active model layers. The upper aquifer layer formed the upper boundary of the model simulated by specified head nodes. Simulated steady-state flows across layers are given in ft/sec.

should be selected. A small error in the water balance is another assurance that the code correctly and accurately solves the mathematical model. MODFLOW, PLASM, and AQUIFEM-1 all have water balance computations.

Finally, the user should consider the code's track record. The two most widely used flow models are MODFLOW and PLASM. Both have been applied to numerous field problems. AQUIFEM-1 has not been used as widely.

Before applying a code to a field problem, the user should become familiar with the logic behind the assembly of input files. Careful study of example problems in the user's manual is recommended, but it is also advisable to apply the code to a simple problem with a known solution that is not in the user's manual. Constructing the data files for this problem will allow the user to test his or her understanding of the structure of the input procedure and determine whether the instructions in the manual have been interpreted correctly.

8.2 Initiating Model Execution

Even though much time has been spent on steps 1–4 of the modeling process, getting the first error-free output will require many additional hours. The first execution of the model requires entry of input data into computer files, execution of the model, and interpretation of results. These steps may appear straightforward, but in practice the process is involved and time consuming.

An overview of the processes involved in synthesizing and screening field data and translating the data to model input was presented in Chapter 3 (Section 3.4). These parameter values become either the initial guesses for a trial-and-error calibration or prior information for an automated calibration (Section 8.3). Preparing the data files necessary for executing the code can be time consuming if the problem is complex or if the modeler is unfamiliar with the code. Some data entry errors should be anticipated; the first few runs will help identify these errors. Preprocessors and codes with a user-friendly screen menu can help in assembling input data, but they may limit flexibility if all options contained in the model are not presented. They may be useful in preparing data files for the first run of the model because they ensure proper formatting of input. On the other hand, a user familiar with the input structure of a particular code will find it tedious to respond to a lengthy list of questions and will prefer to compose the input files directly. An editor is essential for making modifications to existing input files. If complex changes are required, e.g., changing selected entries in a transmissivity array, a preprocessor such as Model Cad™ (Rumbaugh and Duffield, 1989) may prove helpful in manipulating the data files.

Model execution requires that the code be in an executable file. Most codes are written in a version of FORTRAN (e.g., MODFLOW and AQUIFEM-1) or BASIC (e.g., PLASM). Executable files can be produced with an appropriate compiler. Successful model execution also requires a computer system with sufficient available random access memory (RAM) to store input data and arrays created during execution of the model. Until recently, most FORTRAN compilers for IBM or compatible PCs accessed only 640K of RAM, limiting the number of nodes and options that could be used to represent the groundwater system. These limitations are overcome by using computer systems with new processors, larger RAM, and appropriate compilers. Storage and memory needs should be evaluated when selecting a particular code and computer.

The length of time necessary to execute the computer program for a given set of input data is the execution time. Execution time becomes a concern when deadlines are approaching or many executions are needed. The time involved in executing a model depends on (1) the efficiency of the compiler used; (2) the speed at which the computer can execute the program; (3) the size of the model, i.e., the number of nodes and type of governing equation being solved; and (4) the type of output device. Compilers read the source code and generate machine language statements for the computer hardware. Some compilers

Table 8.2

Comparison of MODFLOW Executions of Problem 7.3c on
Selected IBM Compatible Computers with Math CoProcessors

Computer system	Run time
PC	49 min 43 sec
XT	26 min 13 sec
286 (AT)	7 min 7 sec
386 16 MHz	5 min 8 sec
386 25 MHz	5 min 2 sec
486 25 MHz	1 min 40 sec
486 33 MHz	1 min 16 sec

All input-output was from the system hard drive.

route commands and information more efficiently than others, allowing faster execution. The design of the computer hardware also affects the execution time. Results of tests using MODFLOW to solve Problem 7.3c on different computers are shown in Table 8.2. These data give a rough relative comparison of how other models may perform on the listed computers.

The total time required for a simulation will also depend on the speed of the output device. Some models ask for keyboard input during the execution and write to the screen and/or printer frequently, thereby increasing the total run time. Execution can be speeded up by modifying the code to eliminate printer output or by purchasing a faster printer. Many codes write output to a file which may be viewed on screen. It is advisable to review complete print-outs of input and output files regularly, especially in the early stages of model execution when data errors are being identified.

Model executions may be accompanied by error messages that may terminate the execution. Operating system errors usually indicate that the computer was instructed to perform an illegal mathematical operation or that insufficient RAM is available for model execution. Error messages involving illegal mathematical operations can often be traced to incorrect data formats or errors in input values. Errors indicating insufficient memory must be corrected by providing more memory or reducing the size of the model. In MODFLOW, for example, the size of the model is limited by the size of the X array. Messages are printed at the beginning of the simulation to indicate the amount of space used in the X array. If the model is too large for the current dimensions of the X array, it may be redimensioned and the program recompiled, provided the computer has sufficient RAM and processing capability.

An iterative solution that does not meet an error criterion within the maximum number of iterations specified by the user produces a solution that has not converged. Failure to converge sometimes is due to inappropriate initial guesses of heads and parameters. It can also result from errors in typing input into computer files. The best way to isolate such errors is to print out and review input data as read by the code. Nonconvergence may also occur as a result of a poor conceptual model. For example, a model to solve for the steady-state head distribution in an aquifer receiving areally uniform recharge and surrounded by no-flow boundaries is a poorly conceived conceptual model. This model would never reach steady state; thus, the solution would not converge. Finally, nonconvergence may occur when the error criterion is set below the precision of the numerical solution. Considerations involved in selecting the error criterion are discussed in detail below.

ERROR CRITERION

The simulated heads include errors arising from truncation or discretization error and round-off error. The errors caused by discretization of both space and time should be assessed independently in the early stages of calibration. The effect of the size of the nodal spacing on the solution should be determined by varying the cell or element sizes and rerunning the model. Because of the difficulty in changing a finite difference grid, this is seldom done in finite difference modeling. Luckey and Stephens (1987) and Robson (1978) are among the few investigators to report on errors caused by coarse spacing in a finite difference grid. It is much easier to change a finite element grid because nodes can be added very easily without redesigning the entire grid. The effect of time discretization on transient simulation results should also be examined by using successively smaller time steps and examining the effect on the solution. Certain solution techniques have a limit on the size of the time step in order to ensure a stable solution. If the limit is exceeded, cumulative round-off errors will grow and eventually cause numerical instability (Huyakorn and Pinder, 1983).

The set of algebraic equations that represents the flow equation (Section 2.2) may be solved directly using matrix methods, indirectly using iteration, or by a combination of direct and indirect methods. When iteration is involved in the solution, there is an additional source of error known as iteration residual error. In this case, the user specifies an *error criterion* or *error tolerance* to judge convergence. The iteration residual error is reduced during iteration so that the change in the solution between iterations progressively decreases. The more iterations performed, the closer one expects to come to the exact solution, provided that the nodal spacing and time step have been selected to minimize discretization errors and avoid instability. The choice of error criterion controls the size of the iteration residual error and influences the number of iterations performed to achieve a solution within the specified error tolerance. Iteration stops when the change in heads between iterations is less than the error criterion selected by the user. The residual error can be minimized by selecting appropriately small values for error criteria for the head solution and/or the water balance.

Error Criterion for Heads

The value selected for the error criterion will depend on the method used to calculate the change in the solution between iterations. The change in heads between successive iterations is always computed for each node. Some codes then select the largest absolute difference in head as a measure of the residual error and compare it to the user-selected error criterion. This type of error check is used in MODFLOW. As a rule of thumb, the error criterion should be one to two orders of magnitude smaller than the level of accuracy desired in the head results.

Another method of calculating residual error is used in PLASM. The absolute values of the nodal differences in heads between successive iterations are summed for all the nodes in the grid and the sum is compared to the error criterion. PLASM calculates the default value for the error criterion to be 0.01*NR*NC, where NR and NC are the numbers of rows and columns, respectively. Given the different way numerical error is calculated in PLASM and MODFLOW, it is obvious that it is necessary to select a larger error criterion when using PLASM in order to achieve the same accuracy. AQUIFEM-1 allows the user to select one of two options for calculating model error: (1) the average root mean squared change in heads between successive iterations or (2) the maximum absolute change in heads between successive iterations.

Ideally, the error criterion for heads should be small to ensure an accurate solution. However, if the error criterion is set too low, the precision of the numerical solution will be exceeded and the model will approach the criterion value but never reach it. Some codes (Fig. 8.2) monitor the approach to closure and provide a printout of the error after each iteration. If the error criterion has been set lower than the precision of the numerical solution, the residual will oscillate around some value that is higher than the error criterion. Convergence can be obtained by raising the error criterion to this value.

The effect of the choice of error tolerance on results of a specific numerical solution is shown in Fig. 8.2 and also considered in Problem 8.3.

Error Criterion for the Water Balance

Another way of checking the amount of residual error in the solution is to compare the total simulated inflows and outflows as computed by the water balance (Table 8.1). In transient simulations, uptake and release of water to and from storage are also included in the water balance. In some codes (e.g., MODFLOW), cell areas or volumes associated with specified head nodes may not be considered part of the problem domain for the water balance computation. For example, refer to Example 2 in Box 4.2 (Figs. 3 and 4). When using a code that does not include specified head nodes in the water balance, inflow to this problem domain would consist of areal recharge applied to all nodes *except* the specified head nodes. In the water balance computation, the area receiving recharge would be the area of the problem domain *minus* the area of

a U.S. GEOLOGICAL SURVEY MODULAR FINITE-DIFFERENCE GROUND-WATER MODEL

STEADY STATE TOTH PROBLEM
1 LAYERS   6 ROWS   11 COLUMNS
   LAYER AQUIFER TYPE
   ----
      1   0

COLUMN TO ROW ANISOTROPY =   1.000000
DELR =   20.00000   DELC =   20.00000
TRANSMIS. ALONG ROWS =   10.00000   FOR LAYER 1

SOLUTION BY THE STRONGLY IMPLICIT PROCEDURE

HEAD CHANGE CRITERION FOR CLOSURE =   0.10000E+01

----

2 ITERATIONS FOR TIME STEP   1 IN STRESS PERIOD 1
MAXIMUM HEAD CHANGE FOR EACH ITERATION:
HEAD CHANGE LAYER,ROW,COL

-1.561   (1, 6, 11)
0.7873   (1, 6, 1)

HEAD IN LAYER 1 AT END OF TIME STEP 1 IN STRESS PERIOD 1

1	100.00	101.00	102.00	103.00	104.00	105.00	106.00	107.00	108.00	109.00	110.00
2	100.96	101.52	102.30	103.17	104.09	105.03	105.98	106.90	107.75	108.47	108.95
3	101.52	101.90	102.52	103.29	104.14	105.04	105.93	106.78	107.53	108.08	108.37
4	101.87	102.15	102.66	103.35	104.16	105.02	105.88	106.68	107.35	107.82	108.04
5	102.12	102.31	102.74	103.38	104.15	104.99	105.82	106.58	107.22	107.65	107.86
6	102.26	102.40	102.76	103.37	104.13	104.94	105.75	106.50	107.12	107.56	107.78

VOLUMETRIC BUDGET FOR ENTIRE MODEL AT END OF TIME STEP 1 IN STRESS PERIOD 1

         PERCENT DISCREPANCY =   -6.59

Fig. 8.2 Effect of changing the error criterion for the regional groundwater flow problem described in Box 2.1 and Box 4.2. Output is from MODFLOW.
(a) The error criterion equals 1.0. Two iterations are required. The water balance error of 6.59% is unacceptably high.
(b) The error criterion equals 0.001. Ten iterations are required. The water balance error of 0.02% is acceptable.

b   U.S. GEOLOGICAL SURVEY MODULAR FINITE-DIFFERENCE GROUND-WATER MODEL

STEADY STATE TOTH PROBLEM
    1 LAYERS   6 ROWS   11 COLUMNS
     LAYER   AQUIFER TYPE
     ----
      1   0
COLUMN TO ROW ANISOTROPY =   1.000000
DELR =   20.00000   DELC =   20.00000
TRANSMIS. ALONG ROWS =   10.00000   FOR LAYER 1

SOLUTION BY THE STRONGLY IMPLICIT PROCEDURE

HEAD CHANGE CRITERION FOR CLOSURE =   0.10000E-02

----

10 ITERATIONS FOR TIME STEP 1 IN STRESS PERIOD 1
MAXIMUM HEAD CHANGE FOR EACH ITERATION:
HEAD CHANGE LAYER,ROW,COL

-1.561   (1, 6, 11)
 0.7873   (1, 6, 1)
-0.7703   (1, 6, 11)
 0.3431   (1, 6, 1)
-0.7109E-01 (1, 3, 8)
-0.5795E-02 (1, 6, 4)
 0.6747E-02 (1, 6, 11)
-0.5684E-02 (1, 6, 1)
-0.2974E-02 (1, 6, 2)
-0.7823E-03 (1, 5, 6)

HEAD IN LAYER 1 AT END OF TIME STEP 1 IN STRESS PERIOD 1

1	100.00	101.00	102.00	103.00	104.00	105.00	106.00	107.00	108.00	109.00	110.00
2	101.34	101.84	102.54	103.32	104.15	105.00	105.85	106.68	107.46	108.16	108.66
3	102.18	102.49	102.99	103.61	104.29	105.00	105.71	106.39	107.01	107.51	107.82
4	102.72	102.94	103.32	103.82	104.39	105.00	105.61	106.18	106.68	107.06	107.28
5	103.05	103.22	103.54	103.96	104.46	105.00	105.54	106.04	106.46	106.78	106.95
6	103.20	103.35	103.64	104.04	104.50	105.00	105.50	105.96	106.36	106.65	106.80

VOLUMETRIC BUDGET FOR ENTIRE MODEL AT END OF TIME STEP 1 IN STRESS PERIOD 1

         PERCENT DISCREPANCY =   -0.02

Fig. 8.2 (continued)

the specified head cells. Likewise, flows between specified head cells are not normally computed.

In the water balance totals, release of water from storage is counted as inflow and uptake is counted as outflow. The difference between total inflow and outflow is divided by either inflow or outflow to yield error in the water balance. Ideally the error in the water balance is less than 0.1% (Konikow, 1978). An error of around 1%, however, is usually considered acceptable. MODFLOW and AQUIFEM-1 do not allow the user to set a separate error criterion for the water balance. PLASM, however, does allow the user to specify separate error criteria for the head solution and the water balance computation. Examples of water balance calculations performed by MODFLOW, PLASM, and AQUIFEM-1 are given in Fig. 8.3.

In addition to serving as a check on solution accuracy, a water balance is a way of identifying errors made when designing the model. For example, errors in entering transmissivity data may be reflected in unreasonably high or low fluxes to or from the model. Errors in entering the storage parameter will be reflected in the water balance as unreasonably high or low volumes of water entering or leaving storage. Failure of the model to reach a solution or a solution with a high water balance error may also indicate errors in data entry or may indicate that the conceptual model of the system is invalid. A large error in the water balance could also mean that the numerical solution is inaccurate because the error criterion was set too high or that the solution did not converge within the maximum number of iterations set by the user. Finally, a large error in the water balance may occur when using MODFLOW's General Head Boundary (GHB) Package (Box 4.1), but in this case the high error does not necessarily mean that the solution is unacceptable (Box 8.1).

8.3 The Calibration Process

Calibration of a flow model refers to a demonstration that the model is capable of producing field-measured heads and flows which are the *calibration values* (Fig. 8.4). Calibration is accomplished by finding a set of parameters, boundary conditions, and stresses that produce simulated heads and fluxes that match field-measured values within a preestablished range of error (Fig. 8.4). Finding this set of values amounts to solving what is known as the *inverse problem*. In an inverse problem the objective is to determine values of the parameters and hydrologic stresses from information about heads, whereas in the forward problem system parameters such as hydraulic conductivity, specific storage, and hydrologic stresses such as recharge rate are specified and the model calculates heads. Most classroom problems are formulated as forward problems but most field problems require solving an inverse problem. A complication in groundwater problems is that information about the head distribution is always incomplete. A simple type of inverse model that uses a water balance computa-

a U.S. GEOLOGICAL SURVEY MODULAR FINITE-DIFFERENCE GROUND-WATER MODEL
STEADY STATE TOTH PROBLEM
CONSTANT HEAD CELL-BY-CELL FLOWS WILL BE PRINTED
HEAD/DRAWDOWN PRINTOUT FLAG = 1   TOTAL BUDGET PRINTOUT FLAG = 0
CELL-BY-CELL FLOW TERM FLAG = 1

OUTPUT FLAGS FOR ALL LAYERS ARE THE SAME:
 HEAD   DRAWDOWN HEAD DRAWDOWN
PRINTOUT PRINTOUT SAVE   SAVE
----
  1   0   0   0

CONSTANT HEAD PERIOD 1 STEP 1 LAYER 1 ROW 1 COL 1 RATE -13.42507
CONSTANT HEAD PERIOD 1 STEP 1 LAYER 1 ROW 1 COL 2 RATE -8.425144
CONSTANT HEAD PERIOD 1 STEP 1 LAYER 1 ROW 1 COL 3 RATE -5.389025
CONSTANT HEAD PERIOD 1 STEP 1 LAYER 1 ROW 1 COL 4 RATE -3.244348
CONSTANT HEAD PERIOD 1 STEP 1 LAYER 1 ROW 1 COL 5 RATE -1.531670
CONSTANT HEAD PERIOD 1 STEP 1 LAYER 1 ROW 1 COL 6 RATE -0.7129161E-03
CONSTANT HEAD PERIOD 1 STEP 1 LAYER 1 ROW 1 COL 7 RATE 1.530261
CONSTANT HEAD PERIOD 1 STEP 1 LAYER 1 ROW 1 COL 8 RATE 3.242986
CONSTANT HEAD PERIOD 1 STEP 1 LAYER 1 ROW 1 COL 9 RATE 5.387733
CONSTANT HEAD PERIOD 1 STEP 1 LAYER 1 ROW 1 COL 10 RATE 8.423918
CONSTANT HEAD PERIOD 1 STEP 1 LAYER 1 ROW 1 COL 11 RATE 13.42389

VOLUMETRIC BUDGET FOR ENTIRE MODEL AT END OF TIME STEP 1 IN STRESS PERIOD 1

CUMULATIVE VOLUMES   L**3   RATES FOR THIS TIME STEP   L**3/T
-----------------   -----------------
IN: IN:
---   ---
STORAGE =   0.   STORAGE =   0.
CONSTANT HEAD =   32.009   CONSTANT HEAD =   32.009
TOTAL IN =   32.009   TOTAL IN =   32.009

OUT: OUT:
----   ----
STORAGE =   0.   STORAGE =   0.
CONSTANT HEAD =   32.016   CONSTANT HEAD =   32.016
TOTAL OUT =   32.016   TOTAL OUT =   32.016

IN - OUT = -0.71869E-02   IN - OUT = -0.71869E-02
PERCENT DISCREPANCY =   -0.02   PERCENT DISCREPANCY =   -0.02

----

b Flow from storage = 0   Flow into storage = 6.475564E-10
Flow in from leakance = 0   Flow out via leakance = 0
Flow in from withdrawal = 0   Flow out via withdrawal = 0
Flow from constant heads = 32.01813 Flow to constant heads =-32.00371

Total Flow IN = 32.01813   Total Flow OUT =-32.00371
Percent of unaccounted water = -4.505589E-02
WB = 100*(1 - ABS(TOTIN/TOTOUT))

Fig. 8.3 Examples of output showing water balance results.
(a) Output for the MODFLOW simulation of Fig. 8.2b.
(b) Output from PLASM for the same problem as in 8.3a.
(c) Output from AQUIFEM-1 for the same problem as in 8.3a. The grid for this problem with node and element numbers is shown in Box 3.3, Fig. 3.

C

NODE NO	PIEZOMETRIC HEAD		INTERNAL FLUXES (INSIDE THE AQUIFER)		EXTERNAL FLUXES (THROUGH THE BOUNDARIES)				FLUX VECTOR*	
	HEAD	DRAWDOWN	QX	QY	SOURCE/SINK INFLOWS	LEAKAGE INFLOWS	INFLOWS AT BOUNDARIES		R	D
	(L)	(L)	(L2/T)	(L2/T)	(L3/T)	(L3/T)	(L3/T)	ID	(L)	(D)
1	103.340	.000	-.043	.043	.000	.000	.000	0	.061	314.998
2	103.253	.000	-.045	.073	.000	.000	.000	0	.086	328.123
3	102.981	.000	-.055	.172	.000	.000	.000	0	.180	342.255
4	102.472	.000	-.080	.304	.000	.000	.000	0	.315	345.231
5	101.600	.000	-.151	.527	.000	.000	.000	0	.548	343.963
6	100.000	.000	-.341	.641	.000	.000	-12.998	1	.726	331.982
7	103.426	.000	-.071	.041	.000	.000	.000	0	.082	300.003
8	103.347	.000	-.089	.083	.000	.000	.000	0	.122	312.714
9	103.099	.000	-.109	.174	.000	.000	.000	0	.205	327.892
10	102.653	.000	-.158	.290	.000	.000	.000	0	.330	331.401
11	101.963	.000	-.286	.449	.000	.000	.000	0	.532	327.540
12	101.000	.000	-.439	.421	.000	.000	-9.631	1	.609	313.772
13	103.671	.000	-.146	.034	.000	.000	.000	0	.150	283.191
14	103.608	.000	-.166	.067	.000	.000	.000	0	.179	291.965
15	103.414	.000	-.197	.136	.000	.000	.000	0	.239	304.614
16	103.080	.000	-.258	.211	.000	.000	.000	0	.333	309.368
17	102.600	.000	-.363	.285	.000	.000	.000	0	.462	308.153
18	102.000	.000	-.459	.259	.000	.000	-5.995	1	.527	299.419
19	104.043	.000	-.203	.025	.000	.000	.000	0	.205	276.951
20	104.000	.000	-.221	.047	.000	.000	.000	0	.226	281.904
21	103.871	.000	-.254	.091	.000	.000	.000	0	.270	289.680
22	103.653	.000	-.313	.135	.000	.000	.000	0	.341	293.315
23	103.355	.000	-.399	.171	.000	.000	.000	0	.434	293.171
24	103.000	.000	-.469	.146	.000	.000	-3.547	1	.491	287.292
66	104.500	.000	-.239	.014	.000	.000	.000	0	.240	273.412
25	104.479	.000	-.254	.025	.000	.000	.000	0	.255	275.642
26	104.414	.000	-.286	.046	.000	.000	.000	0	.290	279.219
27	104.308	.000	-.341	.066	.000	.000	.000	0	.347	281.014
28	104.166	.000	-.415	.081	.000	.000	.000	0	.423	281.033
29	104.000	.000	-.472	.055	.000	.000	-1.658	1	.476	276.672
30	105.000	.000	-.254	.004	.000	.000	.000	0	.254	270.806
31	105.000	.000	-.264	.004	.000	.000	.000	0	.264	270.775
32	105.000	.000	-.296	.003	.000	.000	.000	0	.296	270.674
33	105.000	.000	-.349	.003	.000	.000	.000	0	.349	270.502
34	105.000	.000	-.419	.002	.000	.000	.000	0	.419	270.264
35	105.000	.000	-.472	-.028	.000	.000	.000	1	.473	266.654
36	105.500	.030	-.246	-.007	.000	.000	.000	0	.247	268.343
37	105.521	.000	-.254	-.018	.000	.000	.000	0	.254	265.977
38	105.586	.000	-.286	-.039	.000	.000	.000	0	.288	262.200
39	105.692	.000	-.339	-.060	.000	.000	.000	0	.344	260.050
40	105.834	.000	-.412	-.076	.000	.000	.000	0	.419	259.513
41	106.000	.000	-.469	-.114	.000	.000	1.658	1	.482	256.283
42	105.957	.000	-.218	-.018	.000	.000	.000	0	.219	265.321
43	106.000	.000	-.221	-.039	.000	.000	.000	0	.225	259.928
44	106.130	.000	-.254	-.083	.000	.000	.000	0	.267	251.956
45	106.347	.000	-.310	-.126	.000	.000	.000	0	.335	247.949
46	106.645	.000	-.392	-.163	.000	.000	.000	0	.425	247.459
47	107.000	.000	-.459	-.218	.000	.000	3.546	1	.508	244.592
48	106.329	.000	-.168	-.028	.000	.000	.000	0	.170	260.482
49	106.392	.000	-.167	-.060	.000	.000	.000	0	.178	250.215
50	106.586	.000	-.198	-.127	.000	.000	.000	0	.235	237.411
51	106.920	.000	-.256	-.198	.000	.000	.000	0	.323	232.298
52	107.401	.000	-.351	-.267	.000	.000	.000	0	.441	232.717
53	108.000	.000	-.439	-.360	.000	.000	5.994	1	.568	230.647
54	106.574	.000	-.099	-.037	.000	.000	.000	0	.106	249.471
55	106.654	.000	-.093	-.078	.000	.000	.000	0	.121	229.946
56	106.901	.000	-.116	-.166	.000	.000	.000	0	.202	214.846
57	107.347	.000	-.164	-.272	.000	.000	.000	0	.318	211.157
58	108.037	.000	-.265	-.398	.000	.000	.000	0	.478	213.658
59	109.000	.000	-.394	-.588	.000	.000	9.631	1	.707	213.839
60	106.660	.000	-.045	-.041	.000	.000	.000	0	.061	227.281
61	106.747	.000	-.051	-.101	.000	.000	.000	0	.113	206.650
62	107.019	.000	-.070	-.204	.000	.000	.000	0	.216	198.812
63	107.528	.000	-.121	-.345	.000	.000	.000	0	.366	199.340
64	108.401	.000	-.288	-.572	.000	.000	.000	0	.641	206.697
65	110.000	.000	-.500	-.800	.000	.000	12.997	1	.943	212.014
TOTALS				.000		.000	-.001			

NET FLOW INTO AQUIFER (L3/T)   -.001

THE TOTAL LEAKAGE INTO THE AQUIFER IS   .000E+00

---

* R and D are the magnitude and direction, respectively, of the flux vector.

Fig. 8.3 (continued)

tion to calculate recharge rates given the water table configuration was discussed in Chapter 5 (Section 5.2). The inverse problem also can be interpreted more broadly to include estimation of boundary conditions, hydrologic stresses, and the spatial distribution of parameters by methods that do not involve consideration of heads (e.g., kriging, Section 3.4).

Model calibration can be performed to steady-state or transient data sets. Most calibrations are performed under steady-state conditions but may also involve a second calibration to a transient data set. Some care is necessary in selecting a representative steady-state water level from a long transient record. Several types of average water levels for a 36-year hydrograph are shown in Fig. 8.5. The mean water level for the period of record, the mean annual water level, and the mean March water level are shown. Any one of these three averages might be selected as a steady-state calibration value, depending on the objective of the simulation. In the absence of a long record, it may be appropriate to let the seasonal average of heads in a given year represent dynamic average steady-state conditions. Or, depending on the problem and the modeling objective, it might be appropriate to assume that water levels measured during a certain period of time represent quasi steady-state conditions under stresses that prevail during that period.

In some hydrogeologic settings it may be inappropriate to assume steady-state conditions owing to large seasonal fluctuations in water levels, or a steady-state data set may not be available. In this case, the model may be calibrated to transient conditions (Fig. 8-6). In a transient calibration, calibration values may be taken from well hydrographs or from water levels during long-term production from the aquifer. The most common type of transient calibration begins the simulation from the calibrated steady-state solution. For example, initial conditions for the transient calibration may represent steady-state conditions prior to development of the aquifer (e.g., Luckey et al., 1986; Hearne, 1985). The model is then calibrated to a time series of water level changes caused by pumping. Alternatively, the model may be calibrated to a particular point in time represented by a field-measured potentiometric surface map or a set of hydrographs (Lines, 1976; Watts, 1989). The initial conditions are set arbitrarily and the model is run until the solution hits the calibration targets. It is assumed that the effects of the initial conditions do not influence the solution. A transient calibration or a transient verification test (Section 8.4) is necessary to calibrate values of the storage parameters, which are needed if a transient prediction is required (Section 8.5).

There are basically two ways of finding model parameters to achieve calibration, i.e., of solving the inverse problem: (1) manual trial-and-error adjustment of parameters and (2) automated parameter estimation. Manual trial-and-error calibration was the first technique to be used and is still the technique preferred by most practitioners. In the late 1970's, the use of automated calibra-

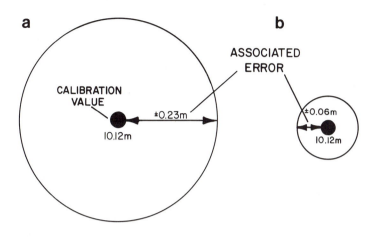

Fig. 8.4 A calibration target is defined as a calibration value and its associated error.
(a) A target for head with a large associated error, 10.12 m ± 0.23 m.
(b) A target for head with a small associated error, 10.12 m ± 0.06 m.

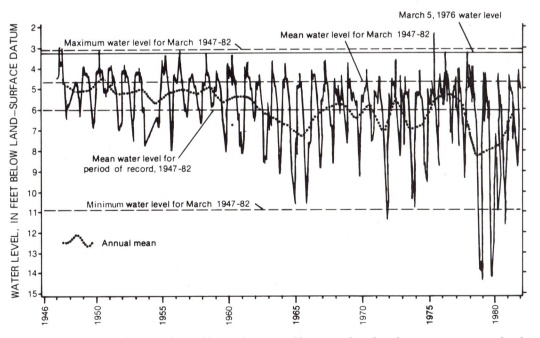

Fig. 8.5 Definition of possible steady-state calibration values based on a 36-year water level record for an observation well at Cortland, New York (Reynolds, 1987). Reynolds (1987) selected the water level for March 5, 1976, which he assumed represented a quasi steady-state. Other possible calibration values are the mean water level for the period of record, the mean water level for March for the period of record, and the mean water level for 1976.

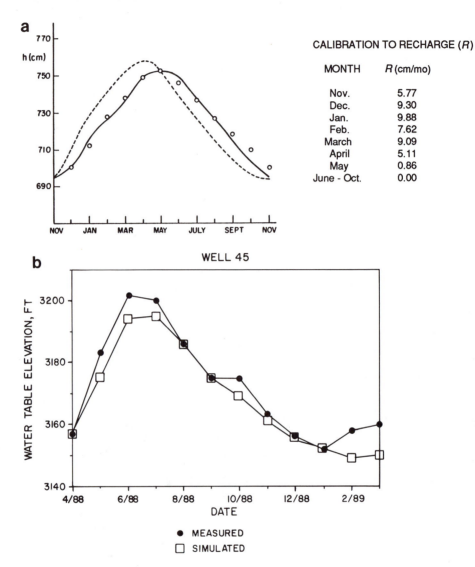

Fig. 8.6 Transient calibration.

(a) Calibration to dynamic cyclic conditions as defined by a well hydrograph. Calibration is achieved by adjusting monthly recharge rates. The solid line is an observed hydrograph for a site in New Jersey. The dashed line is the hydrograph simulated with a groundwater model; the circles represent a hydrograph simulated by means of a groundwater model coupled to a model of the unsaturated zone. (Pikul, Street and Remson, Water Resources Research, 10(2), pp. 295–302, 1974, copyright by the American Geophysical Union).

(b) Transient calibration to a well hydrograph for an unconfined aquifer receiving recharge from adjacent highlands and leaking streams and irrigation ditches located in western Montana. (Makepeace, 1989.) Initial conditions consisted of a calibrated steady-state solution for April 1988.

tion strategies started to be explored mainly in research applications. Efforts to find the best strategy for automated calibration have generated a considerable volume of literature, some of which is reviewed by Carrera (1988) and Yeh (1986). Only now in the early 1990's are codes that perform automated calibration being introduced to modeling practitioners. It may take another 10 to 20 years before the use of automated calibration codes becomes standard practice. This lag time can be attributed in part to the mathematical complexity of the problem and the need to establish a track record of successful applications to a variety of field problems.

The use of an automated calibration model imposes a framework for setting up the calibration strategy and examining and evaluating the calibration results. Modelers who perform trial-and-error calibration, however, have not been forced to adhere to any protocol. Consequently, many trial-and-error calibrations are poorly documented, making their reliability suspect.

Prior to performing the calibration either by trial and error or by using an automated calibration model, it is necessary to assess the head values and fluxes or other calibration data (called *sample information*), as well as the parameter estimates (called *prior information*) that will be used during the calibration process. A framework for assessing each of these two types of information is presented below.

SAMPLE INFORMATION

Field-measured values of heads and fluxes form the *sample information* or *calibration values*. These values always have associated error that must be quantified. The calibration value with its associated error forms the *calibration target* (Fig. 8.4). Calibration targets should be set before attempting to calibrate the model.

Heads

Head values always form part of the sample information. Sources of error in each calibration value must be assessed and the magnitude of the total error quantified. The field-measured heads may reflect the presence of *transient effects* that are not represented in the model. The head values also include *measurement error* associated with the accuracy of the water level measuring device, the operator, and the location and accuracy of the elevation survey point. Under ideal circumstances, measurement error will be on the order of a few hundredths of a foot. A regional study, however, may have larger errors, depending on survey accuracy.

Another source of error is caused by *scaling effects*. For example, heads may be measured in wells with long screens but the model may require point values. Head measurements averaged over long screens may be appropriate for calibrating a two-dimensional areal model but are usually not representative of heads calculated by a three-dimensional model. Gelhar (1986) discussed errors in heads caused by another kind of scaling effect. The cells or elements of the grid represent average aquifer properties within the cell or element. Field-measured heads, however, may be influenced by small-scale heterogeneities that are not captured by the model. Unmodeled heterogeneity causes error in the simulated heads. The magnitude of this error can be quantified if the variance of the log hydraulic conductivity and the horizontal and vertical correlation lengths of the log K field are known. Details are given by Gelhar (1986).

Calibration values ideally should coincide with nodes, but in practice this will seldom be possible. This introduces *interpolation* errors caused by estimating nodal head values. This type of error may be 10 feet or more in regional models. The points for which calibration values are available should be shown on a map to illustrate the locations of the calibration points relative to the nodes. Ideally, heads and fluxes should be measured at a large number of locations, uniformly distributed over the modeled region.

It is desirable to minimize all of these errors in order to minimize the size of the calibration targets (Fig. 8.4), thereby reducing some of the uncertainty inherent in model calibration. In automated inverse models the reliability of the head observation may be represented by a weighting factor associated with each calibration value (Box 8.4, Eqn. 1) or by explicitly stated error terms.

Fluxes

Field-measured fluxes, such as baseflow, springflow, infiltration from a losing stream, or evapotranspiration from the water table may also be selected as calibration values. Estimates of flux have associated errors that are usually larger than errors associated with head measurements. Nevertheless, it is advisable to use estimates of flow as calibration values in addition to heads in order to increase the likelihood of achieving a unique calibration. For example, when calibrating a model an increase in hydraulic conductivity creates the same effect on heads as a decrease in recharge making it possible to calibrate the model to heads by adjusting either hydraulic conductivity or recharge. Calibration to flows gives an independent check on hydraulic conductivity values. Some investigators have used velocities (Duffield et al., 1990) and information on solute distributions (Krabbenhoft et al., 1990; Medina et al., 1990; Kauffmann et al., 1990; Keidser et al., 1990) as additional calibration information for the flow model.

PRIOR INFORMATION

Calibration is difficult because values for aquifer parameters and hydrologic stresses are typically known at only a few nodes and, even then, estimates are influenced by uncertainty. If the parameters used in the model are not consistent with the field-measured heads, an incorrect description of the system will result.

Boundary conditions too are uncertain, particularly if the model boundaries do not correspond to the natural physical boundaries of the aquifer. In general it is true that the use of specified head boundary conditions will help achieve calibration since this type of boundary will supply the model with numerous calibrated points. However, the user should be wary of specified head boundary conditions that may erroneously affect the predictive simulation (Sections 4.3 and 7.3).

Prior information on hydraulic conductivity and/or transmissivity and storage parameters is usually derived from aquifer tests. Prior information on discharge from the aquifer may be available from field measurements of springflow or baseflow. Direct field measurements of recharge are usually not available but it may be possible to identify a plausible range of values. In a Bayesian statistical framework it is possible to base prior estimates of aquifer parameters on hydrogeologic judgment rather than relying on site-specific measurements (Box 8.2).

Uncertainty associated with estimates of aquifer parameters and boundary conditions must also be evaluated. The coefficient of variation (standard deviation divided by the expected value) may be used to quantify the uncertainty associated with each piece of prior information. A plausible range of parameter values and hydrologic stresses should always be determined prior to the calibration.

CALIBRATION TECHNIQUES

Parameter estimation is essentially synonymous with model calibration, which is synonymous with solving the inverse problem. Kriging (Section 3.4) is a method of estimating the spatial distribution of parameters (or heads), but it is generally recognized that kriging should be combined with an inverse solution because the uncertainty associated with estimates of transmissivity can be greatly reduced when information about the head distribution is used to help estimate transmissivities. In other words, better estimates of aquifer parameters can be obtained when both prior information and sample information are used in the analysis.

Solving the inverse problem by manual trial-and-error adjustment of parameters does not give information on the degree of uncertainty in the final parameter selection, nor does it guarantee the statistically best solution. An automated statistically based solution of the inverse problem quantifies the uncertainty in parameter estimates and gives the statistically most appropriate solution for the given input parameters provided it is based on an appropriate statistical model of errors.

Trial-and-Error Calibration

In trial-and-error calibration, parameter values are initially assigned to each

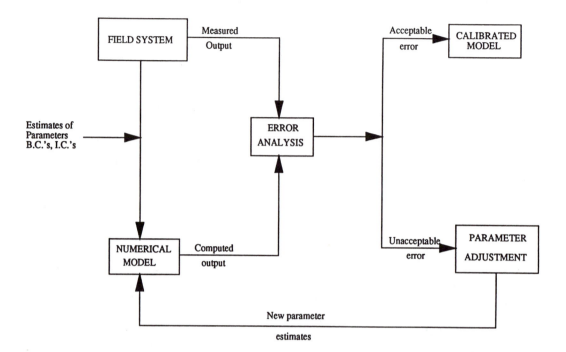

Fig. 8.7 Trial-and-error calibration procedure (modified from Peters, 1987). The field system is converted to a numerical model and calibration targets are set. The model is executed and results are compared to the calibration targets. If the error in the simulated results is acceptable, the model is considered calibrated; if the level of error is unacceptable, parameter values are adjusted and the model is run again until acceptable results are achieved. *B.C.*, boundary condition; *I.C.*, initial condition.

node or element in the grid (Section 3.4). During calibration, parameter values are adjusted in sequential model runs to match simulated heads and flows to the calibration targets.

Prior to calibration, the range of uncertainty in each parameter value is quantified, as discussed above. Some parameters may be known with a high degree of certainty and therefore should be modified only slightly or not at all during calibration. The results of each model execution are compared to the calibration targets; adjustments are made to all or selected parameters and/or boundary conditions, and another trial calibration is initiated (Fig. 8.7). Tens to hundreds of model runs are typically needed to achieve calibration. Maclay and Land (1988), for example, reported 300 simulations to achieve calibration.

Trial-and-error calibration may produce nonunique solutions when different combinations of parameters yield essentially the same head distribution. Freyberg (1988), for example, reported nine different acceptable calibrations to a set of calibration values. A large number of uniformly distributed calibration targets, each having small associated error, will increase the likelihood of obtaining a unique calibration, as will the use of fluxes as calibration targets. Because trial-and-error calibration does not quantify the statistical uncertainty or reliability of the results, calibration should be followed by a detailed sensitivity analysis (Section 8.4). A case study of a trial-and-error calibration is presented in Box 8.3.

The trial-and-error process is influenced by the modeler's expertise and biases. The modeler uses all information about the system to evaluate its response to changes in parameters and boundary conditions and then makes decisions that eventually lead to calibration. Proponents of trial-and-error calibration argue that this method uses information that is "unquantifiable," i.e., the subjective good judgment of the modeler. Detractors stress that modeler bias should be minimized by using automated methods. The truth, as usual, lies in between these two extreme viewpoints.

Automated Calibration

Although the theory involved in automated inverse modeling is beyond the scope of this book, some aspects relevant to calibration are discussed briefly below. The interested reader is referred to Cooley and Naff (1990), Menke (1989), Carrera (1988), Chapter 5 of Peck et al. (1988), Carrera and Neuman (1986a), and Yeh (1986) for more information.

Automated inverse modeling is performed using specially developed codes that use either a direct or indirect approach to solve the inverse problem. In a *direct solution*, the unknown parameters are treated as dependent variables in the governing equation and heads are treated as independent variables. This means that values for head must be input for all nodes. Heads are known only at points where there are observation wells, making it necessary to estimate heads elsewhere in the grid, usually by kriging (Section 3.4). The solution minimizes the nodal mass balance errors caused by using these heads and the computed parameter values. Direct solutions are prone to instability. Furthermore, they do not recognize measurement errors. According to Carrera (1988), it may be this failure to provide a statistical framework in which to view field data that has caused direct methods to be practically abandoned. Direct solutions will not be considered further here.

The *indirect approach* is similar to performing trial-and-error calibrations in that the forward problem is solved repeatedly. However, an inverse code automatically checks the head solution and adjusts parameters in a systematic way in order to minimize an objective function, an example of which would be to minimize the sum of the squared residuals, i.e., differences between simulated and observed heads. Methods used to minimize objective functions are often based on Gauss-Newton (e.g., Cooley, 1977, 1979) or gradient search methods (e.g., Carrera et al., 1984).

Although more stable than direct solutions of the inverse problem, indirect solutions may be unstable and give unreasonable solutions involving negative parameter values. It may be possible to control instability by proper zonation of aquifer parameters (Section 3.4). For example, Garabedian (1986) controlled instability in his simulations by combining zones to form single regression parameters. Instability can also be controlled by using prior information to provide bounds on parameter values. When prior information about parameter values is used, the simulation is said to be *conditional*, whereas in an *unconditional* simulation, the uncertainty in parameter values at or near measurement points is taken to be the same as the uncertainty in areas where measurements are lacking.

Indirect solutions are formulated in a statistical framework in which errors in heads and parameters are quantified. In a *weighted least squares* statistical framework, head measurements and prior information on parameter values are weighted to indicate the relative confidence in the measurement. In this way it is possible to place greater emphasis on measurements that are thought to be of higher reliability. The objective function becomes a weighted sum of the squared differences between observed and simulated heads and between initial and current parameter estimates. In this statistical framework, errors are assumed to be normally distributed and have zero mean. Cooley (1977, 1982) used a least squares framework to develop an indirect solution algorithm for two-dimensional steady-state groundwater flow.

In a *Bayesian* statistical framework, the unknown parameters are random variables described by a probability density function (pdf). In classical statistics, field measurements are used to describe the parameter distribution, but in a Bayesian approach, geologic judgment based on data at other similar sites may be used (Freeze et al., 1990). These prior estimates then form the prior information for an unconditional simulation of the parameter distribution. When field data become available the prior estimates are updated to posterior estimates. Kriging (Section 3.4) may be used to interpolate between measurement points, thereby creating a more detailed data set to be used as prior information (Clifton and Neuman, 1982). A Bayesian framework is also used in stochastic modeling (Box 8.2).

The *Fisherian* statistical framework, which includes the *maximum likelihood (ML)* approach, assumes that the unknown parameters are deterministic rather than random. That is, uncertainty in the model parameters is attributed to insufficient information and corruption by noise, rather than to randomness. While the ML approach assumes that the prior information consists of parameter values influenced by errors, it avoids the subjective procedure of assigning reliability weights to head and parameter measurements. ML theory uses estimates of these errors, called prior errors, and attempts to find a solution that maximizes the likelihood of obtaining the measured values of head. Prior errors in both heads and parameters are assumed to be Gaussian with zero mean;

parameters may need to be transformed in order to meet this assumption; e.g., K is replaced by $Y = \log K$. The criterion used to find the solution involves minimizing residuals in both heads and parameters using what is known as the log-likelihood criterion. Carrera and Neuman (1986a) developed an algorithm for an indirect solution of the inverse problem for steady and transient flow in two dimensions. Carrera et al. (1984) discussed a quasi three-dimensional transient code known as INVERT-3.

To date, there are several inverse models for flow that are documented and readily available. The code MODINV (distributed by Scientific Software Group), is a three-dimensional transient inverse model compatible with MODFLOW. MODFLOWP (Hill, 1990a) is a three-dimensional inverse code supported by the U.S. Geological Survey. It too is linked to MODFLOW. MODFLOWP uses a new solution procedure to solve the groundwater flow equation (Hill, 1990b). An earlier USGS finite element inverse code was developed by Cooley (1977, 1979) and a finite difference version was documented by Cooley and Naff (1990). These two-dimensional codes assume steady-state flow; applications were reported by Cooley (1979), Bradbury (1982), Waddell (1982), Czarnecki and Waddell (1984), Garabedian (1986), Cooley et al. (1986), and Connell and Bailey (1989). A three-dimensional, steady-state inverse model is contained in the code FTWORK (Faust et al., 1990) that was developed for the Savannah River Site. While FTWORK is normally used in the forward mode to solve three-dimensional, transient flow, and contaminant transport problems, an automated inverse package is included to help calibrate steady-state flow problems. Other inverse codes, including INVERT-3 (Carrera et al., 1984), have been developed within academia but are not distributed commercially.

All of the codes mentioned above use an indirect solution to the inverse problem. MODINV, MODFLOWP, and the code that comes with FTWORK use a least squares statistical framework and a Gauss-Newton solution procedure. INVERT-3 uses a maximum likelihood statistical framework and a gradient search solution procedure.

To date, automated inverse models have had limited application. They are criticized because of problems with nonuniqueness and instability, but according to Sampler et al. (1990) nonuniqueness and instability "depend on the problem and the way it is posed, not on the approach used for calibration." Nonuniqueness is especially a problem in the absence of prior information on transmissivities (Neuman et al., 1980). Sampler et al. (1990) believe that the calibrated model produced with automated techniques is not necessarily superior to one produced using manual trial-and-error calibration. Rather, an advantage to using automated calibration codes is that they speed the modeler through the most time-consuming and frustrating part of the modeling process. Carrera and Neuman (1986a) summarize the situation as follows:

> The method of calibration used most often in real world situations is manual trial and error. However, the method is recognized to be labor intensive (therefore ex-

pensive), frustrating (therefore often left incomplete), and subjective (therefore biased and leading to results the quality of which is difficult to evaluate).

Automated inverse models provide a protocol for model calibration, forcing the modeler to formulate the problem properly and solve it in a rational way. They also provide information on uncertainty in the calibration. Case studies of two automated calibrations are presented in Box 8.4. As modelers acquire more statistical expertise and more powerful computers, automated parameter estimation methods will be used more frequently in model calibration. Meanwhile calibrations performed manually by trial and error should be formulated using a protocol for quantifying errors, selecting calibration targets, and defining plausible ranges in parameters and hydrologic stresses as outlined above. Evaluation of the calibration should also follow a standard protocol as described in Section 8.4. Case studies that illustrate features of automated calibration are presented in Boxes 8.3 and 8.4.

8.4 Evaluating the Calibration

The results of the calibration should be evaluated both qualitatively and quantitatively. Even in a quantitative evaluation, however, the judgment of when the fit between model and reality is good enough is a subjective one. To date, there is no standard protocol for evaluating the calibration process, although the need for a standard methodology is recognized as an important part of quality assurance in code application (National Research Council, 1990). Claims such as "The measured and simulated contours compared favorably; therefore, the transient-state model was considered calibrated" are not easily evaluated. Commonly used measures for evaluating trial-and-error calibrations are discussed below. We also suggest a standard protocol for reporting and evaluating model calibration.

TRADITIONAL MEASURES OF CALIBRATION

Comparison between contour maps of measured and simulated heads provides a visual, *qualitative* measure of the similarity between patterns, thereby giving some idea of the spatial distribution of error in the calibration (Fig. 8.8). However, contour maps of field data include errors introduced by contouring and therefore should not be used as the only proof of calibration. A scatterplot of measured against simulated heads is another way of showing the calibrated fit (Fig. 8.9). Deviation of points from the straight line should be randomly distributed.

A listing of measured and simulated heads together with their differences and some type of average of the differences is a common way of reporting calibration results (Table 8.3). The average of the differences is then used to

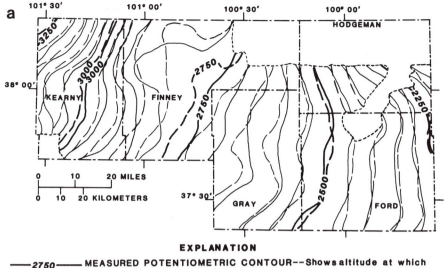

EXPLANATION

—2750——— MEASURED POTENTIOMETRIC CONTOUR--Shows altitude at which
water level would have stood in tightly cased wells, January 1982.
Contour interval 50 feet. National Geodetic Vertical Datum of 1929

— —2750— — SIMULATED POTENTIOMETRIC CONTOUR--Shows altitude at which
water level would have stood in tightly cased wells, December 31,1981.
Contour interval 50 feet. National Geodetic Vertical Datum of 1929

---------- APPROXIMATE LIMIT OF HIGH PLAINS AQUIFER

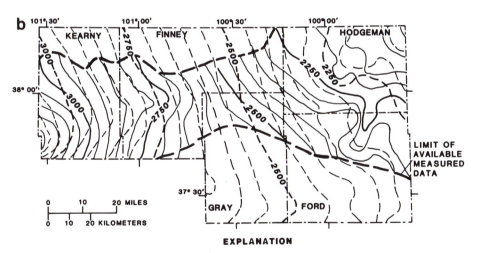

EXPLANATION

—2750——— MEASURED POTENTIOMETRIC CONTOUR--Shows altitude at which water level
would have stood in tightly cased wells, October 1981 to March 1982.
Contour interval 50 feet. National Geodetic Vertical Datum of 1929

— —2750— — SIMULATED POTENTIOMETRIC CONTOUR -- Shows altitude at which water
level would have stood in tightly cased wells, December 31, 1981.
Contour interval 50 feet. National Geodetic Vertical Datum of 1929

Fig. 8.8 Comparison of measured and simulated potentiometric surfaces, January 1982, for aquifers in southwestern Kansas (Watts, 1989).
(a) High Plains aquifer (unconfined), a hydraulically connected mixture of lenticular deposits of silt, clay, sand, and gravel.
(b) Dakota aquifer (confined), a sequence of sandstone and shale.

quantify the average error in the calibration. The objective of the calibration is to minimize this error, sometimes called the *calibration criterion*. Three ways of expressing the average difference between simulated and measured heads are commonly used.

1. The *mean error* (ME) is the mean difference between measured heads (h_m) and simulated heads (h_s).

$$\text{ME} = 1/n \sum_{i=1}^{n} (h_m - h_s)_i \qquad (8.1)$$

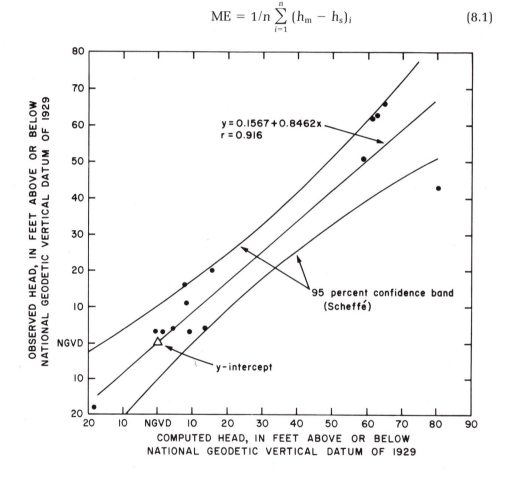

Fig. 8.9 Plot of simulated versus measured heads in a sand and gravel aquifer in western Florida (Trapp and Geiger, 1986). The model was originally calibrated to 1972 pumping conditions. The calibration was tested by simulating three other pumping periods. The results of one of these simulations are shown here. Observed heads for the period were compared to those predicted by the model. Heads were then plotted as shown. Confidence intervals for the linear regression analysis are also shown.

Table 8.3

Measured and Simulated Heads for a model of a portion of Cape Cod, Massachusetts[a]

Node (i, j, k)	U.S. Geological Survey well no.	Water level (feet above sea level)		Node (i, j, k)	U.S. Geological Survey well no.	Water level (feet above sea level)	
		Observed average 1963–76	Calculated at center of grid block			Observed average 1963–76	Calculated at center of grid block
6, 16, 7	WNW 78	2.7	3.4	16, 26, 7	TSW 198	7.6	7.1
11, 26, 7	TSW 216	4.1	5.3	17, 22, 7	TSP 16	7.6	7.9
15, 14, 7	WNW 30	6.6	6.7	18, 17, 7	WNW 34	8.0	8.0
16, 14, 7	WNW 30	6.6	7.4	21, 19, 7	TSP 18	6.4	6.5
16, 21, 7	TSP 17	6.8	7.2				

$$ME = -0.34 \text{ ft}$$
$$MAE = 0.40 \text{ ft}$$
$$RMS = 0.55 \text{ ft}$$

[a] The mean error (ME), mean absolute error (MAE), and root mean squared (RMS) error are given. Modified from Guswa and Le Blanc, 1985.

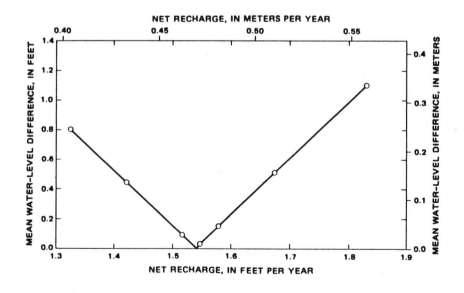

Fig. 8.10 An example showing the use of the ME as a calibration criterion for a model of the unconfined aquifer associated with the Rocky Mountain Arsenal in Colorado. The ME at all nodes in an irrigated area is minimized when a net recharge rate of 1.54 ft/yr is used (Konikow, 1977).

where n is the number of calibration values. The ME is simple to calculate but is usually not a wise choice because both negative and positive differences are incorporated in the mean and may cancel out the error. Hence, a small mean error may not indicate a good calibration. An example of minimizing this measure of calibration for net recharge is shown in Fig. 8.10.

2. The *mean absolute error* (*MAE*) is the mean of the absolute value of the differences in measured and simulated heads.

$$\text{MAE} = 1/n \sum_{i=1}^{n} |(h_m - h_s)_i| \qquad (8.2)$$

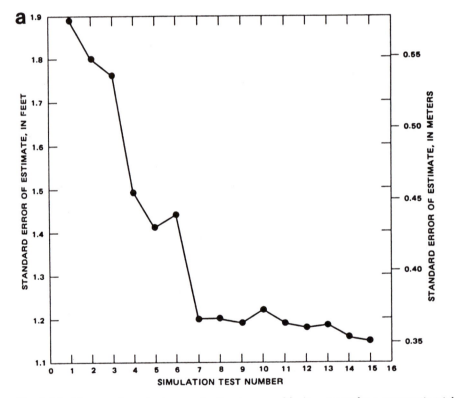

Fig. 8.11 The use of standard error of estimate as a calibration target for a systematic trial-and-error calibration.
(a) Reduction of the standard error of the head during calibration of a steady-state flow model of the unconfined aquifer in the vicinity of the Rocky Mountain Arsenal, Colorado (Konikow, 1977).
(b) The reduction in the root mean squared error of heads by increasing the hydraulic conductivity in layer 3 of a finite difference model describing groundwater flow in south central New York State (Yager, 1986)

3. The *root mean squared* (*RMS*) error or the *standard deviation* is the average of the squared differences in measured and simulated heads.

$$\text{RMS} = \left[1/n \sum_{i=1}^{n} (h_m - h_s)_i^2 \right]^{0.5} \tag{8.3}$$

It is similar to the standard error (SE) of estimate (Box 8.4, Eqn. 1). The systematic reduction of the standard error and RMS error during calibration is illustrated in Fig. 8.11.

The choice of calibration criterion may affect the values of the parameters selected for the calibrated model. For example, the effect of changing the calibration criterion on the calibrated value of recharge for a simulation of the southern High Plains aquifer is shown in Fig. 8.12. Note that the minimum value of each criterion occurs at a different value of recharge. The RMS is usually thought to be the best measure of error if errors are normally distributed. For the simulation in Fig. 8.12, Luckey et al. (1986) selected the ME because it provided the best-defined minimum.

It is important to note that these measures of error can only be used to evaluate the *average* error in the calibrated model. For example, in Fig. 8.11a, the model would be considered calibrated when the SE is 0.35 m. The maximum acceptable error is often set during the calibration but ideally should be established prior to calibration (Woessner and Anderson, 1990). The maximum acceptable value of the calibration criterion depends on the magnitude of the change in heads over the problem domain. If the ratio of the RMS error to the total head loss in the system is small, the errors are only a small part of the overall model response.

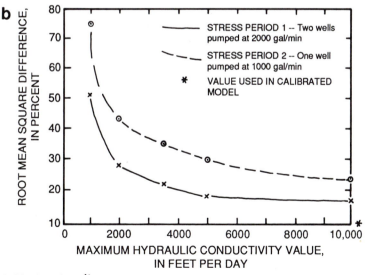

Fig. 8.11 (*continued*)

CALIBRATION LEVEL AND DISTRIBUTION OF ERROR

The three measures of error discussed above quantify the *average* error in the calibration but say nothing about the distribution of error. Such lumped measures of calibration hide poorly calibrated portions of the model by the averaging process. Comparison of head contours (e.g., Fig. 8.8) gives a purely qualitative and subjective indication of the spatial distribution of error. A quantitative analysis of the distribution of error should be part of calibration assessment. The distribution of head residuals may be shown in discrete form (Fig. 8.13a) or using contours (Fig. 8.13b). The error in the residuals should be randomly distributed over the grid. If a trend is evident (e.g., heads in a portion of the model are too high) parameter values or boundary conditions should be adjusted to eliminate the trend. A scatterplot (Fig. 8.9) may also be helpful in detecting trends. From this analysis it may be possible to isolate sources of

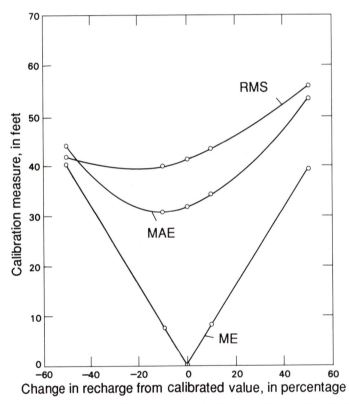

Fig. 8.12   The effect of the choice of calibration measure — root mean squared (RMS); mean absolute error (MAE); mean error (ME) — on the recharge value used to achieve calibration for a model of the southern High Plains aquifer system (modified from Luckey et al., 1986).

a

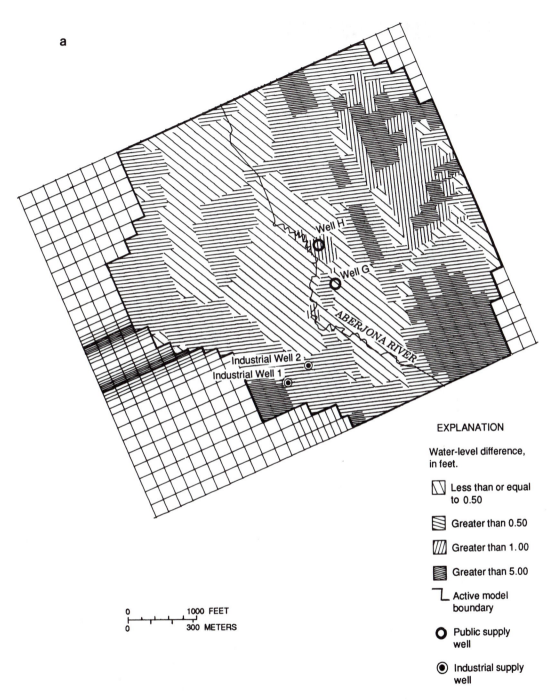

EXPLANATION

Water-level difference, in feet.

Less than or equal to 0.50

Greater than 0.50

Greater than 1.00

Greater than 5.00

Active model boundary

O Public supply well

◉ Industrial supply well

Fig. 8.13 Representation of the spatial distribution of error or residual calculated as the difference between measured (or interpolated) heads and simulated heads.
(a) Discrete intervals of head residual are shown for a unconfined aquifer in Woburn, Massachusetts (de Lima and Olimpio, 1989).

b

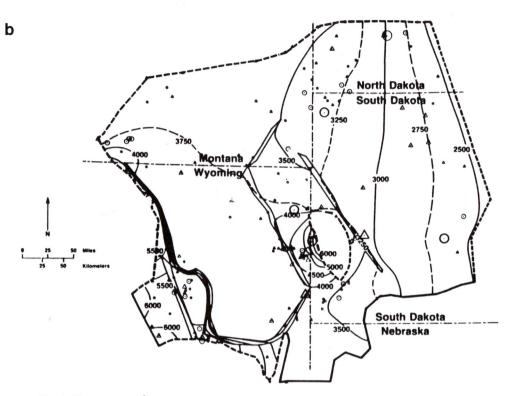

Fig. 8.13 (*continued*)

(b) Residuals of head are plotted as triangles (positive residual) and circles (negative residual) for a simulation of the Dakota aquifer. Large symbols indicate absolute values greater than two standard deviations, medium symbols indicate between one and two standard deviations, and small symbols indicate less than one standard deviation. The potentiometric surface calculated by the calibrated model is shown (Cooley, Konikow and Naff, Water Resources Research, 22(13), pp. 1759–1778, 1986, published by the American Geophysics Union).

Table 8.4
Levels of Calibration

Level 1	Simulated value falls with target (highest degree of calibration)
Level 2	Simulated value falls within two times the associated error of the calibration target
Level 3	Simulated value falls within three times the associated error of the calibration target
Level *n*	Simulated value falls within *n* times the associated error of the calibration target (lowest degree of calibration)

Modified from Woessner and Anderson, 1990.

error and take steps to improve the field information used in the model.

In order to evaluate calibration results it is helpful to know how well the calibrated model matched each of the calibration targets. Then the level of calibration may be quantified for *each* calibration target using the guidelines presented in Table 8.4. Level 1 represents the highest level of calibration where the simulated value lies within the calibration target (Fig. 8.4). Other calibration levels are defined using multiples of the error associated with each calibration target and are used to delineate those portions of the grid where level 1 calibration could not be achieved. Ideally the entire model should be calibrated to level 1; however, in practice the level of calibration may vary spatially (Box 8.3, Fig. 1b).

The distribution of levels of calibration should be shown on a map (Box 8.3, Fig. 1). Steady-state calibrations require one map for each type of calibration value, e.g., heads and fluxes. Transient calibrations require a map for each type of calibration value and for each time step used in the calibration. A single map may display transient results if the same calibration level was achieved at each time step. The match between the simulated values and the calibration targets also should be presented in tabular form (Box 8.3, Table 1). A time series plot of measured and simulated heads (Fig. 7.1a) is also a useful way to present results from a transient calibration.

The reporting of calibration level using the procedure described above shows the spatial distribution of the error in the calibration. It also requires that the number and distribution of calibration targets and corresponding nodes be reported. Reporting the distribution of error allows a more meaningful assessment of the calibration than the measures of average error discussed in the preceding section. Measures of average error serve as additional expressions of the degree of calibration but they should not be used as the sole measure of calibration.

It should be obvious that the proposed method of target setting and calibration assessment does not eliminate the need to describe the process of calibration and the effect of the level of calibration achieved on final model results. In particular, differences between initial parameter estimates and the final calibrated values should be listed (Box 8.4, Table 1).

Although these guidelines for evaluating and reporting calibration results are intended to guide trial-and-error calibrations, such as the one discussed in Box 8.3, they may also apply to automated calibrations. Automated calibration techniques require the modeler to use a code-specific series of tests to select the best model. In so doing many of the types of evaluations discussed above are produced. Two case studies using automated calibration are discussed in Box 8.4.

It should also be noted that there is no guarantee that a model calibrated to level 1 will result in a valid interpretative or predictive model. Trial-and-error calibration procedures do not produce unique solutions and thereby introduce uncertainty in model results. Automated calibration also produces a suite of possible solutions. Garabedian (1986), for example, presented seven possible inverse solutions to a two-dimensional model of a regional basalt aquifer in Idaho. Carrera and Neuman (1986c) presented five possible inverse solutions to a problem involving flow to the Colorado River, and Connell and Bailey (1989) presented three solutions for a profile model through the Oak Ridge Reservation in Tennessee. The possibility of nonuniqueness is enhanced if there is no or little prior information on parameter values, if calibration targets are large, and/or if the targets are few or poorly distributed. The use of clearly defined levels of calibration at least will facilitate interpretation of the spatial distribution of uncertainty associated with the calibrated model.

While the objective of calibration is to demonstrate that the calibrated model can reproduce measured heads and fluxes, the ultimate modeling objective usually is to produce a model that can accurately simulate future conditions for which no head data are available. Given that the calibration may be nonunique, we have no guarantee that the predictive model will produce accurate results when the model is stressed differently from calibrated conditions. In light of this uncertainty, the calibrated model should be subjected to a sensitivity analysis and, if possible, a verification test.

SENSITIVITY ANALYSIS

The purpose of a sensitivity analysis is to quantify the uncertainty in the calibrated model caused by uncertainty in the estimates of aquifer parameters, stresses, and boundary conditions. A sensitivity analysis is an essential step in all modeling applications:

> Not only do we have uncertainty as to the parameter values needed for our design calculations, we even have uncertainty about the very geometry of the system we are trying to analyze. The uncertainties of lithology, stratigraphy, and structure introduce a level of complexity to geotechnical and hydrogeological analysis that is completely unknown in other engineering disciplines. (Freeze et al., 1990)

During a sensitivity analysis, calibrated values for hydraulic conductivity, storage parameters, recharge, and boundary conditions are systematically changed within the previously established plausible range (Section 8.3). The magnitude of change in heads from the calibrated solution is a measure of the

Fig. 8.14 Examples of methods used to present the results of a steady-state sensitivity analysis. (a) A profile model of the aquifer near Barstow, California was tested to determine the sensitivity of the vertical head distribution at a discharge boundary to differences in horizontal and vertical hydraulic conductivities (Robson, 1978).
(b) Sensitivity analysis of the effect of the value of hydraulic conductivity and recharge on the absolute value of mean residual water level (ME) for a predevelopment-period model of the High Plains (Luckey et al., 1986).
(c) The effect of varying recharge, hydraulic conductivity, and stream leakage on the calibrated steady-state water table elevation at nodes in row 22, columns 58 through 70. (Gerhart and Lazorchick, 1988).

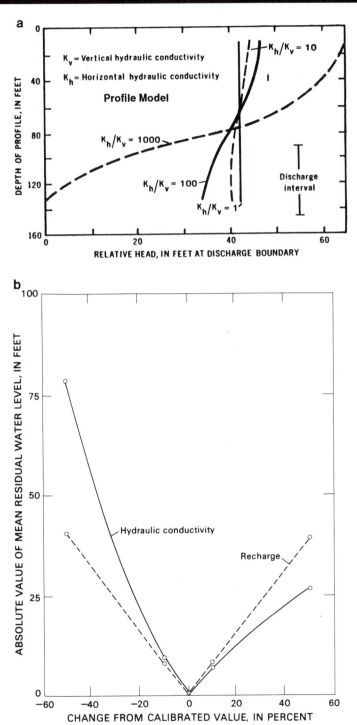

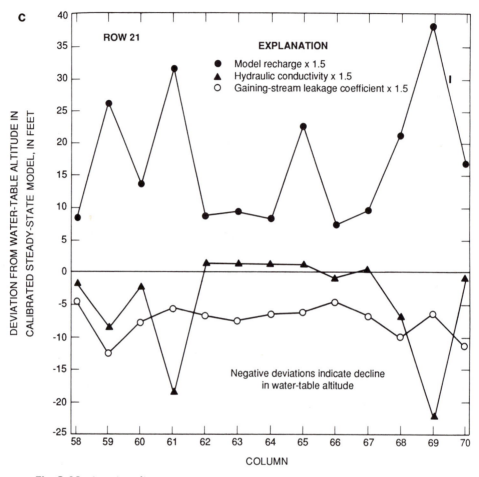

Fig. 8.14   (continued)

sensitivity of the solution to that particular parameter. The results of the sensitivity analysis are reported as the effects of the parameter change on the average measure of error selected as the calibration criterion. Ideally, the effect on the spatial distribution of head residuals is also examined.

Sensitivity analysis is typically performed by changing one parameter value at a time. The effects of changing two or more parameters also might be examined to determine the widest range of plausible solutions. For example, hydraulic conductivity and recharge rate might be changed together so that low hydraulic conductivities are used with a high recharge rate and high hydraulic

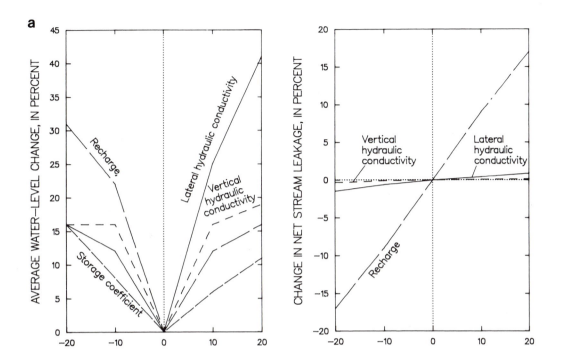

CHANGE FROM CALIBRATED VALUES, IN PERCENT

Fig. 8.15 Examples of additional ways of presenting the results of a sensitivity analysis with examples from transient calibrations.

(a) The left-hand-side figure shows the effect of varying the storage coefficient on the average water level change, plotted with other sensitivity analyses, for a model of aquifers in the Umatilla Plateau and Horse Heaven Hills area, Oregon and Washington (Davies-Smith et al., 1988). The right-hand-side figure shows the effect on stream leakage.

(b) Presentation of the effect of varying leakance and transmissivity by 50% for all aquifers and confining beds in a model of the Potomac aquifers of northern Delaware (Martin, 1984). Results are presented for head in the upper aquifer at node 16, 25.

(c) The residuals in head for a series of sensitivity analyses are compared with the final calibrated model (labeled FINAL along the horizontal axis) using box plots which show the mean and the magnitude of the extreme values (de Lima and Olimpio, 1989). The sensitivity to streamflow loss is also shown.

conductivities are used with a low recharge rate. Mandle and Kontis (1986) reported using a simulation combining low estimates of hydraulic conductivities with low values of leakance to define one end of a range of possible conditions and a simulation with high estimates of hydraulic conductivities and leakance to define the other end. Some of the more common ways of displaying the results of sensitivity analyses are shown in Figs. 8.14 and 8.15 and Table

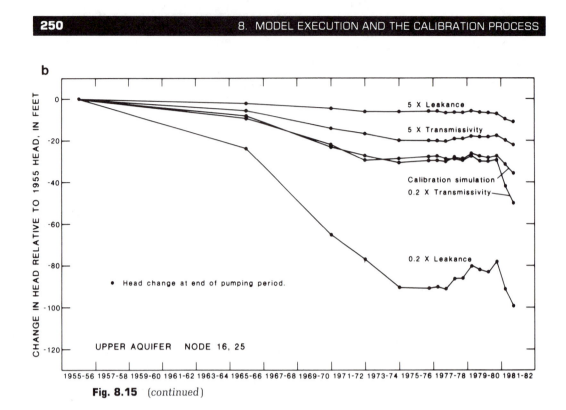

Fig. 8.15 (*continued*)

8.5. Table 8.6 illustrates a qualitative approach to sensitivity analysis in which the result of a sensitivity test is described in words.

A sensitivity analysis may also test the effect of changes in parameter values on something other than head. Figure 8.15a shows the effect of changes in hydraulic conductivity and recharge on stream leakage and Fig. 8.15c shows the effect of changes in several parameters on streamflow. Figure 8.16 shows the effect of changing the hydraulic conductivity of riverbed sediments on the capture zone of a pumping well (Fig. 8.16a) and on the sources of water to the pumping well (Fig. 8.16b). Sensitivity analyses also are performed when using automated calibration codes. For example, Garabedian (1986) tested the effect of horizontal anisotropy on hydraulic conductivity, parameter zonation, and the sensitivity to leakance and recharge estimates.

A more formal sensitivity analysis can be performed by calculating sensitivity coefficients or sensitivities (Box 8.4). Cooley et al. (1986), for example, produced sensitivity maps (Fig. 8.17) to illustrate areas of the model most sensitive to changes in a given parameter value. The procedure of calculating sensitivities for all parameters at all nodes can be automated by formulating adjoint equations (Sykes et al., 1985; Townley and Wilson, 1983, 1985; Carrera and Neuman, 1984). Sensitivity analysis also can be done using stochastic modeling (Box 8.2).

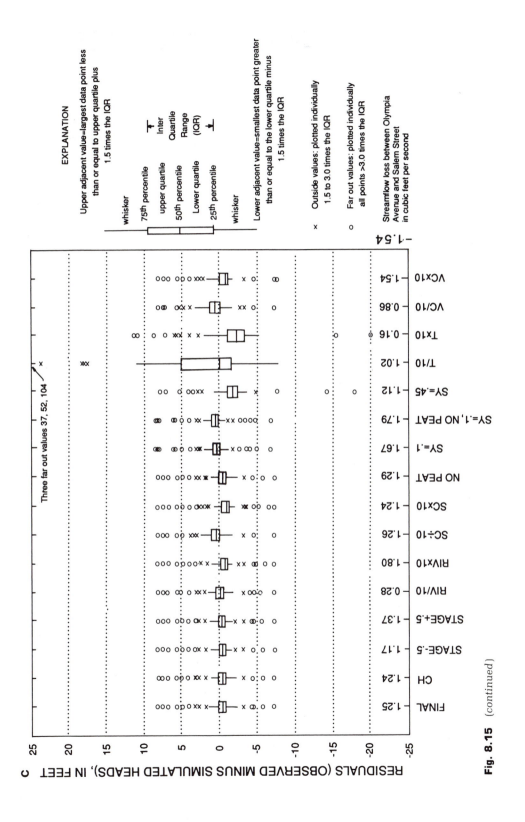

Fig. 8.15 (continued)

Table 8.5

Results of a Transient Calibration Sensitivity Analysis

Aquifer and hydrologic characteristic varied	Hydrologic value simulated	Percent change[a]	Resulting changes			
			Average difference (feet)[b]	Standard deviation (feet)[c]	Average difference (feet)[b]	Standard deviation (feet)[c]
High Plains aquifer			*High Plains aquifer*		*Dakota aquifer*	
Specific yield	0.27	+50	−3.5	15.8	−2.4	53.1
(dimensionless)	0.225	+25	−1.7	16.2	−1.2	53.6
	0.18	0	1.0	17.0	−0.1	54.7
	0.135	−25	5.3	18.8	1.9	55.6
	0.09	−50	13.5	23.7	6.1	60.0
Horizontal	120	+50	1.0	17.6	−0.1	54.1
hydraulic	80	0	1.0	17.0	−0.1	54.7
conductivity	40	−50	0.8	16.4	−0.7	54.8
(feet per day)						
Vertical	1.5	+50	1.0	17.0	−0.1	54.7
hydraulic	1.0	0	1.0	17.0	−0.1	54.7
conductivity	0.5	−50	1.0	17.0	−0.1	54.7
(feet per day)						
Niobrara-Graneros confining unit						
Vertical	0.0000015	+50	1.0	17.0	−1.2	55.3
hydraulic	0.000001	0	1.0	17.0	−0.1	54.7
conductivity	0.0000005	−50	1.0	17.0	0.9	54.1
(feet per day)						
Dakota aquifer						
Specific yield	0.14	+33	1.0	17.0	−1.8	54.9
(dimensionless)	0.105	0	1.0	17.0	−0.1	54.7
	0.07	−33	1.0	17.1	−1.8	54.9
Specific	0.0000075	+50	1.0	17.0	−1.1	56.4
storage	0.000015	0	1.0	17.0	−0.1	54.7
(per foot)	0.0000225	−50	1.0	17.0	−1.1	54.3
Horizontal	10.5	+50	1.0	17.1	−0.8	55.0
hydraulic	7.0	0	1.0	17.0	−0.1	54.7
conductivity	3.5	−50	1.1	17.0	−1.4	56.2
(feet per day)						

[a] Percentage change from calibrated value.

[b] Average difference between simulated hydraulic heads for test and for calibrated model.

[c] Standard deviation of the differences between simulated hydraulic heads for test and for calibrated model.

From Watts, 1989.

MODEL VERIFICATION

Owing to uncertainties in the calibration, the set of parameter values used in the calibrated model may not accurately represent field values. Consequently, the calibrated parameters may not accurately represent the system under a different set of boundary conditions or hydrologic stresses.

Model verification will help establish greater confidence in the calibration. According to Konikow (1978), a model is verified "if its accuracy and predictive capability have been proven to lie within acceptable limits of error by tests independent of the calibration data." In a typical verification exercise, values of parameters and hydrologic stresses determined during calibration are used to simulate a transient response for which a set of field data exists. Examples of transient data sets include pumping test data and changes in water levels in

Table 8.6
Results of a Sensitivity Analysis Displayed in a Qualitative Format

Condition varied	Range tested	Results
Hydraulic conductivity	1000–2670 ft/d	Major effect, more intruded for high-K case. Width of transition zone doubles at aquifer base over range tested.
Anisotropy	10 : 1–500 : 1	Slightly more intruded for high-anisotropy case. Tilt and curvature of front increase with higher anisotropy.
Recharge	6–18 in./yr	Major effect, more intruded for low-recharge case. Width of transition zone doubles at aquifer base over range tested
Lateral influx	5–15 ft^3/d/ft	Minor effect in movement and shape of front for range tested.
Pumping well	0–5 ft^3/d/ft	Similar response as to lateral influx case.
Dispersivity	100–500 ft (α_L) 5–50 ft (α_T)	Insensitive to changes of this magnitude of α_L. Increasing values of α_T makes front more vertical, less intruded at aquifer base.
Layering ($K_{top} : K_{bot}$)	1 : 1–1 : 0.01	Pronounced layering compresses saltwater front, increases curvature. Intrusion is especially limited in upper (permeable) part of aquifer.
Saltwater boundary	Atlantic Ocean–inland saltwater bodies	Transition zone translates laterally depending on concentration at aquifer top.

From Andersen et al., 1988.

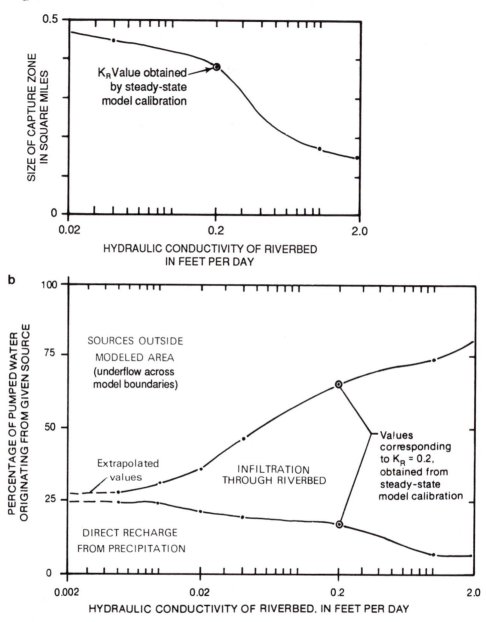

Fig. 8.16 Results of a sensitivity analysis to test the effect of changing the hydraulic conductivity of riverbed sediments on the size of the area contributing water to a well field located near the Susquehanna River in southern New York State (Yager, 1986).

(a) The effect of altering the riverbed hydraulic conductivity (K_R) on the size of the area recharging the well field.

(b) The effect of altering the riverbed hydraulic conductivity on three sources of water to the well field, namely, direct recharge from precipitation, infiltration through the riverbed, and underflow across boundaries.

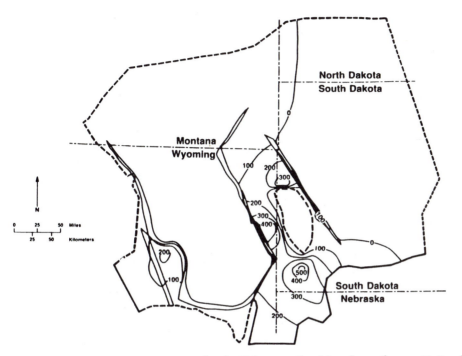

Fig. 8.17 Sensitivity map for the Dakota aquifer. Map shows the sensitivity of the heads to changes in hydraulic conductivity. (See Box 8.4, Eqn. 2.) Contours are in feet of head. (Cooley, Konikow, and Naff, Water Resources Research, 22(13), pp. 1759–1778, 1986, published by the American Geophysical Union.)

response to a drought or to long-term pumping (Fig. 7.1). In the absence of a transient data set, the calibration may be tested using a second independent set of steady-state data (e.g., Bair et al., 1990).

If the model is calibrated to steady-state heads, values for storage parameters will not be calibrated. A transient verification test may be performed using best guesses for the storage parameters but if the parameters established during steady-state calibration are changed during verification, the calibration simulation should be run again using the changed parameter values. This is done to demonstrate that there is still a good match between simulated and measured heads for the original calibration data set. If the calibrated parameters were changed significantly during verification, it may not be possible to match the calibration targets using the new parameter values. In this case it will be necessary to repeat the process until a set of parameter values is identified that produces a good match to both the calibration and what were intended to be verification targets. If it is necessary to adjust parameters during verification, the verification becomes a second calibration and another independent data set is needed to perform the verification. Verification is accomplished when the verification targets are matched without changing the calibrated parameter values. An example of model verification to a transient data set is shown in Fig. 7.1a.

Unfortunately it is often impossible to verify a model because usually only one set of field data is available. That data set, of course, is needed for calibration. If this is the case, the model cannot be verified. A calibrated but unverified model can still be used to make predictions as long as careful sensitivity analyses of both the calibrated model and the predictive model are performed and evaluated. Predictions resulting from calibrated but unverified models generally will be more uncertain than predictions derived from verified models.

SUMMARY

The procedures to use before and after calibration are summarized below.

A. Prior to Calibration
1. Select calibration values from heads, head gradients, flows, or other field data.
2. Estimate the error in the calibration values including measurement error, interpolation error, and errors from scale effects and transient effects. Define calibration targets.
3. Compile the field data needed to set boundary conditions, parameter values, and hydrologic stresses, and estimate plausible ranges in boundary conditions, parameter values, and hydrologic stresses.
4. Assign parameter values to zones in the grid and calculate the coefficient of variation for each zone.
5. Prepare a map showing the location of calibration targets relative to nodes in the grid.
6. Prepare a table showing initial estimates of boundary conditions, parameters, and hydrologic stresses and their coefficients of variation.

B. After Calibration
1. Calculate coefficients of variation (standard deviation divided by mean value) using calibrated estimates of parameter values. A small coefficient of variation indicates a relatively high degree of certainty. The range of acceptable parameter values is determined during calibration and sensitivity analysis.
2. Prepare a table showing differences between calibration targets and simulated values of heads and fluxes.
3. Calculate the ME, MAE, and RMS error in the heads. Also calculate the ratio of the RMS error to the total head loss in the system.
4. Present the spatial distribution of residual in several ways, selecting from the following types of presentation:

    (a) Map of superimposed contours of head.

    (b) Map showing contours of head residuals.

    (c) Map showing location and value of calibration targets and simulated values.

    (d) Plot of calibration values vs. simulated values showing deviation from a straight-line correspondence.

    (e) Box plot of residual head for each important calibration run.

    (f) Plot of ME, MAE, and RMS vs. the calibration run number to show the approach to calibration.

    (g) Plot of ME, MAE, and RMS vs. parameter values to show the sensitivity of the calibration to changes in a parameter value.

5. Prepare a lucid discussion of the calibration procedure and discuss the changes made in initial parameter estimates and the sensitivity of the model to these changes.

6. The calibration procedure will include a sensitivity analysis. The results of the sensitivity analyses should be discussed and results should be summarized using one or more of the presentation forms listed under item 4 above. Presentations b, c, e, and g are especially recommended.

8.5 Prediction

In a predictive simulation, the parameters determined during calibration and verification are used to predict the response of the system to future events. Some environmental problems require predicting the system response many years, perhaps as many as 10,000 years, into the future. An important task in predictive modeling is to determine the length of time for which the model will accurately predict the future. The confidence to be placed in model predictions depends largely on the results of the calibration, sensitivity analyses, and verification tests. The modeler must consider the extent to which the model has been validated (Section 1.4). Faust et al. (1981) suggest that a predictive simulation not be extended into the future more than twice the period for which calibration data are available, but this may not be possible if regulations require longer simulations.

Two major pitfalls are involved in making predictions: uncertainty in the calibrated model and uncertainty about future hydrologic stresses. Each of these requires a different type of sensitivity analysis. Even though the set of calibrated parameters may give close agreement during calibration and verification, the model may not accurately reflect system behavior when the model is stressed in some new way. Therefore, a sensitivity analysis as described in Section 8.4 should be performed for at least one of the predictive simulations in order to test the effect of uncertainty in the calibrated parameters.

Furthermore, many predictive simulations require guesses about the likelihood and magnitude of future hydrologic or human-regulated events such as

Table 8.7

Summary of Results of the Baseline Predictive Simulation and Associated Sensitivity Analysis for Eight Different Scenarios of Pumping from the Dakota Aquifer for a 20-Year Period

Simulation number or name	Minimum distance of well to Dakota subcrop (miles)	Minimum distance between Dakota wells (miles)	Maximum pumping rate (acre-feet per year per well)	Initial pumping rate during projection year 1 (thousands of acre-feet)	Final pumping rate during projection year 20 (thousands of acre-feet)
Baseline[a]	—	—	—	25	25
Case 1	5	2	300	191	191
Case 2	2	1	100	294	294
Case 3	5	2	300	241	241
[b]Case 4	2	1	100	33	191
[b]Case 5	5	2	300	38	294
[b]Case 6	2	1	100	36	241
[b]Case 7	5	4	300	28	78
Case 8	10	2	300	138	138

Case	High Plains aquifer, average difference (feet)[c]	Dakota aquifer, average difference (feet)[c]
1	0.87	124.30
2	1.17	137.12
3	1.14	133.15
4	0.76	119.02
5	0.91	129.59
6	0.88	127.13
7	0.67	92.95
8	0.69	110.30

[a] Continuation at estimated 1982 pumping rates for the 20-year projection period.

[b] Pumpage rates were incrementally increased from the estimated 1982 pumping rate (about 25,000 acre-feet) to attain the maximum pumping rates in year 20 of the projection period.

[c] The average difference is the mean of the differences between the simulated hydraulic heads from the baseline projection with those from the hypothetical cases (cases 1–8) for grid blocks in the study area.

From Watts, 1989.

future recharge events or pumping rates. Because such information is known only with uncertainty, new errors are introduced into the simulation. These errors explain in part why postaudits of some modeling applications demonstrate that models do not give reliable predictions (Chapter 10). In the predictive sensitivity analysis, several variations of a particular scenario are simulated. For example, several different pumping rates may be simulated or the response of the system to different assumed recharge rates may be tested. The resulting heads and drawdowns for each scenario are reported.

An example of results from a predictive simulation with the latter type of sensitivity analysis is discussed briefly below. Watts (1989) used a three-dimensional model to predict the effects of continued pumpage from the High Plains and Dakota aquifers in a portion of Kansas for a period of 20 years under a variety of pumping scenarios. Pumping from the High Plains aquifer was assumed to continue at 1982 rates, but several different rates and distributions of pumpage were assumed for the Dakota aquifer with the baseline projection assuming continuation of 1982 pumping rates. Eight other scenarios of pumpage were devised by the local groundwater management district. Results of the baseline projection and the associated sensitivity analysis using the eight alternative scenarios are shown in Table 8.7.

All scenarios showed that continued pumping will cause relatively small changes in heads in the High Plains aquifer but large changes in heads in the Dakota aquifer. Most of the Dakota aquifer within the study area will convert from confined to unconfined conditions as water is taken out of storage. Other examples of multiple scenarios used in prediction may be found in the literature (e.g., Brown and Eychaner, 1988; Luckey et al., 1988).

Box 8.1
Water Balance Errors in MODFLOW's General Head Boundary Package

An apparent water balance error occurs when MODFLOW's General Head Boundary (GHB) Package, discussed in Box 4.1, is used to simulate specified head boundary conditions. A specified head boundary is simulated by setting the head at a boundary node equal to the specified head value and then using a large value of conductance to simulate low resistance between the boundary node and the aquifer (Box 4.1, Fig. 4). A large conductance (e.g., 1E9) causes the head adjacent to the boundary to be controlled by the assigned head at the boundary, which is the desired effect, but it also may cause a water balance error of as much as

200%.

The water balance error occurs as a result of the way in which numbers are stored in the computer. The error tolerance (Section 8.2) is typically set to around $1E - 3$ to $1E - 5$ feet. The head values are saved in the computer to 13 decimal places or more. As a result, the model will calculate a value for head near the boundary node that is slightly different from the assigned boundary head and there will be a small gradient between the aquifer and boundary node. In the water balance calculation, the flux across the boundary will be calculated using a small gradient and a large conductance, resulting in a large flux of water. If a number of GHB nodes are used to simulate specified head nodes, a large water balance error may result. Whether this error appears as excess water in the inflow or outflow portion of the water balance depends on the relative magnitude between the head in the aquifer and the specified head boundary node.

This type of water balance error does not invalidate the results of the model as long as the GHB is identified as the source of the error. The water balance error can be eliminated by reducing the GHB conductance value until the model produces an acceptable water balance.

Box 8.2
Stochastic Modeling to Evaluate Uncertainty _____

In trial-and-error and automated solutions of the inverse problem discussed in Section 8.3, sensitivity analysis is used to study uncertainty in the solution owing to incomplete knowledge about the head distribution, aquifer parameters, and hydrologic stresses. In stochastic model-

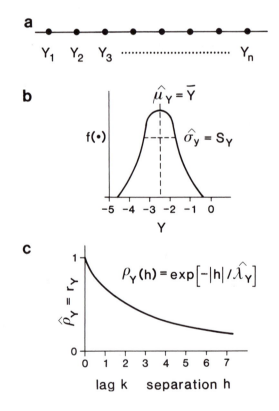

Fig. 1 Probability density function and autocorrelation function for $Y = \log K$ (Freeze et al., 1990).
(a) A one-dimensional sequence of log hydraulic conductivity values, Y.
(b) pdf for Y, with the mean ($\overline{Y}$) and standard deviation (S_Y) indicated.
(c) Autocorrelation function for Y.

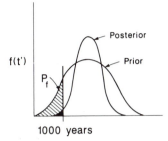

Fig. 2 Bayesian updating (Freeze et al., 1990). Results of stochastic simulations of a steady-state flow field are shown as probability density functions of travel time. Failure occurs prior to 1000 years. Note the reduction in probability of failure (P_f) after updating from prior to posterior estimates.

ing, uncertainty is addressed directly by assuming that the parameters are random variables. For example, hydraulic conductivity is assumed to be a stochastic process (Y) that can be described by a probability density function (pdf). The pdf for hydraulic conductivity (K) is commonly assumed to be log normal, i.e., $Y = \log K$, although a number of investigators have found sets of hydraulic conductivity data that are not log normal (e.g., Connell and Bailey, 1989). The random field of Y values is described by the expected value or mean, the variance or variability about the mean, and the correlation scale or the degree of correlation between points (Fig. 1), defined by an autocorrelation function. A stochastic model produces stochastic output in the form of a pdf of heads.

A stochastic consciousness was introduced into the groundwater literature in the seminal paper by Freeze (1975). Since then a voluminous literature has accumulated on various applications of stochastic analysis to groundwater flow and contaminant transport modeling (Dagan, 1989; Gelhar, 1991). Here we will discuss only a few main points that relate to evaluating uncertainty in the solution of groundwater flow models.

In stochastic modeling, the mean, variance, and correlation length are used to generate a quantitative description (a realization) of the hydraulic conductivity field. The simulation of the hydraulic conductivity distribution pro-

duced in this way is unconditional if the hydraulic conductivity values are not constrained to match point measurements in the field. In classical statistics, field measurements specific to the site being modeled are used as prior information to calculate the mean, variance, and correlation length. Then unconditional realizations of the hydraulic conductivity distribution are produced by kriging. In Bayesian statistics, information from other similar sites and/or geologic and engineering judgment may be used in the absence of site-specific field data. This subjective information can be used to provide the statistical information for the unconditional simulation. In Bayesian terms this step produces a prior estimate of the hydraulic conductivity field. When site-specific field data are not available, the Bayesian approach is philosophically appealing to hydrogeologists and engineers, but not necessarily to statisticians (Freeze et al., 1990). When site-specific field data become available, they can be used to update the mean, variance, and correlation length. This can be done either in the framework of classical statistics with kriging or in a Bayesian framework. Massmann and Freeze (1989) summarize the procedure as follows:

> In the traditional kriging approach, new data are grouped with prior data to form a larger data set. Updated estimates for the parameters are determined from this larger

data set using methods from classical statistics. In the Bayesian approach the prior estimates of the parameters can be determined subjectively and new data can be combined with subjective prior information using Bayes' theorem.

In Bayesian terminology the *prior* estimates are converted to *posterior* estimates. As additional field information is obtained, new posterior estimates can be produced. This is known as Bayesian updating (Fig. 2).

Clifton and Neuman (1982) and more recently Massmann and Freeze (1989), among others, pointed out the advantage of using conditional simulations to help reduce the variance of the output. In conditional simulation, the hydraulic conductivity distribution is constrained to match field data at measurement points. The process of generating the conditional simulation is identical in classical and Bayesian statistics but the answers will be different because the posterior estimates of the mean and variance are different in each case inasmuch as the Bayesian posterior estimates include subjective prior information. The dif-

ferences in mean and variance will be propagated into the conditional simulations (Freeze et al., 1990).

The uncertainty in heads caused by the uncertainty inherent in the parameter field typically is quantified using Monte Carlo simulations whereby many realizations (hundreds to thousands) of the parameter field are produced. Methods for generating random distributions of the log K field include the turning bands technique (Mantoglou and Wilson, 1982: Tompson et al., 1989), matrix methods (Clifton and Neuman, 1982), and nearest-neighbor approximations (Smith and Freeze, 1979). Some of these techniques assume that the stochastic process, $Y = \log K$, is stationary; i.e., the mean, variance, and correlation length are constant in space. Each realization of the parameter distribution is used in a forward model to produce a head solution. The result is a suite of realizations from which the mean head solution and the variance in heads can be computed. Case studies using stochastic models are discussed by Clifton and Neuman (1982), Mercer et al. (1983), and Frind et al. (1988).

Box 8.3
Case Study of a Trial-and-Error Calibration

The modeling study by Thomas et al. (1989) is one of few reported in the literature in which detailed calibration objectives are clearly laid out. They performed a steady-state calibration for a three-layer model of flow in a closed desert basin. The model was considered calibrated when:

1) All simulated heads were within 10 ft of measured heads (averaged over 2 mi^2 model block);
2) mean absolute departure of simulated heads from measured heads was close to zero and the standard deviation was minimal for the 32 model blocks containing wells in layer one;

3) mass balance of water into and out of the system had minimal error;
4) the average of the simulated head differences between model layers one and two was less than 2 ft of head differences derived from field measurements;
5) the simulated areal distribution of evapotranspiration matched the estimated distribution, and the simulated rate of discharge equalled the estimated rate.

The calibration targets are of three types: heads (items 1 and 2, above), head gradients (item 4), and fluxes (item 5). Item 3 in the above list deals with the error in the water balance

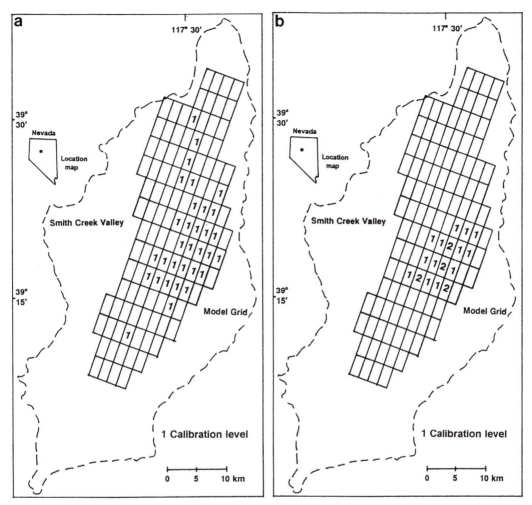

Fig. 1 Calibration levels based on the definitions in Table 8.4 for a simulation of a closed desert basin in Nevada (Woessner and Anderson, 1990).
(a) Calibration levels for heads in layer 1 of the model.
(b) Calibration levels for head gradients between layers 1 and 2.

Table 1
Calibration Levels (ft) for the Simulation Shown in Fig. 1

	Level 1	Level 2	Level 3
Head in layer 1 (32 nodes)	±10 ft 100%[a]	±20 ft —	±30 ft —
Difference in head between layers 1 and 2 (17 nodes)	±2 ft 76%[a]	±4 ft 24%	±6 ft —

[a] Percent of nodes falling within a given level.
Modified from Woessner and Anderson, 1990.

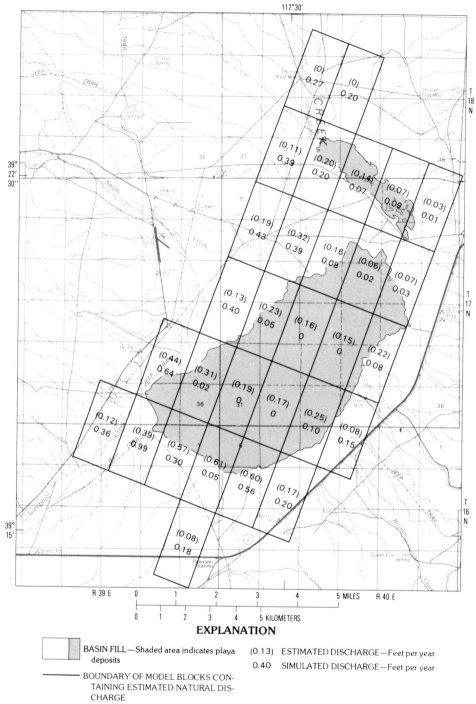

Fig. 2 Comparison of simulated and measured evapotranspiration rates for a simulation of flow in a closed desert basin in Nevada (Thomas et al., 1989). The distribution of discharge into and around a playa near the center of the basin is shown.

and is not so much a measure of calibration as a measure of numerical accuracy and a correctly formulated conceptual model (Section 8.1). Heads measured at 53 sites were used to develop calibration targets for 32 nodes. An error of 10 feet was assigned to field-measured heads and attributed to interpolation errors resulting from interpolation of measured heads to nodes of the block-centered finite difference grid. Vertical head gradients between model layers 1 and 2 formed the second type of calibration target. The error of 2 feet was based on surveyed head differences with an error of a few hundredths of a foot and an error of greater than 1 foot associated with interpolation. The error is smaller than that used for heads in layer 1 because the vertical head differences between layers 1 and 2 showed less variation over the area of the finite difference cells. Woessner and Anderson (1990) used the information given by Thomas et al. (1989) to calculate the spatial distribution of calibration levels (Fig. 1). Table 1 shows the number of nodes used for calibration and the percent of points falling within each calibration level.

The third type of calibration target consisted of fluxes from evapotranspiration (ET). All groundwater discharge occurs via evapotranspiration from a playa in the interior of the basin. ET was measured in several ways but the error associated with the ET measurements was not presented. Although calibration targets cannot be defined or evaluated, Thomas et al. (1989) did present the areal distribution of measured and simulated ET rates (Fig. 2).

Thomas et al. (1989) used the MAE as an average measure of calibration (item 2 above). In the final calibrated model, the MAE was +0.2 ft with a standard deviation of 4.1 ft. Although the discharge rates were considered to be the best known hydrologic parameter, the match between measured and simulated values does not appear to be especially good (Fig. 2). However, sensitivity analysis did not produce a better fit to ET rates and the calibrated model was accepted as the best solution of the inverse problem.

Box 8.4
Case Studies of Two Automated Calibrations _____

Garabedian (1986) used an automated inverse code developed by Cooley (1977, 1979), with modifications to allow the use of interpolated heads where measured heads are not located at nodes. Garabedian calibrated a regional model of two-dimensional, steady-state flow in a basalt aquifer in Idaho. A total of 824 head measurements made in Spring 1980 were used as sample information, i.e., calibration values. Errors in heads were not quantified; all measurements were assumed to be of equal reliability. The quantity and distribution of springflows were used as important calibration data in preliminary simulations. Because no prior information was available for hydraulic conductivity or saturated thickness, transmissivity was treated as a regression parameter and allowed

to vary unrestrained during the automated calibration. Recharge and leakance were used as prior information to constrain head gradients. They were held fixed in preliminary simulations but were treated as regression parameters during a sensitivity analysis.

Garabedian observed, as have many other researchers (e.g., Carrera and Neuman, 1986b), that proper zonation of transmissivity is critical in achieving a stable solution. Zonation was based on rock type and water level gradients, but during initial simulations zonation of transmissivity was adjusted until a stable solution was produced. Then, calibration was achieved by manually adjusting leakance values until the model-calculated springflows approximated measured springflows. Approximately 40 runs were required to achieve satisfactory results. Run number 40 was selected as the base case simulation.

The calibration criterion used to judge the overall fit to measured heads was the standard error (SE):

$$SE = \sqrt{\frac{\sum_{i=1}^{n} (WF)_i (h_m - h_s)_i^2}{n - P}} \qquad (1)$$

where $(WF)_i$ are weighting factors used to indicate the reliability of a head measurement; h_m and h_s are measured and simulated heads, respectively; n is the number of head measurements; and P is the number of regression parameters. The SE was 40.4 ft for the base case simulation. The ratio of the SE to the total head loss over the system was 40.4/3619 or 0.011, indicating that the error in the heads represents a small fraction of the overall model response. The spatial distribution of head residuals showed that there were large residuals in several places, especially along the margins of the model. These errors were attributed to transient effects not included in the steady-state simulation and to coarse discretization. Calculated springflows were within 3% of measured values, which was within the range of the errors estimated for the water balance compiled from field data.

In addition to heads and springflows, the inverse solution calculated values for transmissivity in each of the zones and the standard error (also called coefficient of variation) associated with each estimate. The standard error for parameters is defined to be the calculated standard deviation divided by the mean value. The standard error is a measure of the range over which the parameter may be varied and produce a similar solution. The standard error should be small relative to the absolute value of the parameter.

In a sensitivity analysis, an additional five simulations were performed to test the effect of anisotropy in transmissivity, zonation, and the magnitude of the coefficient of variation of the prior information on leakance and recharge.

Coefficients of variation were set equal to 0.05, 0.1, and 0.2 in successive simulations. A verification test was also done by using a set of 427 heads measured during Summer 1980. As a result of the sensitivity tests, Garabedian concluded that "There is no best simulation presented; instead, results are compared for indications of model improvement. ... the similarity of the results should be noted." Thus, an inverse model does not necessarily give a unique solution.

The sensitivity of the base case solution to changes in parameter values was demonstrated by means of sensitivity parameters or scaled sensitivities (SW) defined as follows:

$$SW = (\partial h_i / \partial a_m)\, a_m \qquad (2)$$

where h_i is the head at location i and a_m is the value of the regression parameter. Large values of SW indicate a large influence on calculated heads in response to a change in parameter value. Garabedian presented values of the sensitivity as defined by SW for each aquifer zone in response to changes in transmissivity, leakance, recharge, and springflow.

One could argue that calibration by an automated inverse solution is a time-consuming and frustrating experience not unlike trial-and-error calibration. The inverse solution was prone to instability and did not give a unique solution. Furthermore, trial-and-error adjustment of leakance in several runs of the model was required to achieve calibration of springflow. The difficulties in this calibration can be largely attributed to the absence of prior information on transmissivity. In a trial-and-error calibration, the modeler would face the same lack of prior information but would usually be convinced that the model is calibrated despite the uncertainty. In an automated calibration, the uncertainty in the solution is clearly documented.

The example discussed above illustrates that a certain amount of user interaction with an automated inverse model will be required even though the calibration is said to be automated. Connell and Bailey (1989) reported an applica-

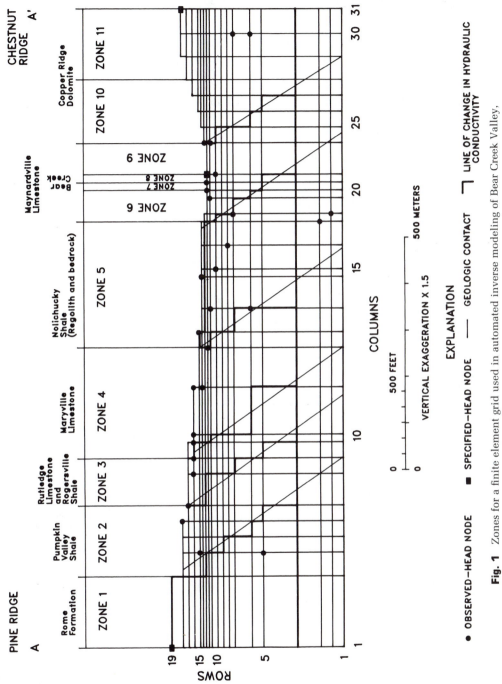

Fig. 1 Zones for a finite element grid used in automated inverse modeling of Bear Creek Valley, Tennessee (Connell and Bailey, 1989).

tion of the Cooley inverse model that required fewer trial runs. The automated inverse model was used to calibrate the profile model of Bear Creek Valley discussed in Box 6.1. The problem description is given in Box 6.1.

For the application of the automated inverse model, the grid was divided into 11 zones as shown in Fig. 1. All geologic units were assumed to be homogeneous and isotropic. Ranges and means in hydraulic conductivities, estimated from field data (Fig. 2), were used to define coefficients of variation for hydraulic

conductivity of each zone. Hydraulic conductivity for each zone was a regression parameter. Three specified head nodes were fixed at the water table (Fig. 1). Coefficients of variation for the field-measured heads were calculated to be 0.0021 for the node near Bear Creek and 0.0055 and 0.0057 for the two nodes on the divides. All other nodes along the upper boundary were specified flux values that were designated to be regression parameters with coefficients of variation ranging from 0.115 to 0.59. The inverse model was run three times (Table 1). The SE for

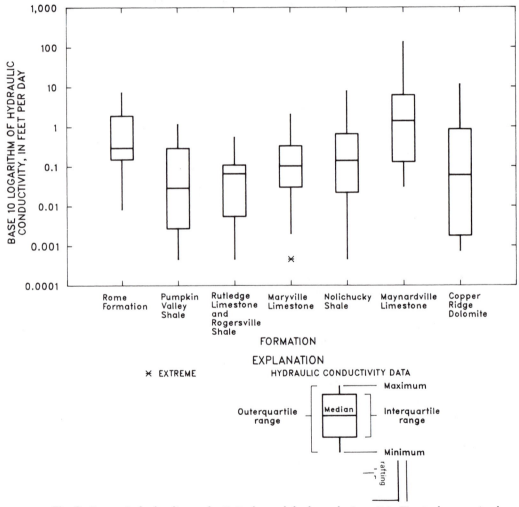

Fig. 2 Ranges in hydraulic conductivity for each hydrogeologic unit in Fig. 1, shown using box plots (Connell and Bailey, 1989).

Table 1
Results of Three Runs of an Automated Inverse Model Applied to a Profile Simulation of Bear Creek Valley, Shown in Fig. 1

Geologic unit	Initial estimate of hydraulic conductivity,[a] in feet per day	Run 1		Run 2		Run 3	
		Regression estimate of hydraulic conductivity, in feet per day	Percentage difference[b]	Regression estimate of hydraulic conductivity, in feet per day	Percentage difference[b]	Regression estimate of hydraulic conductivity, in feet per day	Percentage difference[b]
Rome Formation	0.44	0.30	22	0.30	22	0.27	30
Pumpkin Valley Shale	.019	.016	16	.016	16	.015	19
Rutledge Limestone and Rogersville Shale	.035	.037	.6	.037	.6	.034	3
Maryville Limestone	.11	.034	68	.034	68	.027	75
Nolichucky Shale							
Regolith	.59	—	—	—	—	.51	.3
Bedrock	.138	.059	57	.059	57	.011	79
Maynardville Limestone	1.08	.039	96	.039	96	.041	96
Copper Ridge Dolomite	.042	.031	26	.031	26	.032	24
Deep Bedrock	.00078	—	—	.00078	0	—	—

Modeled zone	Initial recharge rate,[c] in inches per year	Regression estimate of recharge rate,[c] in inches per year		
		Run 1	Run 2	Run 3
1	25.0	28.03	28.03	25.4
2	-12.5	-12.79	-12.79	-11.96
3	-9.38	-10.91	-10.91	-10.99
4	-1.00	-1.02	-1.02	-1.03
5	.88	.99	.99	-1.12
6	-3.52	-4.38	-4.38	3.89
7	14.06	-16.21	-16.21	-14.89
8	28.12	-24.53	-24.53	-23.2
9	49.3	-38.54	-38.54	-38.1
10	5.7	2.19	2.19	2.50
11	20.0	14.89	14.89	15.77

Location of specified-head node	Initial estimate of specified head, in feet above sea level	Regression estimate of head, in feet		
		Run 1	Run 2	Run 3
Pine Ridge	1045.0	1044.2	1044.2	1044.0
Bear Creek	886.4	886.14	886.14	886.31
Chestnut Ridge	1005.0	1002.0	1002.0	1002.3

[a] Initial estimates of hydraulic conductivities are median values.
[b] Percentage difference = |(simulated value − initial value)/initial value| × 100. Negative value indicates regression estimate is greater than the initial estimate.
[c] Negative indicates discharge.
From Connell and Bailey, 1989.

heads was 3.21 ft for runs 1 and 2 and 2.91 ft for run 3. The ratio of SE to total head loss was 0.21 for runs 1 and 2 and 0.19 for run 3. Analysis of the standard errors (coefficients of variation) in the parameter values suggested that the model could be improved most by improving the initial estimate of the hydraulic conductivity of the Copper Ridge Dolomite and by collecting more water level data for the Copper Ridge Dolomite, particularly in zone 10.

The application reported by Connell and Bailey (1989) required fewer trial runs of the inverse code then Garabedian's (1986) application because of the use of prior information on the hydraulic conductivities and on water table fluxes and heads. These case studies demon-strate that an automated inverse model forces the modeler to adopt a protocol for calibration. Such protocol is notably absent from most trial-and-error calibrations. Prior to running an auto-mated inverse code, the modeler is required to quantify the reliability in head measurements and to establish the plausibility range in param-eter estimates by specifying coefficients of vari-ation. The solution calculates the ME, MAE, and SE for the heads and the SE for the calcu-lated parameters, thereby providing quantita-tive measures from which to judge the calibra-tion. Scaled sensitivities allow systematic examination of the sensitivity of the solution to changes in parameter values.

Problems

8.1 Review the documentation for PLASM, MODFLOW, or AQUIFEM-1 and outline the procedures used for code verification. Comment on the criteria used to assess how well the model reproduced verification data sets.

8.2 A water balance is important in model conceptualization and calibration because it provides information on fluxes that is especially useful when attempting to calibrate a steady-state model. Produce a water balance for the simulation performed in Problem 4.1a and calculate the steady-state outflow to the swamp. If you were told that a water balance estimated from field data yielded a discharge to the swamp 30% lower than your model-derived value, which parameters would you suspect may be poorly approximated in the model?

8.3 The error tolerance for numerical closure should be small enough to produce heads that will not change within a specified range. It should also yield an acceptably small error in the simulated water balance and yet be large enough to minimize computer run time.

(a) Illustrate the effect of changing the error tolerance on calculated heads and computational time for Problem 7.2. Compare and contrast the resulting heads and computational time (the number of iterations can be used for iterative solvers) needed for closure.

(b) Use the same error tolerances as in Problem 8.3a in your model for Problem 5.1c and calculate the head at the end of 10 days of pumping. Explain why the heads are different. Outline a method to determine the most effective error tolerance to use in a given model situation.

8.4   Sensitivity analysis is used to evaluate the effect of parameter uncertainty on model results.

(a) Run your model for Problem 5.5 to calculate the heads in the aquifer after 5 days of mine dewatering. Vary the original transmissivity value of 150 m²/day by plus and minus 20 and 50%. (You will need to run the model four more times.) Keep all other parameter values the same.

(b) Select a single node located about 100 m from the mine shaft and plot the absolute value of the difference between the head calculated after 5 days of dewatering for the base case of 8.4a ($T = 150$ m²/day) and each of the four heads produced during sensitivity analysis vs. the percentage change in transmissivity. Construct a second plot by comparing 10 nodal values for each of the four model runs with the results using $T = 150$ m²/day and plot the mean absolute difference in head vs. the percentage change in transmissivity. Comment on how the number and distribution of water level measurements will affect the sensitivity analysis. How could the net change in mine shaft inflow calculated from these analyses be used to describe sensitivity?

8.5   Chapter 8 described the process of calibration and verification of groundwater models. This problem presents a hydrogeologic setting for which you are asked to develop calibrated two-dimensional steady-state and transient models. A third set of data is provided to use in verifying your model.

The model area encompasses a 1500 by 1500 m unconfined sand and gravel aquifer (Fig. P8.1). The east and west boundaries correspond to bedrock. The south boundary is represented by a 100-m-wide gravel-bottomed ditch which slopes to the east. The ditch leaks large quantities of water continuously. This leakage and leakage from many other such ditch systems south of the model area result in a constant flux boundary corresponding to the ditch location. The northern boundary is formed by a 100-m-wide eastward-flowing river. The average stage is plotted as meters above sea level at measured points. The river has an average depth of 2 m and a bottom composed of 2 m of sand and fine gravel with a vertical hydraulic conductivity of 30 m/day. The entire area receives an average daily recharge of 0.0001 m/day. The driller's logs for the aquifer show river sand and gravel with isolated lenses of silt and clay over much of the aquifer. The geologic logs for two wells, N and E in Fig. P8.1, show over 50% silt and clay, overbank and oxbow sediments. The hydraulic conductivity of clean sand and gravel is assumed to range from 30 to 120 m/day. Aquifer tests at wells A and M yielded an hydraulic conductivity of 75 m/day. The storage coefficient for the sand and gravel was estimated to be about 0.10.

(a) Calibrate a two-dimensional model to the steady-state heads in Table P8.1. Refer to Section 8.4 and select at least three ways to illustrate your calibration results. Use average measures of calibration to judge your calibration and also document the spatial distribution of error. Produce a water table map showing field and model results and construct other

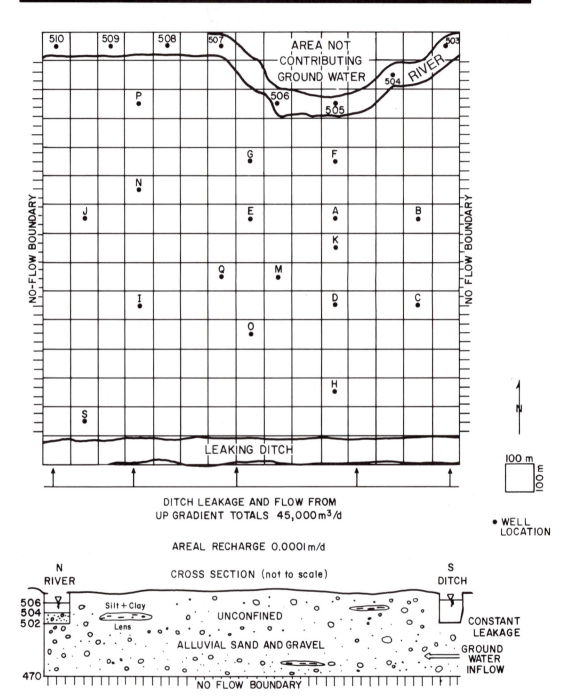

DITCH LEAKAGE AND FLOW FROM
UP GRADIENT TOTALS 45,000 m³/d

AREAL RECHARGE 0.0001 m/d

• WELL
LOCATION

CROSS SECTION (not to scale)

Fig. P8.1

Table P8.1

Field-Measured Heads for Model Calibration and Verification

| Well | Row | Column | Head (m)[a] | | |
			I	II	III
P	3	4	509.12	509.11	508.53
G	5	8	508.19	507.99	506.84
F	5	11	508.71	507.79	506.25
N	6	4	512.83	512.83	509.78
J	7	2	515.71	515.71	511.62
E	7	8	513.17	513.04	507.68
A	7	11	512.22	508.80[b]	507.27
B	7	14	511.95	511.29	507.63
K	8	11	513.88	512.21	507.76
Q	9	7	518.32	518.18	510.00
M	9	9	517.12	516.68	504.36[b]
I	10	4	519.28	519.25	512.82
D	10	11	516.71	516.17	509.44
C	10	14	516.03	515.66	510.12
O	11	8	519.02	518.86	511.08
H	13	11	519.70	519.55	512.94
S	14	2	521.96	521.95	515.85

[a] I, Steady-state heads to be used in calibration; II, heads after 3 days of pumping at A, to be used in transient calibration; III, heads after 1 year of pumping at well M, to be used in verification. Assume head measurement error is ±0.001 m, survey error ±0.02 m and no measurable error resulting from transient effects.

[b] Head is the average within a 100 by 100 m finite difference cell with a pumping well.

graphs as needed to justify your calibration. (Note: Calibration will require some adjustment of the hydraulic conductivity distribution.)

(b) A second phase of calibration includes matching a set of transient head data. If your model is sufficiently calibrated it should reproduce the 3-day continuous pumping test data for well A without additional parameter adjustment. It is more likely that you will need to make additional modifications in hydraulic conductivity and specific yield values to obtain a calibrated transient model.

Simulate the head distribution at the end of 3 days of continuous pumping of well A at a constant discharge of 20,000 m³/day (Table P8.1). Make additional runs as necessary for steady-state and transient calibration. Describe the analyses you used to decide when the model is calibrated. Justify your calibration criteria and distribution of aquifer parameters.

(c) Next you should examine the sensitivity of your model to parameter uncertainty by performing a series of analyses, as in Problem 8.4, for hydraulic conductivity, specific yield, recharge, and river seepage. However, to reduce the time required to complete this problem, we will assume you have done such analyses of your calibrated model and find the results satisfactory. Now you have the opportunity to verify the model against a second set of transient data which were not used to calibrate the model. Simulate the head distribution that results from pumping a well located at site M for 1 year. The well was pumped continuously at 30,000 m^3/day. Do not alter any parameter values. Make only one run of the model. Compare your head data with Table P8.1 and examine the water balance calculated by the model. Discuss the criteria you used to decide if the model is verified.

(d) Based on all your model development efforts, how would you respond to suggestions that this model be used to predict the magnitude of well interference from the operation of two newly proposed city well fields in the vicinity of Q and J on existing wells in the model area? Support your response.

8.6 Read the paper by Gilham and Farvolden (1974). Why do they warn that the procedure of model calibration that consists of adjusting hydraulic conductivities until computed head values agree with field data "must be applied with extreme caution"?

8.7 Read the paper by Gomez-Hernandez and Gorelick (1989). Discuss how the assignment of a uniform hydraulic conductivity based on either the harmonic or geometric mean affects the results of simulating flow in an unconfined aquifer influenced by pumping.

9

DOCUMENTING AND REPORTING YOUR MODELING STUDY

"The writer does the most who gives his reader the most knowledge, and takes from him the least time."
—*Sydney Smith*

A good report is essential to the effective completion of a modeling study. To help in report preparation, it is advisable to keep a journal during the modeling project. Documenting the modeling effort by means of a journal will also allow the model to be revived at some later time by another investigator. Some provision for archiving the modeling journal and associated input/output files should be made.

9.1 Documenting the Modeling Effort

During a typical modeling study there will be many changes in parameter values and boundary conditions and possibly even in modeling strategy between the initial runs of the model and the final runs. Maps of the study area, including maps showing the grid and assignment of parameter values to each node, are essential in model design, and these maps should be preserved as part of the documentation of the modeling effort. Of course, the modeling effort could not be reproduced from these maps alone. The modeler typically will start modeling with good intentions of recording each change in the original design and the consequences of the adjustment, but when modeling gets under way and the deadline approaches it is easy to become careless. Because of the ease of editing input on the computer, note taking commonly is abandoned and replaced with, at best, numbered input and output files. Several weeks after the completion of the study, it will require lengthy efforts by the original modeler to reconstruct justifications for final selection of model parameters and boundary conditions. It may be impossible for another modeler to reconstruct the original modeler's thought process.

Keeping a journal or a log during the modeling study is well worth some extra time because it will facilitate report preparation, allow reconstruction of the model at a later time, and also reduce calibration time. An infinite number of combinations of parameter values can be changed during model calibration. Therefore, it is easy to become confused over what should be adjusted next. Keeping track of what you have done and its effect on model output will allow you to plan a logical sequence of parameter adjustments.

The journal should chronicle the purpose of each model run, the changes in input files, the rationale for the changes, and the effects of the changes on the results. A good example of the type of information that should be recorded in the journal was given by Maclay and Land (1988), who presented a summary of over 300 runs of a model of the Edwards aquifer (Table 9.1). In their work each of the ten series of runs listed in the first column of Table 9.1 had a specific purpose, which is listed in the second column of the table. Modifications to the code and the model input are described in the third and fourth columns. The best simulation of the series is described in the last two columns.

9.2 The Modeling Report

The following discussion focuses on the application of an existing code rather than on code development. If a new code is developed, a separate report should be prepared to describe the code. This report should include details of the governing equation, how boundary conditions are simulated, data input requirements, the numerical method used, code verification results as discussed in Chapter 8, and a user's manual. A listing of the code should be included in an appendix if the code is to be in the public domain. If the code is proprietary, a listing of the code should be provided to the client, if requested.

We assume you are familiar with the standard elements of a report, including an abstract or executive summary at the beginning and a list of reference citations at the end. Our intent is to highlight the portions of the report that describe model-related information.

The report should include the purpose of the model, formulation of the conceptual model, governing equation, boundary and initial conditions, aquifer parameters, the grid of the numerical model with locations of boundaries and internal sources and sinks, calibration results and sensitivity analyses, and results of predictive simulations. A list of assumptions and the field data used to formulate the conceptual model, to set reasonable ranges of parameter values, to calculate initial conditions, to calibrate the model, and to estimate the

Table 9.1

Excerpt from a Modeling Journal

Series	Purpose of series	Modification of model	Simulation condition	Input	Best simulation of series Results
I. Steady state; eastern half of the San Antonio region.	To improve original estimates of transmissivities.	Rearrangement of output matrices for readability.	Used recharge and discharge data for 1972. Initial heads were winter of 1972–73 data. Underflow to Bexar County is represented by simulated injection wells along western Bexar County line. Varied regional anisotropy in z-direction (row) for the entire area by multiplication factors that range from 1 to 10. Varied estimates of transmissivity for individual cells. Original estimate of maximum transmissivity was about 12 ft^2/s. For numerical purposes, steady state was approximated by a very low storage coefficient and a 2-year simulation.	Anisotropy multiplication factor of transmissivity in z-direction (row) was 1.2. Varied transmissivity at individual cells up to 50 percent from original estimates. In general, the model transmissivities were about 20 percent higher than those originally estimated for cells in the confined zone of the aquifer in Bexar County.	Simulated heads were much too high along eastern and western boundaries, too low in most of confined zone, and much too low in unconfined zone of the aquifer.
II. Steady state; western half of the San Antonio region.	Same as series I.	None.	Outflow from Medina County was represented by simulated pumping wells along eastern Medina County line. Varied anisotropy in the z-direction for the entire area from 1 to 10. Varied estimates of transmissivity for individual cells.	Regional transmissivity increased by 20 percent over original estimate. Varied cell estimates of transmissivity up to 100 percent.	Simulated heads were 5 to 20 ft too high in confined zone and about 20 to 60 ft too low in unconfined zone of the aquifer.
III. Steady state; entire San Antonio region.	Same as series I.	None.	Initial experiment used transmissivity derived from previous best experiments.	Anisotropy multiplication factor for transmissivity in z-direction (row) was 1.4. Varied transmissivity in individual cells up to a factor of 10.	Not very satisfactory. Simulated heads were 100 ft too high in the west. No combination of transmissivities produced reasonable results.
IV. Transient, 1972–73.	To refine estimates of transmissivity and to improve estimates of storage coefficients.	Features to change all transmissivities in any of the 26 subareas by a subarea multiplier, to input recharge by basin, and to change recharge and discharge by separate multipliers. Subdivided the San Antonio region into 26 areas.	Simulated hydrology of calendar years 1972 and 1973. To reduce model start irregularities, continued to start the model simulation using winter of 1972–73 water-level data and repeated the simulation of 1972. Simulated results were compared with winter of 1972–73 measurements. Subdivided 1973 into 150-, 65-, 60-, and 30-day simulation periods. The recharge distribution included identifying the cells in each of the drainage basins, estimating the percent of recharge for those cells overlying streams, and inputting recharge by drainage basin.	No satisfactory solution. The following parameters were varied: transmissivity and anisotropy for each of the 26 subareas, transmissivity value of individual cells, and storage coefficients for unconfined and confined zones of the aquifer.	Excessive springflows were computed for 1973. Simulated Comal Springs discharge usually was several times too high. Simulated heads were too high in the western part of the San Antonio region. Could not adequately control simulated storage within the recharge area using transmissivity and anisotropy multipliers by subarea.
V. Transient, 1972–73; barrier faults.	To test the significance of barrier faults on water levels and springflow.	Feature to allow anisotropy in y-direction to be changed cell by cell. Value must be greater than zero.	Delineated significant barrier faults. The anisotropy factor is equivalent to the fraction of water that would flow across the cell for an expected hydraulic gradient and transmissivity if the barrier fault did not exist. Varied anisotropy	Subarea multipliers for transmissivity (T) and anisotropy in y-direction: T Anisotropy T Anisotropy A 0.60 0.80 C 0.40 0.60 B 0.40 0.50 D 1.00 0.60	For 1972, the simulated heads were about 10 ft below measured heads in the confined zone of the aquifer and 10 to 50 ft above them in the recharge area in the west (Uvalde and Kinney Counties).

Table 9.1 (continued)

Series	Purpose of series	Modification of model	Simulation condition	Input T	Input Anisotropy	Best simulation T	Best simulation Anisotropy	Results
V.—Continued			factor by subareas from 0.4 to 1.0 in the y-direction. Varied transmissivity values of individual cells in accordance with revised concepts of possible flow patterns. Redistributed cell-by-cell recharge percentage to add more water near the downstream part of the recharge area. Varied specific yields for unconfined zone of aquifer from 0.01 to 0.10.	E 0.60 F 0.60 G 0.50 H 2.00 I 0.85 J 1.50 K 2.00 L 3.00 M 2.00 N 1.50 O 1.50	0.60 0.60 0.60 1.00 0.75 0.75 0.85 0.75 0.75 0.75 0.50	P 1.50 Q 1.00 R 1.00 S 2.00 T 2.00 U 1.00 V 1.00 W 1.00 X 0.50 Y 1.00 Z 1.00	0.50 0.75 0.75 1.00 0.75 1.00 1.00 1.00 1.00 1.00 1.00	

Storage coefficient of aquifer in unconfined zone of aquifer was 0.05.

Series	Purpose of series	Modification of model	Simulation condition	Input	Results
VI. Sensitivity tests.	To test the sensitivity of computed springflow and heads to changes in transmissivity and anisotropy in subareas and to the occurrence and length of selected barrier faults.	None.	Used the results from transient 1972-73, with fault barrier series as base run for sensitivity studies. All parameters except one were held constant. The varied parameter was then investigated by making changes of the values and then studying the computed results.	--	When isotropy was assigned to subareas representing the confined zone of the aquifer, the match of computed and measured springflow and water levels improved. Computed very excessive springflow when simulated barrier fault at Haby Crossing fault was removed. The additions of a simulated subsurface spring near Uvalde (Leona Springs) lowered the discharge at Comal Springs in April 1973 from 367 to 348 ft³/s. This alteration also lowered the head in western Uvalde County. Simulated springflows are very sensitive to presence and effectiveness of barrier faults.

Series	Purpose of series	Modification of model	Simulation condition	Input
VII. Transient, 1972-73; barrier faults included, anisotropy varied by subareas in unconfined zone and set to 1.00 in confined zone of the aquifer.	To test need for anisotropy away from faults in unconfined zone of the aquifer.	Feature to print fluxes across cell boundaries and to plot velocity vectors.	Adjusted anisotropy for cells representing barrier faults. Added additional minor barrier faults in the recharge area. Storage coefficient of recharge area set at 0.05.	Subarea multipliers for transmissivity (T) and anisotropy in y-direction:

T	Anisotropy	T	Anisotropy
A 0.60	0.80	N 1.50	1.00
B 0.40	0.50	O 1.50	1.00
C 0.40	0.40	P 1.50	1.00
D 0.60	0.60	Q 1.00	1.00
E 0.60	0.60	R 2.00	1.00
F 0.60	0.60	S 1.00	1.00
G 0.50	0.60	T 2.00	1.00
H 2.00	1.00	U 1.50	1.00
I 0.85	1.00	V 1.00	1.00
J 1.50	1.00	W 1.00	1.00
K 2.00	1.00	X 1.00	1.00
L 3.00	1.00	Y 1.00	1.00
M 2.00	1.00	Z 1.00	1.00

Results:

Spring	Springflow (ft³/s) Simulated	Measured
	End of 1972	
Comal	265	310
San Marcos	139	161
	May 1973	
Comal	336	365
San Marcos	178	211

For 1972 the simulated heads were 5 to 10 ft too low throughout the confined zone and 10 to 30 ft too high in the unconfined zone of the aquifer.

Series	Purpose of series	Modification of model	Simulation condition	Input	Results
VIII. Transient, 1972–73; isotropic except for barrier faults.	To test the significance of leakage across barrier faults on springflow and water levels. To locate the most significant barrier faults.	Feature to allow y-direction anisotropy to be set at 0.00.	Sets anisotropy for individual cells along mapped barrier faults between 0.0 and 0.5 and isotropic elsewhere. Storage coefficient of aquifer held at 0.05 for the recharge area. Tested the effects of barrier faults in northern Uvalde County on springflow. Changes were made in simulated length of selected faults.	--	Spring — Springflow (ft³/s) Simulated / Measured **End of 1972** Comal 269 / 310 San Marcos 138 / 161 **May 1973** Comal 343 / 365 San Marcos 177 / 211 Major convergences of flow were noted in the Nueces area and in the vicinity of Sabinal. In 1972, water levels were about 5 to 10 ft too low in the confined zone of the aquifer.
IX. Transient, 1972–76; isotropic except for barrier faults, specific yield set at 0.08.	To refine calibration of transmissivity and anisotropy of faults using a storage coefficient of 0.08 for the recharge area.	None.	Set discharge at Leona Springs to 75 ft³/s in 1973. Varied transmissivity estimates in vicinity of Comal Springs to increase simulated flow to San Marcos Springs.	Transmissivity subarea and its multiplier: A 0.40 N 1.25 B 0.40 O 1.50 C 0.40 P 1.50 D 1.00 Q 1.50 E 1.40 R 1.00 F 1.40 S 2.00 G 0.50 T 2.00 H 3.00 U 1.50 I 0.85 V 1.00 J 1.50 W 1.00 K 2.00 X 1.50 L 3.00 Y 1.00 M 1.25 Z 1.00	Spring — Springflow (ft³/s) Simulated / Measured **End of 1972** Comal 260 / 310 San Marcos 160 / 161 **May 1973** Comal 375 / 365 San Marcos 210 / 211 Barrier fault along row 10 is a very significant control on springflow. Simulated water levels within confined zone of aquifer are within a few feet of measured heads. Simulated water levels in northern Uvalde are about 10 to 30 ft too high.
X. Transient, 1972–76; final selection of transmissivity, anisotropy, and storage coefficient.	To achieve best simulation.	None.	Any combination of minor changes to any or all aquifer properties.	Subarea multipliers for transmissivity (T): A 0.15 N 2.50 B 0.15 O 1.00 C 0.50 P 2.50 D 1.00 Q 1.50 E 1.40 R 1.00 F 1.40 S 2.00 G 0.20 T 2.00 H 4.00 U 1.50 I 5.00 V 1.00 J 1.30 W 1.00 K 2.00 X 0.50 L 4.00 Y 1.00 M 1.00 Z 1.00	Described in text.

From Maclay and Land, 1988.

water balance are also necessary. The modeling study may be one of several methods used to assess a hydrologic system or it may be the only method. In either case, all of the information described below should be included when reporting the modeling study. Furthermore, your report, along with the modeling journal and other archived information, should include sufficient data to allow a reader familiar with groundwater modeling to reproduce your results.

TITLE

In selecting a title for the report it must be remembered that successfully constructing a model is never the purpose of a modeling study. A title like "The Construction of a Hydrogeologic Model of the Black Mesa Area" doesn't tell the reader why a model was needed or what was done with it. The title should reflect the project goals or specific modeling objectives. For example, the modeling objective may be to assess the impact of five different pumping scenarios on the aquifer in the Black Mesa area. In this case, a good report title would be "Simulation of Five Groundwater Withdrawal Projections for the Black Mesa Area" (Brown and Eychaner, 1988) or "Evaluation of the Consequences of Continued and/or Expanded Pumping on the Sandstone Aquifer of the Black Mesa Area." You should select an informative title, not one that masks the contents of the report.

INTRODUCTORY MATERIAL

The introductory material should include a discussion of the importance of the problem being considered, long-term project goals and specific modeling objectives, their relation to previous work, and the general approach to accomplishing the goals and objectives. A discussion of the relationship of the project to previous research including earlier modeling efforts may be part of the discussion of the importance of the problem or may follow the discussion of goals and objectives.

The long-term goals of the hydrogeologic study should be identified. The purpose of the modeling project should be clearly stated following the guidelines given in Chapter 1. Succinctly stated modeling objectives should be listed in the general order in which they would logically be completed to reach the project goals. This ordering sets the direction of the rest of the report and will guide the reader through the report. The success of the study will be judged by the degree to which the objectives are met.

The final section of the introductory material should include a brief paragraph on the overall approach or strategy used to accomplish the modeling objectives. That is, the steps in the modeling protocol used in the study should be reviewed. The modeling protocol should follow the steps in Fig. 1.1 as closely as possible.

HYDROGEOLOGIC SETTING AND CONCEPTUAL MODEL

This section should present what was known about the hydrogeologic system prior to this modeling effort. It should contain information on the geologic setting, aquifer parameters, and a conceptual model of the system along with the simplifying assumptions used to construct the conceptual model. A location map of the study area with surface topography and surface water bodies indicated should be presented at the beginning of this section.

The description of the geologic setting should include the areal distribution of geologic units, their composition, the depositional environment, and the site stratigraphy. A geologic map can be used to show the distribution of formations at the land surface. Geologic cross sections showing the stratigraphy and limits of the materials exposed on the surface and their presence at depth should also be included. A discussion of depositional environments will assist in conceptualizing the spatial variation of geologic units and their hydraulic properties.

The definition of hydrostratigraphic units and a discussion of hydrogeologic parameters should follow the description of the geologic setting. You should use previously defined terminology when possible. If your study is the first hydrogeologic description of an area, presentation of hydrogeologic parameter values and head data may be required to justify your definition of hydrostratigraphic units (Section 3.1). If your work has redefined the hydrostratigraphy, the original classification should be outlined and your revisions presented and supported by the new data. You should present information on hydrogeologic parameters such as porosity, storativity, hydraulic conductivity, and transmissivity and discuss a rationale for spatial variation of these parameters in the study area. You should also discuss the methods used to estimate parameter values and the estimated ranges in parameter values for the materials described. If the properties of a unit are poorly known, it should be so stated. All the parameter values presented in this section should be clearly distinguished from parameter values estimated from model calibration and verification. These premodeling data are derived independently of the modeling effort and form the basis for the conceptual model. Changes made in parameter values during model calibration and verification should be discussed later in the report when discussing model design and results.

The field-measured distribution of hydraulic head should be presented in map view and, if sufficient data are available, in cross section. The directions of groundwater flow and the locations of recharge and discharge areas should be presented, including rivers, spring, drains, irrigated areas, and pumping wells.

A conceptual model of the study area should be presented based on the discussion of geologic setting, hydrogeologic parameters, and the three-dimensional flow system. Hydraulic and physical boundaries of the system are identified from flow system interpretation and the geologic setting. A figure showing hydrostratigraphic units and a schematic picture of the flow system in either one, two, or three dimensions should be presented following the examples in

Fig. 3.1. A steady-state and, if appropriate, a transient water balance should be prepared based on the available field data with a description of how each component of the budget was calculated or estimated. Water budget information should be summarized in tabular form. Error bars should be derived from field data or otherwise estimated for each component of the water budget.

MODEL DESIGN AND RESULTS

This section should include a brief description of the code, the relation between the parameter values used in the numerical model and those used to formulate the conceptual model, calibration targets and procedures, modeling results, and a demonstration of model sensitivity. In short, you should describe the tool you chose, the results of its application, and the degree of success in accomplishing your objectives.

The Code

This section describes why you selected a particular code. Only a reference citation is needed if you used a well-documented public-domain code without modifying it. However, if the code was modified for your application, the modifications and the effects of the changes should be described. If extensive recoding was done you should include a copy of the modified portions of the code in an appendix. A summary of the changes may be included in the body of the text. Typically, modification in data input or output (e.g., reformatting output or creating head files for graphic packages) requires only a line or two of text simply stating that such changes were made.

The Relationship between Conceptual and Numerical Models

This section describes how the conceptual model was translated to the grid of the numerical model. You should describe how space and time were discretized, how model boundaries were defined, and how parameter values were assigned to each node in the grid. The selection of a two-, three-, or quasi three-dimensional approach should be justified based on the conceptual model of the flow system.

If only a portion of the study area is to be modeled, the rationale for selecting the modeling subarea should be presented. The grid should be superimposed on a figure of the study area showing surface water bodies and geologic features such as fault traces or fracture sets (Fig. 3.22b). You should state the dimensions of the grid blocks or elements and the number of nodes in the grid. The relationship between hydrostratigraphic units and model layers as well as locations of model boundaries should also be shown.

The method and rationale used to assign initial parameter values and stresses to each cell or node should be discussed. Figures showing the distribution of parameters on the grid (Fig. 3.23) may be used to display data input.

Assignment of internal stresses is usually indicated by symbols (Fig. 4.13). The types of stresses and assumptions made in locating the stresses should be described in the text.

This section of the report should also include a discussion of the uncertainty in model parameters and stresses and a statement of the range of values that will be accepted as valid during calibration (Box 8.4, Fig. 2). You should also indicate if some of the parameters or stresses are known with such a high degree of uncertainty that they will be estimated solely by model calibration.

Model Calibration and Verification

This section should describe the calibration process. The calibration targets for both head and flows should be defined and justified. Sources and magnitudes of errors associated with each calibration value should be described. Figures like Fig. 8.11 may be used to show the approach to calibration. If both steady-state and transient calibrations were done, both should be discussed. Revisions of parameter values, boundaries, and stresses should be briefly described in the context of how such changes led to model calibration or to the conclusion that calibration was not possible. A list of parameter values and stresses for the final calibrated model should be presented.

The match to the calibration targets is typically shown in one or more figures comparing field-measured and simulated heads (Figs. 8.8, 8.9). This information may also be shown in tabular form (Table 8.3). Portions of the grid that appear to contribute to the overall error more than others should be identified (Fig. 8.13). You should also speculate why such errors occur. The simulated water balance should be presented (Table 8.1, Fig. 8.1) and compared to the field-estimated water balance. Discrepancies should be identified and discussed. If possible, a second set of field data should be used for model verification. The match between field and simulated heads and flows should be presented following the format used to describe the calibration results.

Sensitivity Analysis

The sensitivity analysis recognizes that the calibrated model may not represent a unique match to the calibration target. This section documents the sensitivity of the results to variations in parameter values, grid size, boundary conditions, and calibration criterion. A discussion and justification of the scope of the sensitivity analysis and a set of figures and/or tables showing the results of the analyses are needed (Section 8.4).

Predictions

If you have demonstrated that the model is calibrated and if you have sufficient information to estimate both natural and human-induced future alterations to the system, you may be able to use the model to predict the consequences of changes in internal stresses or water management practices. This section of the report requires additional qualifiers stating the assumptions, uncertainties, and limitations in the predictions. A sensitivity analysis should be performed on the predictive model to quantify the effect of uncertainties in estimating the timing and magnitude of future changes to the system. A number of possible scenarios should be simulated to reflect the uncertainty in future events (Table 8.7). The results of predictive modeling may be presented graphically or using contour maps.

MODEL LIMITATIONS

This section should clearly state the limitations of your modeling effort and the assumptions made. The reliability of the calibration and the sensitivity of the model to various parameters should be discussed in the context of the assumptions used in constructing the model. You should also discuss the degree of uncertainty in your predictive simulations. The appropriate use of the modeling results should also be discussed. For example, is there sufficient certainty in the model calibration and predictive simulations to use the modeling results to make management decisions?

SUMMARY AND CONCLUSIONS

This section should include a brief summary of the modeling results including important information learned from the modeling effort. Conclusions should be listed. If you were not able to achieve an acceptable calibration, you should list the field data that should be collected to aid model calibration. Specific ways in which the conceptual model should be redesigned to improve calibration should also be discussed. You should follow the conclusions with a short discussion of the questions left unanswered and suggestions for future work.

APPENDICES

Appendices contain additional or supplemental information such as documentation of modified computer code or listing of code. Compilations of geologic logs, well inventories, water level measurements, determinations of hydrogeologic parameters, and details of water budget calculations should also be included to support your discussion in the text. If data input files are not included in the appendices, the accessibility and location of these files must be mentioned in the text. Typically, these files are extensive and increase report production costs. If you do not include complete data input files in the report, you should archive them together with the results of calibration and sensitivity simulations. Directions on how to retrieve archived information should be included in the report. Archiving your data and your modeling journal will ensure that your model can be revived by future investigators when new field information becomes available or when new management questions arise.

Problems

9.1 This chapter stresses the importance of recording your modeling efforts. Describe the method of record keeping you used when modeling Problem 8.5. Could a colleague familiar with groundwater modeling take your records and notes from Problem 8.5 in their current condition and reconstruct the steps you used to derive a calibrated transient model? If not, do the problem again and this time document your efforts properly.

9.2 Using your notes and the results from Problem 8.5, prepare an abbreviated modeling report using the outline recommended in this chapter. The four- or five-page report should state what information was supplied by the problem. It should also contain your interpretations, results, conclusions, and limitations. Be sure to title the report.

9.3 Review the reports you read for Problems 1.5 and 7.4. Critique the report title, organization, and completeness against suggestions presented in this chapter.

10

POSTAUDITS: HOW GOOD ARE PREDICTIONS?

"We can only reason from what is;
we can reason on actualities, but not on possibilities."
—Henry St. John Bolingbroke

Pioneering studies of regional flow systems by Toth (1962, 1963) led the way to widespread mathematical simulation of regional-scale groundwater systems in the 1960's. The objective of many of these modeling studies was to predict the long-term response to some applied stress. Some 20 years later, we are in a position to ask: "Can groundwater flow models accurately predict the future?" Concern over the accuracy of predictions has led workers into two different research areas: (1) efforts to incorporate uncertainties into modeling studies by using automated inverse models (Section 8.3) and stochastic simulation (Box 8.2) and (2) postaudits.

The type of postaudit considered here consists of examining the accuracy of a prediction made at least 10 years prior to the postaudit. Verifications of short-term predictions such as described by Wong et al. (1987) are certainly useful but do not provide as rigorous a test of predictive ability as an analysis of long-term predictions because they do not allow sufficient time for the model to move far from the calibrated solution. To date, four postaudits have been reported in the literature. Two of these include postaudits of solute transport modeling, which of course also included a flow model. In all four cases the model did not accurately predict the future. Considering the findings of the four postaudits, two conclusions emerge:

1. Inaccurate predictions were partly caused by errors in the conceptual model of the hydrogeologic system.

2. Inaccurate predictions resulted from a failure to use appropriate values for assumed future stresses. Predictive simulations require the modeler to estimate the magnitude of future stresses such as recharge, pumping, and contaminant loading rates. Uncertainties involved in estimating these future stresses are often large and the modeler's failure to use accurate future stresses is understandable. We tend to reason "on actualities," not on "possibilites," and will tend to extrapolate current trends into the future when the current trends may not apply to future events. In order to build this type of uncertainty into model prediction it is imperative that several different scenarios be simulated using different assumed trends in the applied stresses in order to define a range in the predicted values (Section 8.5).

10.1 Importance of the Conceptual Model

Formulation of the conceptual model was discussed in Section 3.1. The postaudits reported in the literature make it clear that a valid and complete conceptual model is essential for making accurate predictions. Each of the postaudit case studies is reviewed below in order to highlight the type of error in the conceptual model that caused the prediction to fail.

The first two case studies discussed below involve postaudits of electric analog models used to simulate the response of an aquifer to pumping. Electric analog models were used extensively in the 1960's before being superseded by digital models. An electric analog model consists of electrical circuitry (resistors and capacitors) mounted on boards. Because Ohm's law for the flow of electrical current is analogous to Darcy's law, the flow of groundwater can be simulated by monitoring the flow of electricity. The principles used in translating the conceptual model to an electric analog model are the same as those used in translating the conceptual model to a digital model. A prediction made by an electric analog model will be the same as a prediction made with a digital model if both are based on the same conceptual model. The third and fourth case studies discussed below involve postaudits of solute transport predictions made with digital models of flow and transport.

SALT RIVER AND LOWER SANTA CRUZ RIVER BASINS, ARIZONA

Konikow (1986) performed a postaudit of a two-dimensional electric analog model that was calibrated against 40 years of record (1923–1964) and then used to predict water level changes during the following 10 years (1965–1974). During the postaudit, analysis of observed water level changes in 77 wells during 1965–1974 showed that the model consistently predicted lower water levels than actually occurred.

The errors in the prediction can be accounted for, in part, by the failure to use accurate future pumping stresses in the simulation. In other words, the future pumping scenario assumed by the modeler did not match the actual pumping records that occurred during the modeled period. The modeler used the record of historical pumping shown in Fig. 10.1 to estimate the probable future pumping trend. Figure 10.1 shows that pumping rates increased consistently until the last few years of the calibration period, when they leveled off. For the predictive simulation it was assumed that future pumping would continue at the 1964 rate. In fact, pumping declined after 1965 (Fig. 10.1). During the postaudit, examination of the distribution of pumping and the predicted errors in water levels suggested that incorrect assumed pumping rates were partly responsible for the erroneous prediction but that other sources of error were also present. Because the analog model had been disassembled, it was not possible to run the model again to isolate the sources of error. However, it is likely that improvements in the conceptual model including incorporation of land subsidence and a three-dimensional representation of the system would improve the model's predictive ability by better representing changes in aquifer storage and transmissivity.

BLUE RIVER BASIN, NEBRASKA

Alley and Emery (1986) examined predictions of 1982 water level declines and streamflow depletions made in 1965 with an electric analog model. Declines in water levels and streamflow were predicted to occur as a result of increases in pumpage for irrigation. The postaudit showed that the model overestimated the

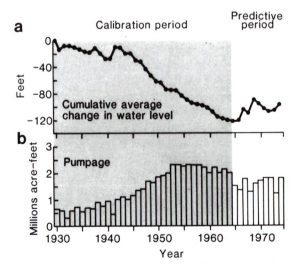

Fig. 10.1   Postaudit of a modeling study of the Salt River Valley (Konikow, 1986).
(a) Average change in groundwater levels in part of the Salt River Valley.
(b) Historical record of pumpage in part of the Salt River Valley.

decline in groundwater levels and underestimated the amount of streamflow depletion.

Net groundwater withdrawals in agricultural areas are difficult to estimate because it is usually necessary to infer net withdrawals from estimates of irrigated acreage, groundwater recharge, and consumptive use of irrigation water (irrigation efficiency). Analyses performed by Alley and Emery (1986) suggested that net groundwater withdrawals used in the analog simulation were too low. The model overestimated groundwater level declines because it assumed that all of the net groundwater withdrawals would come from storage in the aquifer when in fact some water comes from induced recharge from the stream. Furthermore, Alley and Emery (1986) speculated that storage coefficients used in the model were too low. They concluded that "Considerable uncertainty about the basic conceptualization of the hydrology of the Blue River basin greatly limits the reliability of groundwater models developed for the basin."

ARKANSAS RIVER VALLEY, SOUTHEASTERN COLORADO

Konikow and Bredehoeft (1974) used an early version of the USGS Method of Characteristics model (Konikow and Bredehoeft, 1978) to predict concentrations of dissolved solids in the aquifer adjacent to a portion of the Arkansas River in southeastern Colorado. Salinity is a problem in this area owing to recycling of irrigation water. Konikow and Bredehoeft calibrated the flow portion of the model to a transient flow field determined from data collected during the 1971–1972 study period. The solute transport portion of the model was also calibrated to data obtained during 1971–1972. The model predicted that dissolved solids concentrations would increase steadily through 1982 (Fig. 10.2).

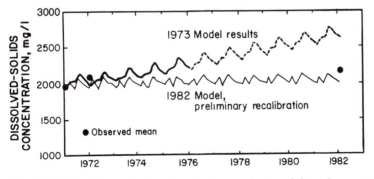

Fig. 10.2 Prediction of salinity trends in groundwater of the Arkansas River valley, Colorado, showing 1973 results and results based on a 1982 recalibration of the original model (Konikow and Person, Water Resources Research, 21(11), pp. 1611–1624, 1985, published by the American Geophysical Union).

Statistical evaluation of the historical data set including data collected in 1982 showed that dissolved solids had not increased in the aquifer above 1971–1972 levels. This suggested that the aquifer is in dynamic equilibrium with respect to salinity (Konikow and Person, 1985). If true, this would mean that current irrigation practices could be continued indefinitely without causing further groundwater salinity degradation within the study area.

Konikow and Person (1985) showed that the error in the original prediction was due to calibration to a period of decreasing river discharge. During the 1971–1972 calibration period, discharge of the river was on a declining trend after a record high in 1966. Concentration of dissolved solids in the river is inversely proportional to discharge; during 1971–1972, river water recharging the aquifer was increasing in salinity. The model propagated this trend into the future. Statistical tests showed that this short-term trend, although statistically significant, was not representative of the long-term salinity trend. The post-audit showed that the flow portion of the model was adequately calibrated.

Person and Konikow (1986) recalibrated the model using an improved regression equation to relate salinity to measured specific conductance. Data from 1971, 1972, and 1982 were used in calculating the new regression equation. Calibration of the model is sensitive to this relationship because about half of the irrigation water is diverted from the river. They also improved the conceptual model of the system by incorporating a lag time for solutes to travel through the unsaturated zone. The recalibrated model successfully simulated the observed long-term trend in salinity from 1971–1982 (Fig. 10.2). Finally, they used the recalibrated model to demonstrate that the system is in dynamic equilibrium with current irrigation practices.

Although the recalibrated model could accurately simulate the observed long-term trend in salinity, it should be noted that the recalibration used field data collected in 1982 in order to simulate 1982 conditions. Only three years of detailed data, including 1982, were available for the 1971–1982 simulation. It is therefore relevant to ask whether the model could have predicted the long-term trend in the absence of the 1982 data. Through statistical analysis of temporal salinity trends using a 32-year record of estimated stream salinities, Person and Konikow (1986) demonstrated that a four-year sampling period was needed to calibrate the solute transport portion of the model to within 10% of the observed mean salinity, while one year's worth of data was sufficient to calibrate the flow portion of the model. The implication is that the long-term trend could have been predicted without the 1982 data, as long as a four-year record of salinity trends was available for calibration. This finding suggests that evaluation of the conceptual model should include not only the spatial properties of the system but also the hydrologic response time and temporal trends that characterize the system.

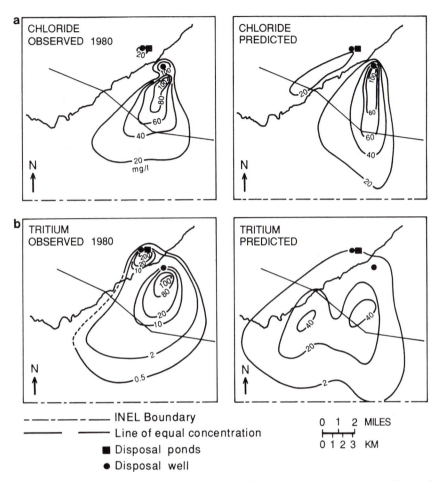

Fig. 10.3 Comparison of observed and simulated concentrations in groundwater beneath the Idaho National Engineering Laboratory (INEL) site (modified from Lewis and Goldstein, 1982). The simulated contaminant plumes on the right are anomalous because the calibration required that longitudinal dispersivity be less than transverse dispersivity.
(a) Chloride concentrations.
(b) Tritium concentrations.

IDAHO NATIONAL ENGINEERING LABORATORY

Robertson (1974) used a two-dimensional version of an early USGS ground-water flow model (Trescott et al., 1976) to simulate flow in a basalt aquifer beneath the Idaho National Engineering Laboratory (INEL). Robertson calibrated the model to an assumed steady-state flow field. He coupled the flow model to a solute transport model solved using the method of characteristics and calibrated the model to the observed concentrations of chloride in groundwater in 1958 and 1969. Tritium and strontium-90 plumes were also simulated and compared with observed data.

Robertson (1974) then used the calibrated model to predict chloride and tritium concentrations in 1980 (Fig. 10.3). Lewis and Goldstein (1982) performed a postaudit of those predictions and concluded that the contaminant plumes predicted by the simulation extended farther downgradient than the actual plumes because of conservative worst-case assumptions in the model input and inaccurate approximations of subsequent waste discharge and aquifer recharge conditions. The model assumed that waste disposal through the disposal well south of the river (Fig. 10.3) would continue at 1973 rates, whereas disposal rates actually increased. The model also assumed that the Big Lost River would recharge the aquifer in odd-numbered years, when in fact there were high flows from 1974 to 1976 followed by four years of low flows when no recharge occurred.

Lewis and Goldstein (1982) pointed out that the conceptual model used by Robertson (1974) was highly simplified. It was not unusual in the 1970's for contaminant transport modeling studies to use simplistic conceptual models that assumed two-dimensional flow along with homogeneous and isotropic aquifer properties. We now know that these assumptions are usually inappropriate for simulations of complex contaminant plumes such as those at the INEL. Recent modeling of this system by Goode and Konikow (1990a) demonstrated that the inclusion of transient effects in the model does not explain some of the anomalies in the Robertson simulation. Another possibility is the likelihood of fracture flow in the basalt aquifer. The conceptual model used by Robertson (1974) viewed the aquifer as a continuous porous medium. In reality, it is likely that flow in this aquifer would be better approximated using a dual porosity model that included flow through the fractures as well as matrix diffusion (Section 12.6). The failure of the prediction can be attributed to an overly simplified conceptual model of the flow system and to inaccurate estimates of future hydrologic stresses and contaminant loading rates.

10.2   Lessons from Postaudits

The answer, then, to the question "How good are predictions?" is that predictions have been inaccurate because of failures of the modeler, not the model. In some cases, the failures of the modeler are understandable and unavoidable. The modeler is called upon to simulate "possibilites." For example, future applied stresses (e.g., pumping rates, contaminant loading rates) must be estimated prior to running the model. The postaudits discussed in Section 10.1 demonstrate that sometimes the "possibilities" the modeler assumes turn out not to be "actualities." Postaudits clearly point to the need for simulations of many different possible future scenarios. In light of these postaudits, the modeler can no longer be content to make one predictive simulation but now must produce a suite of predictions.

All of the postaudits reported in the literature make another point—namely that the failure of the model to predict the future was not due to numerical or theoretical deficiencies in the model. Rather, errors in predictions are attributed to errors in the conceptual model. Continual improvement of the conceptual model by collection of new field data will improve the numerical model. That is, the model should be used in management mode as described below.

10.3 Crisis Mode versus Management Mode

It is likely that there have been hundreds of predictive modeling studies since the 1960's. The fact that only four postaudits have been reported in the literature suggests that models traditionally have been used in a crisis mode rather than a management mode. In other words, a model is constructed to answer some pressing question so that a management decision can be made. After the model has served this purpose, it is "shelved" and forgotten or discarded. In most cases, apparently no effort is made to use the model for management of the groundwater system on a day-to-day, month-to-month, or even year-to-year basis.

Ideally, models should be archived as discussed in Chapter 9 so that the model can be resuscitated years later when new data become available and/or a new modeling objective is defined. A few examples of using models in this way are reported in the literature. Jorgensen (1981) described a succession of three increasingly sophisticated models used to predict drawdowns in the aquifer system underlying Houston, Texas, and surrounding area. The first model was an electric analog model constructed in the early 1960's. The model accurately simulated observed water level declines in and adjacent to the city of Houston but did not reliably simulate drawdowns in outlying areas. The failure of the model was attributed to the lack of sufficient field data to formulate an adequate conceptual model of the system. Between 1965 and 1975, the conceptual model of the system was improved following the acquisition of new field data. A second analog model was constructed in 1975 using a four-layer representation of the system and including the effects of vertical leakage across clay units and the release of water from storage owing to compaction of clays. The simulated clay compaction was used to assess the ability of the model to predict land subsidence. (The volume of land subsidence is approximately equal to the volume of clay compaction.) Although the model accurately simulated drawdown, except near the boundaries, it did not accurately simulate the distribution of observed land subsidence. In the late 1970's, as part of the Regional Aquifer System Assessment (RASA), a five-layer finite difference model was used to simulate the Houston area once again. This model simulated a larger area than the 1975 analog model and thereby eliminated boundary effects that had been a problem with the analog simulation. The finite difference model used essentially the same conceptual model as the 1975 analog model but incorporated a revised distribution of clay layers and time-dependent storage coefficients for the clays. This model accurately simulated both drawdowns and land subsidence.

The example described by Jorgensen (1981) is not a postaudit in the sense of the examples described in Section 10.1 because a long-term prediction of the model was not evaluated. Rather, successive improvements in the model were made in an effort to achieve a better calibration to observed conditions. This example illustrates that accurate model performance depends upon careful evaluation of the conceptual model. If necessary, the conceptual model should be improved by augmenting the data base. Changes in the conceptual model then should be incorporated into the numerical model.

Jorgensen's example also illustrates the iterative way in which a model may be improved as new information is obtained. Ideally, this is the way all models should develop. The initial motivation for constructing a model may be an immediate need for a management decision (crisis mode). When a model is to be used in the predictive mode the modeling effort ideally should continue past the crisis stage and into the management phase. For example, a model may be constructed to determine whether a particular site is suitable for a landfill. If the landfill is to be constructed, the model should next be used to help in landfill design and design of a monitoring network. New information about the site will be acquired during landfill construction and well installation. This new information should be used to improve the conceptual model. As a result predictions made by the numerical model will improve.

Problems

10.1   Read the paper by Konikow (1986). What criteria were used to assess the accuracy of original model predictions?

10.2   Read the paper by Alley and Emery (1986). The postaudit showed that declines in groundwater levels were overestimated and streamflow depletion was underestimated. What were the magnitudes of these errors?

PARTICLE TRACKING OF GROUNDWATER FLOW AND ADVECTIVE TRANSPORT OF CONTAMINANTS

"As many fresh streams meet in one salt sea;
As many lines close in the dial's centre;
So may a thousand actions, once afoot,
End in one purpose, and be all well borne
Without defeat."
—Henry V, Act I

11.1 Introduction

Particle tracking is used to trace out flow paths, or pathlines, by tracking the movement of infinitely small imaginary particles placed in the flow field. Particle tracking codes are postprocessors to flow models because they accept the head distribution from a flow model and use it to calculate a velocity distribution, which is then used to trace out pathlines. Particle tracking is used in two ways: to help visualize the flow field and to track contaminant paths.

Particle tracking analysis should be used routinely with flow modeling to detect conceptual errors that cannot be identified solely by examining the head distribution. For example, tracking the movement of particles placed in every cell around the perimeter of the model helps in evaluating the effects of different boundary conditions. Particle tracking analysis shows the location of recharge and discharge areas more clearly than results from flow models (Fig. 6.5). Particle tracking is also helpful in assessing the effects of partially penetrating wells and streams.

Contaminants are transported in groundwater by advection, i.e., the movement of a solute at the speed of the average linear velocity of groundwater ($\mathbf{v}$):

$$\mathbf{v} = -\overline{K}/n_e \; (\mathbf{grad} \; h) \tag{11.1}$$

where $\mathbf{v}$ is a vector; $\overline{K}$ is the hydraulic conductivity tensor (Eqn. 2.17) and n_e is the effective porosity. There are, however, two other processes that affect contaminant movement, namely dispersion and chemical reactions. Consideration of all three processes requires solving a solute transport model for concentrations in space and time. As we shall see in Chapter 12 (Section 12.5), solute transport modeling requires input parameters that are difficult to measure. Furthermore, the process of dispersion and the processes involved in chemical reactions in the subsurface are not yet fully understood. Hence, solute transport modeling is fraught with uncertainty. For some objectives, particle tracking of advective transport is an attractive alternative to solving a solute transport model because it does not involve the complicating effects and uncertainties associated with quantifying dispersion and chemical reactions, but still allows solute travel paths and discharge points to be estimated. Average travel times can also be computed. Advective transport modeling may be used to delineate capture zones and wellhead protection areas based on a time-of-travel criterion (U.S. Environmental Protection Agency, 1987) for chemically conservative contaminants or those that can be simulated using a retardation factor (Section 12.5).

11.2   Tracking Methods

Pathline is a general term used to refer to a flow path. For steady-state problems in two dimensions, pathlines become *streamlines*. Plots of streamlines and equipotential lines form flow nets (Section 12.2). In transient problems the path of a particle is dependent on a changing velocity field. Bear (1972) calls these transient pathlines *streaklines*. The movement of a contaminant can be simulated by tracking one or more imaginary particles. A solution that tracks particles influenced by both advection and dispersion in effect is a solution of the advection-dispersion equation (Section 12.5, Eqn. 12.11). Particles may be tracked with advection only using a particle tracking code.

Nelson (1978) was an early advocate of particle tracking and his work led to the development of the computer program S-PATHS (Oberlander and Nelson, 1984), which uses analytical solutions formulated in terms of the streamfunction (Section 12.2) to compute streamlines and calculate travel times of contaminants. Bear and Verruijt (1987), Kinzelbach (1986), and Javandel et al. (1984) discuss analytical particle tracking along streamlines; Newsom and Wilson (1988) used an analytical model to determine capture zones around a pumping well located near a fully penetrating stream. Analytical particle tracking codes are limited to two-dimensional steady-state problems.

Numerical particle tracking codes, which are discussed in the rest of this chapter, are more general than analytical codes. Particles are introduced into the flow field and moved in a continuous spatial domain according to a velocity distribution calculated from the heads generated by a numerical groundwater flow model. Retardation of a contaminant owing to adsorption (Section 12.5, Eqn. 12.10a) can be included by dividing velocities by a retardation factor (R_d) so that $v_c = v/R_d$, where v_c is the retarded velocity of the contaminant and R_d has a value greater than one. Mercer et al. (1982) and Fetter (1988) give information that allows computation of the retardation factor for a number of compounds. The code keeps track of the locations of all the particles, removing particles that enter sinks or reach discharge boundaries.

It is important to remember that grid discretization in both the horizontal and vertical dimensions is critical in determining accurate pathlines, particularly when simulating partially penetrating streams or wells. The accuracy of computed pathlines also depends on the accuracy of the head distribution computed by the flow model. The accuracy of the particle tracking code itself depends on the interpolation scheme used to calculate velocities and the method used to move particles. An interpolation scheme is necessary because particles move in a continuous spatial domain but velocities calculated from the head solution are known only at discrete points. Particle tracking codes use interpolation to calculate velocities at particle locations. Differences among particle tracking codes arise owing to the interpolation scheme used and the method of moving particles during a tracking step.

INTERPOLATION SCHEMES

Interpolation schemes used in particle tracking are generally linear or bilinear, although other schemes such as bicubic interpolation are sometimes used (Shafer, 1990). A linear interpolation formula for v_x is a function of changes in v_x in the x direction only:

$$v_x = (1 - f_x)v_{x(i-\frac{1}{2},j)} + f_x v_{x(i+\frac{1}{2},j)} \qquad (11.2)$$

$$\text{where } f_x = (x_p - x_{i-\frac{1}{2},j})/\Delta x_{i,j}$$

and x_p is the x coordinate of the particle (Fig. 11.1). Similar equations can be written for v_y and v_z.

Bilinear interpolation as used in two-dimensional codes considers linear changes in velocity in both directions for each velocity component (Goode, 1990). Each cell is divided into four quadrants (Fig. 11.1). The usual bilinear interpolation formula for v_x for a point in quadrant 1 is

$$v_x = (1 - F_y)[(1 - f_x)v_{x(i-\frac{1}{2},j-1)} + f_x v_{x(i+\frac{1}{2},j-1)}]$$
$$+ F_y[(1 - f_x)v_{x(i-\frac{1}{2},j)} + f_x v_{x(i+\frac{1}{2},j)}] \qquad (11.3)$$
$$\text{where } F_y = (y_p - y_{i,j-1})/\Delta y_{i,j-\frac{1}{2}}$$

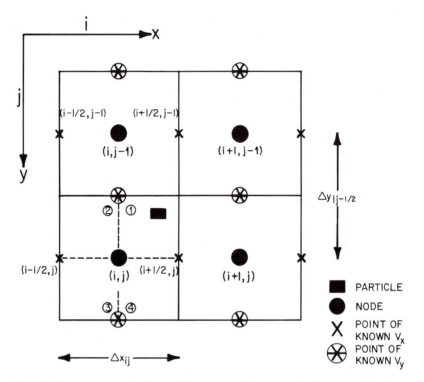

Fig. 11.1 A portion of a finite difference grid showing the locations of nodes and internodal positions where velocity components v_x and v_y are calculated. The quadrants (circled numbers) associated with node (i,j) for use in bilinear interpolation are also shown.

Equation 11.3 can be understood more clearly if the calculation is done in three steps.

Step 1. An intermediate value of velocity is calculated using the two v_x values closest to the particle (Fig. 11.1).

$$(v_x)_1 = (1 - f_x)v_{x(i-\frac{1}{2},j)} + f_x v_{x(i+\frac{1}{2},j)}$$

Step 2. Another intermediate value of velocity is calculated using the next two nearest v_x values.

$$(v_x)_2 = (1 - f_x)v_{x(i-\frac{1}{2},j-1)} + f_x v_{x(i+\frac{1}{2},j-1)}$$

Step 3. The final value of v_x is calculated using the values from steps 1 and 2.

$$v_x = F_y(v_x)_1 + (1 - F_y)(v_x)_2$$

An equation similar to 11.3 is written for v_y. Equations for the other three quadrants are analogous.

In a linear interpolation scheme v_x is continuous along the x-axis but discontinuous along other axes. Similarly v_y and v_z are continuous along their respective axes but discontinuous along other axes. Bilinear interpolation generates a fully continuous velocity field that does not preserve discontinuities in velocity at boundaries between units of different hydraulic conductivity. Furthermore, bi-linear interpolation does not conserve mass within each finite difference cell. Linear interpolation does satisfy the continuity equation (Eqn. 2.15) thereby conserving mass locally. Goode (1990) observed that the ideal interpolation scheme should give smoothly varying velocities where hydraulic properties vary smoothly and discontinuous velocities at media boundaries. He proposed an interpolation scheme that combines linear and bilinear interpolation.

Franz and Guiguer (1990) used an interpolation scheme based on the reverse distance algorithm. Like bilinear interpolation (Eqn. 11.3), it incorporates changes in both directions of a two-dimensional problem domain:

$$v_x = \frac{\sum_{m=1}^{4} (v_x)_m/(r_x)_m}{\sum_{m=1}^{4} (1/r_x)_m} \qquad v_y = \frac{\sum_{m=1}^{4} (v_y)_m/(r_y)_m}{\sum_{m=1}^{4} (1/r_y)_m} \qquad (11.4)$$

where $(r_x)_m$ and $(r_y)_m$ are the distances of the particle from the four nearest locations of known velocity (Fig. 11.2).

The interpolation schemes discussed above can be used to calculate velocities using output from either finite difference or finite element flow models. Other interpolation schemes are also possible.

TRACKING SCHEMES

Particles are tracked along pathlines by solving

$$dx/dt = v_x \qquad (11.5a)$$

$$dy/dt = v_y \qquad (11.5b)$$

$$dz/dt = v_z \qquad (11.5c)$$

Four integration methods are commonly used to solve Eqn. 11.5: semianalytical, Euler, Runge-Kutta, and Taylor series expansion.

A semianalytical solution of Eqn. 11.5 is possible if a linear velocity interpolation scheme is used. Then the solution of Eqn. 11.5a is

$$x_p = x_1 + (1/A_x)[(v_x)_p \exp (A_x \Delta t) - (v_x)_1] \qquad (11.6)$$

$$\text{where } A_x = [(v_x)_2 - (v_x)_1]/\Delta x_{i,j}$$

and $(v_x)_1$ and $(v_x)_2$ are the velocities at either end of the cell. In Eqn. 11.6, x_1 is the x coordinate of the edge of the cell and x_p is the x coordinate of the particle. In this way, the new particle location (x_p) is calculated directly from Eqn. 11.6. The semianalytical approximation is well suited to steady-state problems. In

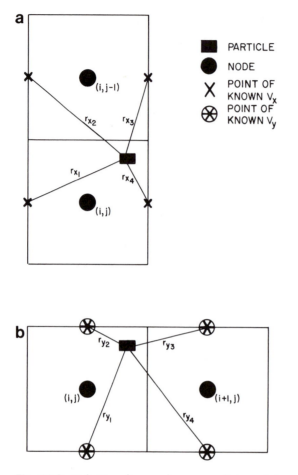

Fig. 11.2 Definition diagram for reverse distance interpolation as used by Franz and Guiguer (1990).
(a) Points used in the calculation of v_x.
(b) Points used in the calculation of v_y.

transient simulations velocities may change within the cell during the tracking step. That is, the time step of the flow problem may be smaller than the tracking step (Δt in Eqn. 11.6) needed to move the particle all the way through a cell. In this case, it may be necessary to take very small tracking steps in order to update the velocity distribution.

The Euler integration formulas are as follows:

$$x_p = x_0 + (v_x)_0 \, \Delta t \tag{11.7a}$$

$$y_p = y_0 + (v_y)_0 \, \Delta t \tag{11.7b}$$

where (x_0, y_0) is the initial position of the particle. Numerical errors tend to be large unless small tracking steps (Δt) are used. For this reason the Runge-Kutta method is preferred over Euler integration. A fourth-order Runge-Kutta method involves calculating the velocity of the particle at four intermediate points for each tracking step: once at the initial position of the particle, twice at midpoints, and once at a trial end point (Fig. 11.3). The final position of the particle is calculated using an average involving all four velocities. The equation for calculating the final x coordinate of the particle at the end of the tracking step is

$$x_p = x_0 + \tfrac{1}{6}(k + 2l + 2m + n) \tag{11.8}$$

$$\text{where } k = \Delta t \, v_{x_0}$$
$$l = \Delta t \, v_{x_1}$$
$$m = \Delta t \, v_{x_2}$$
$$n = \Delta t \, v_{x_3}$$

$(v_x)_0$ is the velocity in the x direction at the initial position of the particle; $(v_x)_1$ is the velocity at a position halfway between the initial position and the trial location $x_k = x_0 + k$; $(v_x)_2$ is the velocity at a position halfway between the initial position and the trial location $x_l = x_0 + l$; $(v_x)_3$ is the velocity at $x_m = x_0 + m$.

In a Taylor series expansion the new position of the particle is calculated from

$$x_p = x_0 + (dx/dt) \, \Delta t + (d^2x/dt^2)(\Delta t^2/2) \tag{11.9a}$$

$$y_p = y_0 + (dy/dt) \, \Delta t + (d^2y/dt^2)(\Delta t^2/2) \tag{11.9b}$$

These are essentially the Euler formulas with an additional higher-order term that represents the time rate of change of velocity or the acceleration. Details of this method as applied to particle tracking are given by Kincaid (1988).

11.3 Particle Tracking Codes

Solute transport codes that solve the advection-dispersion equation (Section 12.5, Eqn. 12.11) by particle tracking include USGS MOC (Konikow and Bredehoeft, 1978) and RNDWALK (Prickett et al., 1981). The MOC code uses stan-

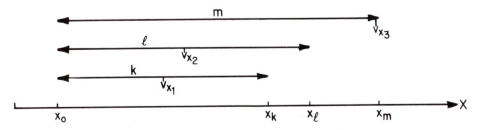

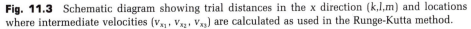

Fig. 11.3 Schematic diagram showing trial distances in the x direction (k,l,m) and locations where intermediate velocities ($v_{x_1}, v_{x_2}, v_{x_3}$) are calculated as used in the Runge-Kutta method.

dard bilinear interpolation (Eqn. 11.3). RNDWALK uses Chapeau functions borrowed from finite elements. RNDWALK can function as an advective particle tracking code when dispersivities (Section 12.5) are set to zero.

MODPATH (Pollock, 1988, 1989) and PATH3D (Zheng, 1989) are three-dimensional tracking codes that are designed to use heads from MODFLOW, but they could be modified to accept a head solution from any other block-centered finite difference code. Both MODPATH and PATH3D use linear interpolation (Eqn. 11.2). MODPATH uses a semianalytical solution (Eqn. 11.6), thereby avoiding errors associated with numerical integration. It also computes cell-by-cell mass balances, which is a useful feature when analyzing flow around sources and sinks. The current version of MODPATH (Pollock, 1989) is not designed for transient simulations. PATH3D uses a fourth-order Runge-Kutta approximation (Eqn. 11.8). Numerical errors are minimized by automatic adjustment of the size of a tracking step according to an error criterion set by the user. The position of the particle is calculated using a full tracking step and two half steps (Fig. 11.4a). If the discrepancy in particle position (Δs) is greater than the error criterion (Δs_0), the time step is reduced and the tracking step is repeated. If Δs is less than Δs_0, Δt for the next tracking step is increased to speed up the solution. PATH3D is equally efficient for steady-state and transient problems.

FLOWPATH (Franz and Guiguer, 1990) includes a finite difference flow code for two-dimensional, steady-state problems with particle tracking. The flow code is similar to PLASM but modified to simulate mesh-centered boundary conditions (Box 4.2). FLOWPATH uses reverse distance interpolation (Eqn. 11.4) with Euler integration (Eqn. 11.7) to move particles along pathlines. The size of a tracking step is controlled by an error criterion. During the error check the particle is moved backward from its new location (x_p, y_p) for the length of time represented by the tracking step (Fig. 11.4b). The discrepancy (d) between the initial position of the particle (x_0, y_0) and the position achieved during backward tracking (x_{back}, y_{back}) is calculated from

$$d = \sqrt{(x_0 - x_{back})^2 + (y_0 - y_{back})^2} \qquad (11.10)$$

If d is greater than the error tolerance, the tracking step is reduced by 50% and the particle is again moved from its initial position (x_0, y_0) using a smaller time period for the tracking step. The error tolerance (E) is defined by

$$E = 0.05 \ \Delta s \qquad (11.11)$$

where here Δs is the average nodal spacing.

The particle tracking code GWPATH (Shafer, 1987, 1990) uses the Runge-Kutta method (Eqn. 11.8). The 1990 version uses bicubic interpolation. It requires input from a flow model. WHPA (Blandford and Huyakorn, 1990) is a collection of programs developed for the U.S. EPA for delineating wellhead protection areas. It has three options for calculating a flow field using analytical solutions. It also includes a two-dimensional particle tracking code (GPTRAC)

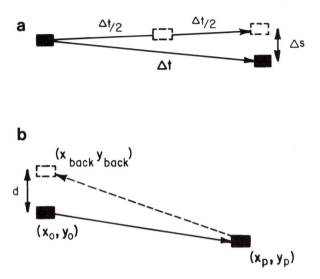

Fig. 11.4   Error checks used to control the tracking step in numerical integration.
(a) Use of two half-tracking steps ($\Delta t/2$) in PATH3D (Zheng, 1989).
(b) Backward tracking used in FLOWPATH (Franz and Guiguer, 1990).

which accepts a head solution from a user-supplied flow code, either block- or mesh-centered finite differences or rectangular finite elements. It uses linear interpolation of velocity and an analytical solution for moving particles. ETUBE (Kincaid, 1988) is a two-dimensional particle tracking code that is part of the FASTCHEM package of codes developed by the Electric Power Research Institute. ETUBE accepts a steady-state head distribution produced by a finite element flow code (EFLOW). Particle movements are calculated using the Taylor series method (Eqn. 11.9).

11.4   Applications

FLOW SYSTEM ANALYSIS

Particle tracking codes should be used routinely as postprocessors to flow modeling. Placement of particles around the perimeter of the model will yield a picture of the flow field (Fig. 6.5b, Box 11.1, Fig. 2) that is particularly useful in engineering design and groundwater management. In analysis of regional flow systems, particle tracking makes it possible to delineate local, intermediate, and regional flow systems together with their associated recharge and discharge areas (Buxton et al., 1990).

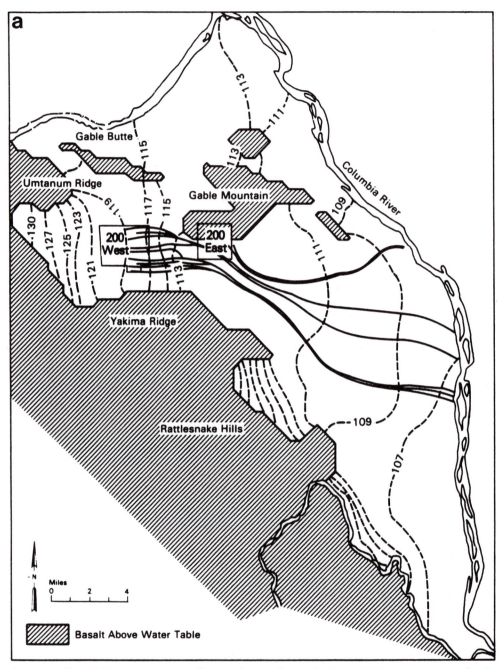

Fig. 11.5 Contaminant pathline analysis at the Hanford Site (DOE, 1987; as reported in Freshley and Graham, 1988). Heads (dashed lines) are in meters above mean sea level. Solid lines are streamlines (m²/day). Boxes labeled 200 West and 200 East indicate locations of waste sites.

(a) Low recharge scenario of 0.5 cm/yr.
(b) High recharge scenario of 5.0 cm/yr.

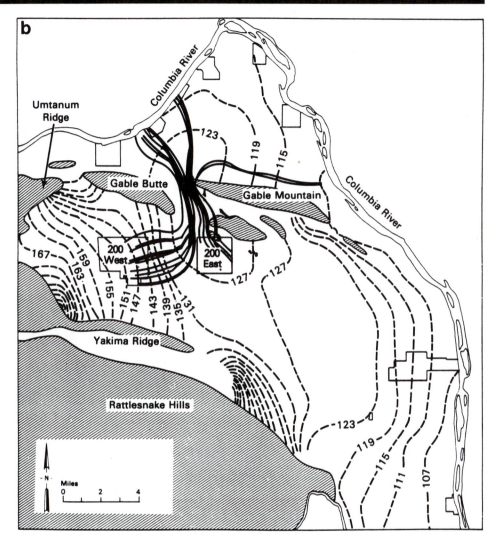

Fig. 11.5 (continued)

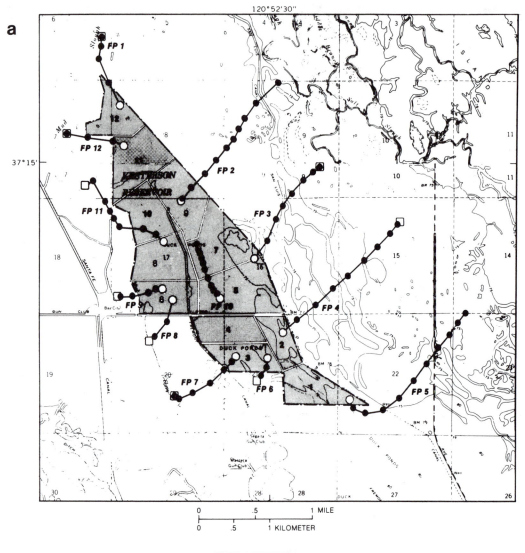

a

EXPLANATION

FP 1   FLOW PATH AND NUMBER - Each data point is equivalent to
10 years. Simulation used high values of hydraulic conductivity

1   KESTERSON POND AND POND NUMBER

○   STARTING LOCATION FOR SIMULATION

□   POINT AT WHICH FLOW PATH TERMINATES AT LAND SURFACE

—·—   BOUNDARY OF KESTERSON PONDS

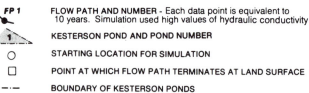

Fig. 11.6 Contaminant flow path analysis at the Kesterson Reservoir, California (Mandle and Kontis, 1986).

(a) Flow paths (labeled FP) indicate the probable flow direction of contaminants leaking from the reservoir into the groundwater system. Most flow paths terminate at streams or wetland areas. The distance between dots along each flow path represents 10 years of travel time.

(b) Selected flow paths shown in cross section. All flow paths terminate at the land surface, except for flow path 2, which did not reach a discharge point within the 100 years of simulated time.

b

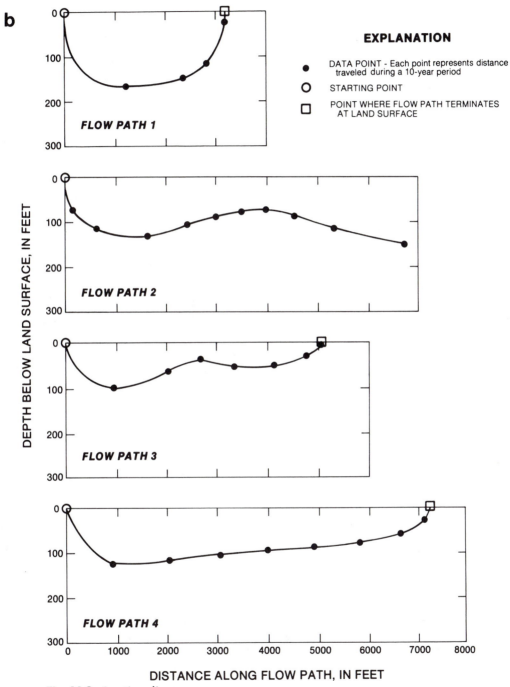

EXPLANATION

● DATA POINT - Each point represents distance traveled during a 10-year period

○ STARTING POINT

□ POINT WHERE FLOW PATH TERMINATES AT LAND SURFACE

FLOW PATH 1

FLOW PATH 2

FLOW PATH 3

FLOW PATH 4

DEPTH BELOW LAND SURFACE, IN FEET

DISTANCE ALONG FLOW PATH, IN FEET

Fig. 11.6 (continued)

a

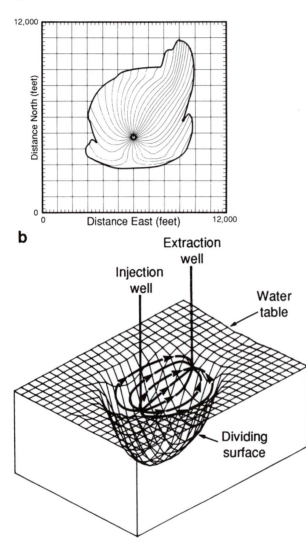

b

Fig. 11.7 Delineation of capture zones.

(a) Twenty-year capture zone for a pumping well in a heterogeneous aquifer calculated using a reverse pathline calculation technique (Shafer, 1987).

(b) Capture zone for an injection/extraction system (Zheng, 1989).

(c) Three capture zones delineated by dividing streamlines (indicated by circled numbers) around an irrigation ditch are shown for a portion of a sandy aquifer in central Wisconsin (Zheng et al., 1988b). Streamlines were calculated by modifying MODFLOW input to produce streamfunctions. ψ is the dividing streamline in m²/day. The three capture zones have the following characteristics: (1) $K_1 = K_2$; (2) $K_1/K_2 = 100$; (3) $K_2/K_1 = 100$.

In all cases the ratio of horizontal to vertical conductivity is 20.

(d) Capture zone of a lake (Townley and Davidson, 1988).

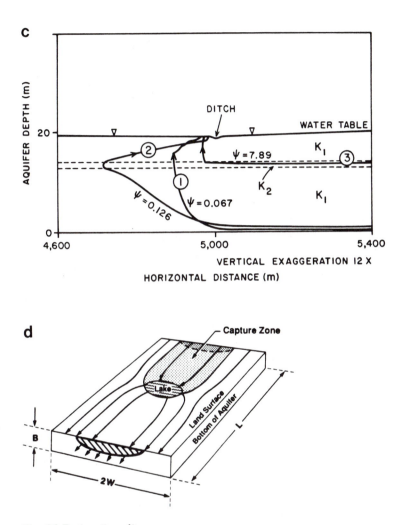

Fig. 11.7 (continued)

CONTAMINANT TRACKING

In a contaminant tracking simulation, one or more particles are introduced at a source location and tracked through the flow field. The results of the analysis include particle pathlines with discharge locations and average travel times (Figs. 11.5, 11.6). Reverse particle tracking may be used to help identify contaminant sources. In reverse particle tracking, particles are introduced where contaminants have been detected. Particles are then tracked backward along pathlines to their source (Shafer, 1987). Schafer-Perini *et al.* (1991) discussed particle tracking algorithms for evaluating breakthrough curves. Zheng *et al.* (1991) demonstrated the use of particle tracking to estimate the time required for aquifer remediation by flushing of the contaminated portion of the aquifer. Because particle tracking codes do not include dispersion (Section 12.5), they are not appropriate when it is necessary to compute the first arrival of measurable contamination.

DELINEATION OF CAPTURE ZONES

Recent interest by the U.S. EPA (1987) in defining capture zones around pumping wells has motivated the development of particle tracking codes specifically designed for wellhead protection analysis (Blandford and Huyakorn, 1990; Shafer, 1987). A capture zone refers to the portion of the flow system that contributes water to a well or a surface water body such as a river or ditch (Zheng et al., 1988a,b) or lake (Townley and Davidson, 1988). Capture zones are best delineated using reverse particle tracking whereby particles are introduced into the well and tracked backwards along pathlines to their source. Some examples of the use of particle tracking programs in capture zone analysis are shown in Fig. 11.7.

Box 11.1
Case Study of a Particle Tracking Application _____

Bair et al. (1990) used a particle tracking code to delineate flow paths of contaminants that may enter groundwater as a result of accidental spills of hazardous cargo being transported on highways in the vicinity of a municipal well field in Columbus, Ohio. The well field is comprised of four radial collector wells situated in a sand and gravel aquifer and located along the Scioto River and Black Walnut Creek (Fig. 1). Some contaminants were expected to be diverted north of the well field to a quarry, which is being actively dewatered (Fig. 1).

Eberts and Bair (1990) described the use of a three-dimensional model to study the flow system in this area including the effects of quarry and mine dewatering operations and pumping of municipal wells. The head solution calcu-lated by MODFLOW was input to the particle tracking code STLINE (GeoTrans, 1987) which computed velocities using linear interpolation. The velocity field, computed for conditions of estimated maximum pumping, is shown in Fig. 2. Maximum pumping rates would be used during periods of prolonged drought. These rates also reflect probable future increases in pumping as a result of population growth. Figure 2 illustrates how a particle tracking code can be used to help visualize the flow field.

Particles were placed at ten potential spill sites shown in Fig. 3a and flow paths were delineated to determine contaminant exit points. Contaminants entering the aquifer at spill sites 1, 8, 9, and 10 exited in the quarry (Fig. 3a), while contaminants from all the other sites en-

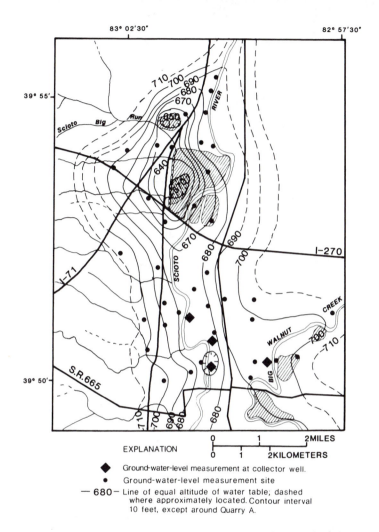

Fig. 1 Location of collector wells and quarries in relation to highways in Columbus, Ohio (Bair et al., 1990). Contours indicate the potentiometric surface in the sand and gravel (glacial drift) aquifer in March 1986. Quarries are defined by shaded areas.

tered one of the collector wells. A particle tracking analysis was also done for 1986 pumping conditions, revealing a number of differences between normal 1986 conditions and maximum pumping conditions. Under 1986 pumping conditions, contaminants from spill site 2 were diverted to the quarry, contaminants from site 5 discharged to the Scioto River, and contaminants from site 7 entered collector well

104 rather than collector well 103. The particle tracking analysis was also used to delineate capture zones for each of the collector wells under maximum pumping conditions (Fig. 3b). The Scioto River and Black Walnut Creek are included in the capture zones because they lose water to the aquifer when the collector wells are pumped.

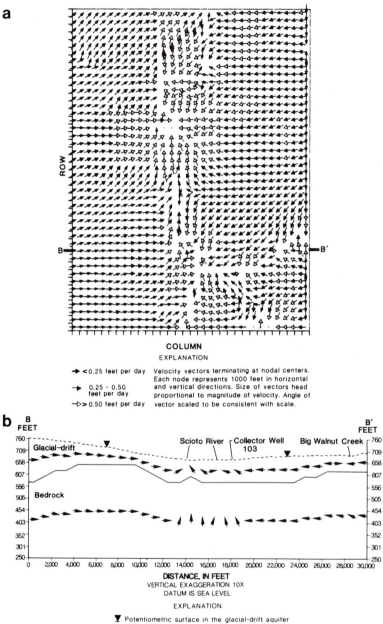

Fig. 2 Plan view (a) of velocity vectors in the sand and gravel aquifer under conditions of maximum pumping (Bair et al., 1990). The flow model consisted of 32 columns, 42 rows, and two layers (b). A no-flow boundary condition was assigned to the bottom of the second layer. Specified flow boundary conditions were assigned along the east, west, and western part of the northern boundary. Flow rates were calculated using the map of the potentiometric surface (Fig. 1). The southern boundary was represented as a no-flow boundary, except for a constant-head node at the Scioto River. The no-flow conditions here were based on the potentiometric surface contours, which suggest that flow is parallel to the southern boundary of the modeled area. Similarly, a no-flow boundary was assigned to the eastern part of the northern boundary. The Scioto River and Black Walnut Creek were simulated as head-dependent stream nodes using MODFLOW's River Package. Also see Fig. 5.8b.

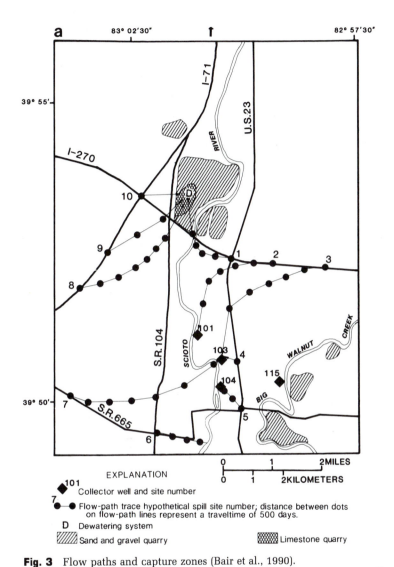

Fig. 3 Flow paths and capture zones (Bair et al., 1990).

(a) Flow paths of contaminants under conditions of maximum pumping. Contaminants enter the sand and gravel aquifer at potential spill sites numbered 1 through 10.

(b) Capture zones for collector wells 101, 103, and 104 and for collector well 105. The dashed line indicates the location of the groundwater divide separating the two capture zone areas. Note that capture zones within each of the two areas are defined based on travel times.

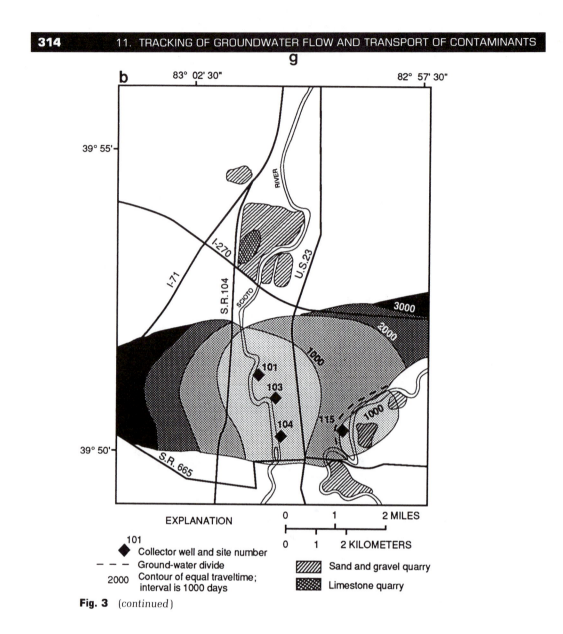

Fig. 3 (*continued*)

Problems

11.1 The town of Hubbertville wants to site a landfill upgradient from the proposed water supply well described in Problem 4.1. The landfill would be located on a few acres owned by the town located 5000 m due south and 500 m due west of the well site shown in Fig. P4.1. If the landfill should leak, leachate could enter groundwater flowing beneath the landfill.

(a) Use a particle tracking program to determine the path that contaminated groundwater would most likely follow under steady-state conditions described in Problem 4.1c, part (i). Initiate tracking from the water table directly below the landfill site. Would the supply well be affected?

(b) If leachate reaches the water table, how many days will it take for leachate to travel 1000 m downgradient from the area beneath the landfill?

(c) Use a particle tracking program that allows backward tracking to define the capture zone around the supply well under steady-state conditions. Comment on the limitations of this method in predicting water quality impacts on water pumped from the well.

11.2 Review the description of the steady-state confined aquifer system in Problem 6.3.

(a) Calculate the travel time required for a contaminant introduced at the midway point on the left-hand boundary to traverse the 490 m length of the cross section under the following conditions:

(i) Assume isotropic and homogeneous conditions with $T = 140$ m^2/day.

(ii) Introduce the zone of low hydraulic conductivity as described in Problem 6.3.

(b) Place several particles along the entire left-hand boundary and estimate the percentage of the total flow which is affected by the presence of the low-K zone used in Problem 6.3b.

11.3 Examine the documentation of the particle tracking programs you have used and summarize the interpolation scheme and tracking methodology used in the code.

12

ADVANCED TOPICS

"I have yet to see any problem, however complicated, which when you looked
at it in the right way did not become more complicated."
—*Poul Anderson*

12.1 Introduction

In preceding chapters, we discussed principles of modeling groundwater flow
and the advection of contaminants. Although these models are adequate for
solving many practical groundwater problems, there are also many situations,
including flow through unsaturated porous media, multiphase flow, and flow
in fractured rock, for which these models are inappropriate. In this chapter we
introduce some of these complex models. Most applications of standard
groundwater flow models can be executed on a personal computer, but some of
the modeling techniques in this chapter require a work station or a mainframe
computer.

The chapter is arranged roughly in order of increasing complexity. We start
with computer generation of flow nets (Section 12.2), which is facilitated by the
availability of easy-to-use graphics packages for personal computers. Unsaturated flow models (Section 12.3) have been used as long as groundwater flow
models but are more difficult to apply owing to sophisticated data requirements
and numerical problems in solving the highly nonlinear governing equation.
These models are also subject to more uncertainty than saturated flow models.
Multiphase flow models (Section 12.4) were first introduced to consider air and
water movement in the unsaturated zone but recently have been adapted to
simulate the movement of non–aqueous phase liquids (NAPLs) in water. Solute transport models (Section 12.5) were introduced in the early 1970's and,

although widely used, are still subject to considerable uncertainty. Much attention is currently focused on simulating flow and solute movement through fractured media (Section 12.6) but difficulties involved in characterizing the system at the field scale limit application of these models. Models that consider density-dependent flow of miscible fluids (Section 12.7) were first introduced to consider saltwater intrusion into coastal aquifers but have since been applied to a wide range of problems. Finally, Section 12.8 discusses several other types of models used in groundwater hydrology, including optimization models and geochemical models.

Detailed treatment of the complex models introduced in this chapter is beyond the scope of this book. We provide information that will help in determining which type of model is most appropriate for a particular need. Effective application of these models requires familiarity with the physics specific to each type of application and with the specific code to be used. More information about complex codes of the kind discussed here can be found in Anderson et al. (1992) and van der Heijde et al. (1988).

The following conclusions of a panel of modeling experts convened by the National Research Council (1990) are relevant to this chapter.

> Ground water models do and should vary in complexity. The complexity of the model used to analyze a specific site should be determined by the type of problem being analyzed. While more complex models increase the range of situations that can be described, increasing complexity requires more input data, requires a higher level and range of skill of the modelers, and may introduce greater uncertainty in the output if input data are not available or of sufficient quality to specify the parameters of the model. (p. 14)

The NRC panel stressed the importance of proper conceptualization and field verification of groundwater models, a task that becomes more difficult as the complexity of the flow system increases.

> There is a range of capability in modeling fluid flow in geologic media. Modeling saturated flow in porous media is straightforward with few conceptual or numerical problems. At the present time, conceptual issues and/or problems in obtaining data on parameter values limit the reliability and therefore the applicability of flow models involving unsaturated media, fractured media, or two or more liquids. (p. 7)

> Numerical solute transport models were first developed about 20 years ago. However, the modeling technology did not have a long time to evolve before a great demand arose for its application to practical and complex field problems. Therefore the state of the science has advanced from theory to practice in such a short time (considering the relatively small number of scientists working on this problem at that time) that a large base of experience and hypothesis testing has not accumulated. It appears that some practitioners have assumed that the underlying theory and numerical methods are further beyond the research, development and testing stage than they actually are. (pp. 113–114)

12.2   Flow Nets

In Chapter 11, we discussed how contaminants can be tracked by calculating groundwater flow paths or pathlines. For steady-state problems in two dimensions, the pathlines are streamlines, which can be computed directly by means of the *streamfunction*. A streamline is plotted as a curve tangent to the groundwater velocity vector at every point in the flow field (Fig. 12.1). This concept is expressed in vector notation as the cross product of the specific discharge ($\mathbf{q}$) and an incremental distance ($d\mathbf{r}$): $\mathbf{q} \times d\mathbf{r} = 0$. In two dimensions this implies that

$$dx/q_x = dy/q_y \qquad\qquad (12.1a)$$

or

$$q_y\,dx - q_x\,dy = 0 \qquad\qquad (12.1b)$$

Plots of streamlines and equipotential lines are called *flow nets*. Flow nets are useful in depicting groundwater flow paths and in calculating flux through the system. Two adjacent streamlines form a *stream tube*; for steady-state flow, the flux through the stream tube is constant (Fig. 12.1). In a homogeneous and isotropic medium, streamlines and equipotential lines intersect at right angles and form curvilinear squares (Fig. 12.2a). When the medium is anisotropic, streamlines intersect equipotential lines at right angles only where flow is

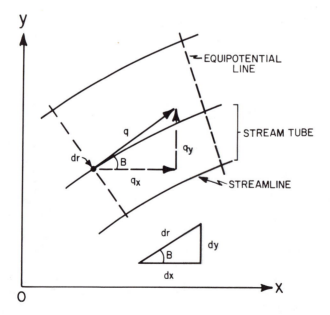

Fig. 12.1   Definition of streamlines and stream tubes; $\mathbf{q}$ is the specific discharge (after Bear, 1972).

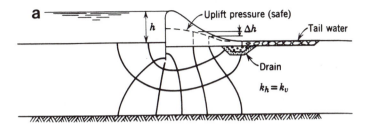

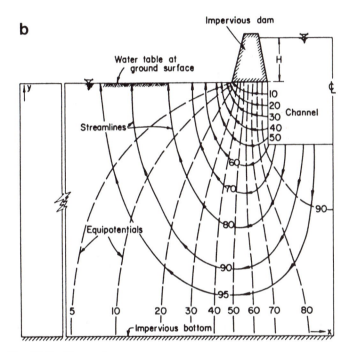

Fig. 12.2 Flow nets.
(a) Isotropic medium (Cedergren, 1967).
(b) Anisotropic medium (Todd and Bear, 1959).

parallel to one of the principal directions of hydraulic conductivity (Fig. 12.2b). For isotropic media, flow nets may be constructed by drawing streamlines by hand at right angles to equipotential lines. For anisotropic media, coordinate transformation (Van Everdingen, 1963; Freeze and Cherry, 1979) is required before drawing streamlines by hand. Flow nets for anisotropic (and isotropic) media may be constructed more easily by generating equipotential lines and streamlines numerically (Fig. 12.3).

The streamfunction is analogous to the equipotential function in that head

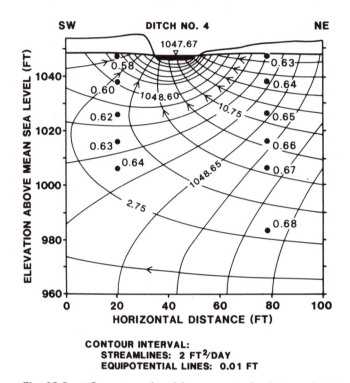

CONTOUR INTERVAL:
STREAMLINES: 2 FT²/DAY
EQUIPOTENTIAL LINES: 0.01 FT

Fig. 12.3 A flow net produced from numerical solutions of the head and streamfunction equations for a sand aquifer in central Wisconsin (Zheng et al., 1988a). Dots show locations of piezometers with head measurements.

is constant along an equipotential line and the streamfunction is constant along a streamline. The streamfunction (U) is defined from Eqn. 12.1 by realizing that $dU = \partial U/\partial x \; dx + \partial U/\partial y \; dy$, so that

$$q_x = -\partial U/\partial y \qquad (12.2a)$$

$$q_y = \partial U/\partial x \qquad (12.2b)$$

from which it is clear that the units of the streamfunction (U) are L^2/T. A stream tube is defined by two streamlines with streamfunction values U_1 and U_2 where $U_2 = U_1 + \Delta U$. The flux through the stream tube is equal to ΔU. The governing equation for two-dimensional flow under steady-state conditions is derived by writing curl $\mathbf{q}/\overline{K} = 0$ (Bear, 1972) and substituting Eqn. 12.2 for $\mathbf{q}$, so that

$$\frac{\partial}{\partial x}\left(\frac{1}{K_y}\frac{\partial U}{\partial x}\right) + \frac{\partial}{\partial y}\left(\frac{1}{K_x}\frac{\partial U}{\partial y}\right) = 0 \qquad (12.3)$$

Note the similarity in form of Eqn. 12.3 to the governing equation for two-dimensional steady-state flow with head as the dependent variable:

$$\frac{\partial}{\partial x}\left(K_x \frac{\partial h}{\partial x}\right) + \frac{\partial}{\partial y}\left(K_y \frac{\partial h}{\partial y}\right) = 0 \tag{12.4}$$

Because of the similarity between Eqns. 12.3 and 12.4, standard ground-water flow codes can be used to calculate streamfunctions by replacing K_x with $1/K_y$ and K_y with $1/K_x$ provided boundary conditions are treated properly. Details of defining boundary conditions for streamfunction problems and of using standard flow codes to calculate streamfunctions are presented in Box 12.1.

Contouring streamfunctions yields a set of streamlines that can be superimposed over a set of equipotential lines to create a flow net. Fluxes in stream tubes are calculated directly from the streamfunction solution. In addition, the average linear velocity, v, can be calculated from the streamfunction solution as follows:

$$v = q/n_e = \Delta U/n_e \, \Delta s \tag{12.5}$$

where Δs is the distance between two streamlines. These velocities can be input to solute transport models that solve the advection-dispersion equation (Section 12.5) or to particle tracking programs (Chapter 11). Frind and Matanga (1985) calculated the velocity distribution in an aquifer with a small head gradient using both head and streamfunction solutions. They demonstrated that the velocities calculated using streamfunctions were less prone to numerical round-off error than velocities estimated from the head solution. Senger and Fogg (1990) found that the streamfunction solution allowed more precise determination of flow patterns than the head solution for a cross-sectional simulation of a groundwater system in the Palo Duro Basin, Texas.

Numerical streamfunction modeling has been used by a number of other investigators. Christian (1980) seems to have been the first to demonstrate the method. Fogg and Senger (1985) used the finite element model FREESURF I to produce complex flow nets for a cross section in the Palo Duro Basin, Texas. Frind et al. (1985) used streamfunctions to define the flow system at the Borden landfill site in Ontario, Canada. Hendry and Buckland (1990) used streamfunctions to identify groundwater recharge areas and calculate recharge rates. Zheng et al. (1988a,b) used MODFLOW to generate flow nets for a stream-aquifer system. Although all the applications discussed above were for cross-sectional models, flow in two-dimensional plan view can also be represented provided that areal recharge and discharge are not included in the model.

12.3 Unsaturated Flow

The customary division of the subsurface into two zones separated by the water table is artificial because there is continuous movement of water between the saturated and unsaturated zones. Furthermore, the demarcation between saturated and unsaturated porous material is not distinct in situations where a thick capillary fringe, or tension-saturated zone, forms at the water table. The porous material in the capillary fringe is saturated with water, yet the water is under tension. The tension-saturated zone tends to be thicker in clayey soils than in sandy soils.

In view of the above, it should not be surprising that water movement through the entire subsurface can be represented by one general governing equation. This equation may be written in terms of total head (h), moisture content (θ), or pressure head (ψ). Pressure head is taken to be positive in the saturated zone and negative in the unsaturated zone so that $\psi = 0$ at the water table. The general governing equation incorporates the dependence of hydraulic conductivity on moisture content. In the saturated zone, the moisture content is always equal to saturation and the governing equation can be simplified to the forms presented in Chapter 2.

The three-dimensional form of the general governing equation written with total head (h) as the dependent variable is

$$\frac{\partial}{\partial x}\left(K(\theta)\frac{\partial h}{\partial x}\right) + \frac{\partial}{\partial y}\left(K(\theta)\frac{\partial h}{\partial y}\right) + \frac{\partial}{\partial z}\left(K(\theta)\frac{\partial h}{\partial z}\right) = C(\theta)\frac{\partial h}{\partial t} + W^* \quad (12.6)$$

where $K(\theta)$ is the unsaturated hydraulic conductivity, which is a function of the moisture content (θ); $C(\theta) = d\theta/d\psi$ is the specific moisture capacity; and W^* is a sink/source term which may include uptake of water by plant roots (Hopmans and Guttierez-Rave, 1988). This form of the governing equation assumes that flow of the air phase is neglected. If this assumption is inappropriate, a two-phase model is required (Section 12.4).

A Neumann condition, i.e., specified flux of infiltration or evapotranspiration, is the typical boundary condition at the top (land surface) boundary. When simulating only the unsaturated zone, the water table may form a specified head condition at the lower boundary. Solution of Eqn. 12.6 gives the distribution of head throughout the problem domain. The flux from the unsaturated zone is the groundwater recharge rate, which may be calculated from the change in moisture content in the system. Unsaturated simulations are usually performed under transient conditions. Hence, initial conditions (Box 12.2) are also required.

Modeling unsaturated flow is more complex than modeling saturated flow for the following reasons:

1. Equation 12.6 requires that the modeler specify the relationship between unsaturated hydraulic conductivity and pressure head (K vs. ψ) as well as the soil moisture characteristic curve, which defines the relationship between moisture content and tension (θ vs. ψ) (Fig. 12.4). These functional relationships are not easily obtained. Techniques for measuring these curves in the laboratory or in the field are time consuming and require special equipment. Published curves are available for a

number of soil types. Fitting functions to these curves has led to empirical formulas that some researchers use in the absence of site-specific information on soil characteristics. Tyler and Wheatcraft (1990) suggested that a fractal model may provide a physical basis for the commonly used empirical formulas.

2. The relationships between K and ψ and between θ and ψ show hysteresis. In other words, the relationship is different depending on whether the soil is draining or wetting. Pickens and Gillham (1980) modeled postinfiltration moisture redistribution in a fine sand with and without hysteresis. Their results show that the maximum error in moisture content caused by using a drying curve instead of one that allowed for hysteresis was about 3% and the maximum error in tensions was about 20 cm.

3. The unsaturated form of the governing equation is highly nonlinear and difficult to solve numerically (Lappala et al., 1987; Celia et al., 1990) because the functional relationships discussed in item 1 above are sensitive to small changes in tension. This sensitivity may cause instabilities to develop during the numerical solution. Instabilities can be minimized by using small nodal spacing (on the order of centimeters or

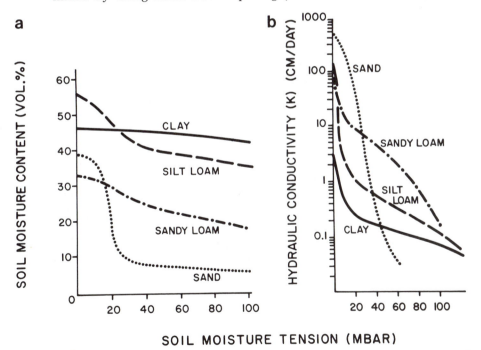

SOIL MOISTURE TENSION (MBAR)

Fig. 12.4 Representative characteristic curves showing the relation (a) between moisture content and soil moisture tension and (b) between hydraulic conductivity and soil moisture tension (adapted from Bouma et al., 1974).

decimeters) and small time steps (on the order of seconds). The USGS code VS2D (Lappala et al., 1987) is designed to handle highly, non-linear unsaturated flow problems.

4. The necessity of using small nodal spacing in the unsaturated portion of an unsaturated/saturated model causes problems in grid design because such small nodal spacing is inappropriate for the saturated portion of the grid. Use of small spacing throughout the grid can result in a prohibitively large number of nodes. Furthermore, there is a discrepancy in time scale between the two zones. The unsaturated zone model requires time steps on the order of seconds, whereas the saturated model may require a total simulated time on the order of years. This discrepancy in time and space scales led Frind and Verge (1978) to conclude that it may not be feasible to model field-scale problems with a general subsurface model. They suggested that one-dimensional unsaturated soil columns linked to a two- or three-dimensional saturated model may be a better alternative (Pikul et al., 1974), an approach that is being pursued by the U.S. EPA in the EPACMS model (HydroGeoLogic, Inc., 1989).

12.4 Multiphase Flow

The flow of fluids of different densities may involve *miscible fluids*, which mix and combine readily, or *immiscible fluids*, which do not. Miscible fluids are discussed in Section 12.7; in this section we discuss immiscible fluids. Multiphase flow refers to the movement of water and one or more immiscible fluid phases. For example, the movement of air and water in the unsaturated zone forms a two-phase system. The movement of an immiscible liquid, i.e., a non–aqueous phase liquid (NAPL), in the saturated zone is also a two-phase system. NAPLs may be either lighter (LNAPL) or denser (DNAPL) than water. Flow of an NAPL in the unsaturated zone may be represented as either two- or three-phase flow depending on how the air phase is treated.

The governing equations that describe multiphase flow are formulated in terms of the pressure of each of the phases. The degree of saturation of the porous material with respect to a given fluid and the relation between the intrinsic permeability and the degree of saturation must also be defined. The governing equations describing multiphase flow can be written following Parker (1989):

$$\frac{\partial}{\partial x_i} \rho_p K_{spij} \left(\frac{\partial \psi_p}{\partial x_j} + \rho_{rp} \frac{\partial z}{\partial x_j} \right) = n \frac{\partial \rho_p S_p}{\partial t} \tag{12.7}$$

where ψ is pressure head, z is elevation head, and the subscript p refers to the phase. One equation of the form of Eqn. 12.7 is written for each phase present. In Eqn. 12.7, ρ_p is the density of phase p; K_{spij} is the saturated conductivity of phase p, which depends on the relative intrinsic permeability (Demond and Roberts, 1987) to account for the reduction in permeability owing to the presence of other phases. Also, n is the porosity, ρ_{rp} a normalized density, and S_p, the saturation, is the fraction of the pore space containing phase p.

Solution of Eqn. 12.7 requires the relations between the pressures of the co-existing phases and both fluid saturations and permeabilities. These relations are nonlinear, influenced by hysteresis, and difficult to measure. Numerical solution of the set of coupled equations represented by Eqn. 12.7 requires an iterative, fully implicit scheme that is computer intensive. Furthermore, the extreme non-linearity of these equations requires fine space and time discretization.

Although this type of modeling is still in the research and development stage, a few applications have been reported. Kueper and Frind (1991a,b), Parker (1989), Abriola and Pinder (1985a,b), and Abriola (1984) reviewed the theory and presented selected applications. Applications involving hydrocarbons were presented by Yazicigil and Sendlein (1981), Hochmuth and Sunada (1985), Baehr and Corapcioglu (1987), Hossain and Corapcioglu (1988), and Corapcioglu and Hossain (1990). Kaluarachchi and Parker (1990), Kuppusamy et al. (1987), and Pinder and Abriola (1986) simulated three-phase flow in cross sections of hypothetical aquifers. Osborne and Sykes (1986) simulated solute migration from a landfill and Faust et al. (1989) used the finite difference code SWANFLOW to simulate transport from a hazardous waste site.

The set of governing equations represented by Eqn. 12.7 assumes that dissolution occurs slowly so that solute transport can be neglected. When this is not the case, or for problems involving miscible fluids, it is necessary to solve the solute transport equation.

12.5 Solute Transport

In Chapter 11 we discussed contaminant movement considering only the process of advection by which contaminants move at the same speed as the average linear velocity of groundwater. Complete description of the transport of a solute requires consideration of two other processes: dispersion and chemical reactions.

Dispersion refers to spreading of the contaminant caused by the fact that not all of the contaminant actually moves at the same speed as the average linear velocity. Groundwater flow modeling is based on the concept of the equivalent homogeneous porous medium by which it is assumed that the real heterogeneous aquifer can be simulated as homogeneous porous media within cells or elements. Then, the specific discharge (**q**) or average linear velocity (**v**) (Eqn. 11.1) is defined using a bulk average hydraulic conductivity for each cell or element. Contaminant movement, however, is strongly influenced by the presence of local heterogeneities that cause deviations from the average linear velocity. These deviations are typically assumed to be represented by a relation similar in form to Fick's law of diffusion, which yields the following expression:

$$\frac{\partial c}{\partial t} = \frac{\partial}{\partial x_i}\left(D_{ij}\frac{\partial c}{\partial x_j}\right) \tag{12.8}$$

where D_{ij} is the dispersion coefficient and c is concentration. The dispersion coefficient is usually calculated from

$$D_{ij} = \alpha_{ijmn}\left(\frac{v_m v_n}{v}\right) + D_d \tag{12.9}$$

where all components of α_{ijmn} are zero, except for $\alpha_{iiii} = \alpha_L$, $\alpha_{iijj} = \alpha_T$, and $\alpha_{ijij} = \alpha_{ijji} = \frac{1}{2}(\alpha_L - \alpha_T)$ for $i \neq j$; D_d is the coefficient of molecular diffusion; α_L, α_T are coefficients known as *dispersivities* that supposedly represent mixing, i.e., dispersion. There is much debate in the literature regarding dispersion. The dispersivity values in Eqn. 12.9 are in a sense correction factors to account for the fact that it is not practical, and perhaps even impossible, to delineate the velocity distribution in detail. Some investigators advocate better resolution of the velocity distribution in order to minimize errors associated with estimating dispersivity values. Others propose that theoretically derived formulas involving parameters that describe the statistics of the hydraulic conductivity distribution be used to calculate dispersivities. Dispersivities have traditionally been estimated from trial-and-error model calibration and from tracer tests (Gelhar et al., 1985).

A complicating factor in quantifying dispersion is the so-called scale effect, whereby dispersivity seems to increase with the size of the contaminant plume; i.e., dispersivity seemingly increases as the plume moves downgradient. Another complication in quantifying dispersion is caused by channeling of contaminants along paths of high hydraulic conductivity or preferential flow paths (Desbarats, 1990; Anderson, 1991). Dispersion influenced by preferential flow cannot be described by a Fickian model and requires additional theoretical work (Silliman and Wright, 1988). Goode and Konikow (1990b) addressed the apparent dispersion caused by transient effects, which also complicate dispersion calculations.

There are also conceptual problems in quantifying chemical reactions in the subsurface. Ideally, all the chemical and biochemical reactions expected to occur should be included in the mathematical formulation. In practice, chemical reactions typically used in transport models are limited to adsorption, described by a retardation factor (R_d), and hydrolysis and decay, described by a first-order rate constant (λ) These terms are defined as follows:

$$R_d = v/v_c = 1 + K_d(\rho_b/n) \tag{12.10a}$$

$$dc/dt = \lambda c \tag{12.10b}$$

where v is the average linear groundwater velocity from Eqn. 11.1, v_c is the velocity of the contaminant, K_d is the distribution coefficient, ρ_b is the bulk density of the porous material, and n is the porosity. In Eqn. 12.10b, $\lambda = \ln(2)/t_{1/2} = 0.693/t_{1/2}$, where $t_{1/2}$ is the half-life.

It is becoming clear that such simple concepts as the retardation factor and a first-order rate constant are adequate for describing only simple contaminant problems. Moreover, there are uncertainties in identifying the nature of the chemical reactions that occur in the subsurface as well as selecting parameters to quantify these processes. Although a few attempts have been made to model reactions among multiple contaminant species (e.g., Lewis et al., 1986; Cederberg et al., 1985; Miller and Benson, 1983; Hostetler et al., 1988), most modeling applications have been limited to single chemical species. Models that incorporate more sophisticated chemistry are currently being developed and tested.

The governing equation for solute transport, known as the advection-dispersion equation, may be derived by writing a mass balance equation using Eqn. 12.8 to represent dispersive flux and Eqns. 12.10 to represent chemical reactions.

$$\frac{\partial}{\partial x_i}\left(D_{ij}\frac{\partial c}{\partial x_j}\right) - \frac{\partial}{\partial x_i}(cv_i) = R_d\frac{\partial c}{\partial t} + \lambda c R_d - \frac{C'W^*}{n_e} \qquad (12.11)$$

where c is concentration and C' is a known source concentration; v_i represents the components of the velocity vector; W^* is a source/sink term and n_e is the effective porosity. The code for a solute transport model typically consists of two submodels: a model to solve the flow equation and another to solve the advection-dispersion equation. The solution of the flow equation yields the distribution of head, from which the velocity field is calculated. Velocities are input to the transport submodel, which predicts the concentration distribution in time and space.

Equation 12.11 is difficult to solve numerically, causing both finite difference and finite element solutions to be affected by numerical errors, including a phenomenon known as "numerical dispersion," which refers to artificial dispersion caused by errors associated with the discretization of the problem domain. To minimize such errors, the grid should be designed so that the Peclet number ($P_e = \Delta l/\alpha$, where Δl is a characteristic nodal spacing and α is a characteristic dispersivity) is less than or equal to one, although acceptable solutions may be obtained with P_e as high as 10 (Huyakorn and Pinder, 1983). It is usually recommended that the grid be designed so that $\Delta l < 4\alpha$. Likewise, care should be taken when discretizing time so that the Courant number ($C = v$ $\Delta t/\Delta l$) is less than or equal to one. That is, the time step should be selected so that $\Delta t < \Delta l/v$, or less than the time it takes solute to move the distance Δl. Finite-element codes that simulate solute transport include SEFTRAN (GeoTrans, 1988), CFEST (Gupta et al., 1987), HST3D (Kipp, 1987), and SUTRA (Voss, 1984b). A recently developed finite-difference code, FTWORK (Faust et al., 1990), uses a flow code similar in structure to MODFLOW.

To avoid the numerical problems associated with finite difference and finite element solutions of the advection-dispersion equation, a number of investigators prefer to use particle tracking solutions. Particle tracking methods without dispersion were discussed in Chapter 11. The USGS MOC model (Konikow and Bredehoeft, 1978) combines particle tracking for advection with a finite difference solution of the dispersion portion of Eqn. 12.11. The codes RNDWALK (Prickett et al., 1981) and RD3D (Koch and Prickett, 1989) use a random number generator to reproduce a Gaussian distribution of particles to simulate dispersion. In some versions of these codes other types of distributions can also be selected. The code MT3D (Zheng, 1990) is a particle tracking code with dispersion that is compatible with MODFLOW. Other solution techniques for avoiding numerical dispersion are discussed by Sudicky (1989) and Leismann and Frind (1989).

The reader is referred to Bear and Verruijt (1987), Grove and Stollenwerk (1987), and Kinzelbach (1986) for additional theoretical discussions of transport modeling and to Anderson et al. (1992) for discussion of well-documented transport codes that are in the public domain. Numerous applications of solute transport models have been reported, some of which were discussed by Anderson (1979). More recent applications include those reported by Anderson et al. (1992). Chapelle (1986), El-Kadi (1988), and Holmes et al. (1989) reported on applications of the USGS MOC code.

12.6 Fractured Media

The models discussed so far are intended for simulations of flow in porous media having continuous interconnected pore space. Yet flow and contaminant transport in fractured rocks have always been and will continue to be of interest to hydrogeologists. Fractured carbonate rocks such as limestone and dolostone form regionally important aquifers that may become contaminated by rapid movement of chemicals in fractures. Moreover, the recognition that clayey sediments, such as some glacial tills and river alluvium, are fractured has generated concern over the siting of waste disposal facilities in these materials. The potential for siting repositories for high-level radioactive waste in fractured media has also prompted interest in simulating flow in fractured rocks.

Forming a conceptual model of a fractured system requires either a gross simplification or a detailed description of the aquifer properties controlling flow. A fractured medium consists of solid rock with some primary porosity cut by a system of cracks, microcracks, joints, fracture zones, and shear zones that create secondary porosity (Fig. 12.5) and form a network for flow when interconnected. Description of flow through a fractured medium requires information on the primary permeability of the rock matrix and the secondary permeability created by the network of fractures. Igneous and metamorphic rock and clay deposits typically have primary intrinsic permeabilities that range from 10^{-12} to 10^{-16} cm^2 or hydraulic conductivities of 10^{-8} to 10^{-11} cm/sec (Freeze and Cherry, 1979). Secondary permeability can increase the effective hydraulic conductivity of a fractured rock system up to five orders of magnitude (Gale,

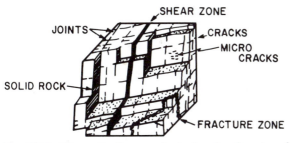

Fig. 12.5 Schematic diagram of a fractured rock system showing the development of secondary porosity and permeability (modified from Gale, 1982).

1982) depending on the type of material and the number, width, and interconnection of the fractures. Extreme alteration of the hydraulic properties of rock can occur when fracture networks in soluble rock like limestone are enlarged by solution. In the case of karst development, conduit flow may occur. Weathering also enhances fracturing and may increase hydraulic conductivity in clay (d'Astous et al., 1989) and rock (Davis and De Wiest, 1966). With the exception of conduit flow in karst, fracture flow models assume that both fracture apertures and flow velocities are small so that Darcy's law applies.

Fractured systems are typically modeled using one or more of the following conceptual models: (1) equivalent porous medium; (2) discrete fractures; (3) dual porosity. Each of these is discussed in detail below.

EQUIVALENT POROUS MEDIUM

Fractured material is represented as an equivalent porous medium (EPM) by replacing the primary and secondary porosity and hydraulic conductivity distributions with a continuous porous medium having so-called *equivalent* or *effective* hydraulic properties (Fig. 12.6b). The parameters are selected so that the flow pattern in the EPM is similar to the flow pattern in the fractured system. An EPM approach assumes that the fractured material can be treated as a continuum and that a representative elementary volume (REV) of material characterized by effective hydraulic parameters can be defined. Simulation of flow in fractured systems using this conceptual model requires definition of effective values for hydraulic conductivity, specific storage, and porosity. Values for effective parameters are derived from aquifer testing (e.g., Sauveplane, 1984; Gingarten, 1982), estimated from water balances or inverse models (Section 8.3), and/or calculated from field description of fracture apertures, lengths and interconnections, and unfractured rock volumes and permeabilities (Cacas et al., 1990a,b; Berkowitz et al., 1988; Hsieh et al., 1985). When effective parameters can be defined, standard finite difference and finite element codes used for continuous porous media may be applied to the EPM that represents the

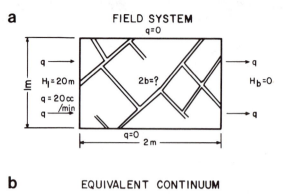

a FIELD SYSTEM

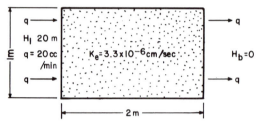

b EQUIVALENT CONTINUUM

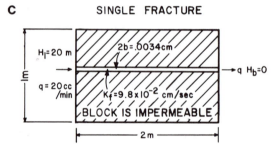

c SINGLE FRACTURE

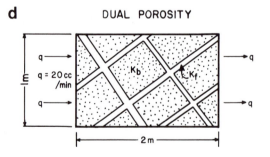

d DUAL POROSITY

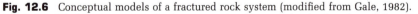

Fig. 12.6 Conceptual models of a fractured rock system (modified from Gale, 1982).
(a) A simplified fracture network of aperture 2b with groundwater flow from left to right.
(b) Equivalent porous medium model of (a).
(c) Discrete fracture model of (a).
(d) Dual porosity medium model of (a).

fractured system (e.g., Gerhart, 1984).

The difficulty in applying the EPM approach arises in determining the appropriate size of the REV needed to define equivalent hydraulic properties (Cacas et al., 1990a,b). When fractures are few and far between and the unfractured block hydraulic conductivity is low, the EPM method may not be appropriate even with a large REV (Gale, 1982). Berkowitz et al. (1988) concluded that if estimates of equivalent hydraulic properties are determined by field testing, including field-scale tracer tests, a reasonable flow model based on EPM concepts can be constructed. Other researchers, however, have concluded that an EPM representation is generally not valid for fractured systems (Long and Billaux, 1987; Cacas et al., 1990a,b). They note that the EPM approach may adequately represent the behavior of a regional flow system, but poorly reproduces local conditions.

DISCRETE FRACTURES

A discrete fracture (DF) model assumes that water moves only through the fracture network (Fig. 12.6c). The DF approach is typically applied to fractured media with low primary permeability such as crystalline rocks. Flow through a single fracture may be idealized as occurring between two parallel plates with a uniform separation equal to the fracture aperture ($2b$). Then the flow rate (Q_f) is calculated from

$$Q_f = 2bwK_f(dh/dl) \qquad (12.12)$$

where w is the width and K_f the hydraulic conductivity of the fracture, h is hydraulic head, and l is the length over which the hydraulic gradient is measured. The hydraulic conductivity of the fracture may be calculated from

$$K_f = \rho g(2b)^2/12u \qquad (12.13)$$

where ρ is fluid density, u is viscosity, and g is the constant of acceleration. By substituting Eqn. 12.13 into Eqn. 12.12, it is evident that Q_f is proportional to the cube of the fracture aperture. Use of a model based on Eqns. 12.12 and 12.13 requires a description of the fracture network, including fracture apertures and geometry. These data are extremely difficult to collect or estimate. Some descriptions of fracture networks have been obtained from mapping done in underground mines, from bore hole tests, and from tracer experiments (Cacas et al., 1990a,b; Dverstorp and Andersson, 1989).

Further complications in using a DF model arise when fracture widths are less than 10 μm and when portions of the fracture surfaces touch or are rough (Witherspoon et al., 1987; National Research Council, 1990). Under these conditions the cubic law for flow through a fracture may not be valid. Furthermore, increases in stress with depth and with decreases in pore pressure (e.g., from dewatering) cause a decrease in fracture aperture. Hence, the relative orientation of the fractures and the stress field in relation to the groundwater flow field must also be considered (Schmelling and Ross, 1989). Huyakorn and Pinder (1983) present governing equations for discrete fracture flow with deformation.

Models based on the DF approach are computationally intensive. To date, applications have been mainly to research problems (Piggott and Elsworth, 1989; Tsang and Tsang, 1987, 1988; Long et al., 1982, 1985; Smith and Schwartz, 1984). Cacas et al. (1990a,b) reported an application of a DF flow and transport model to a granite uranium mine in France. Dverstorp and Andersson (1989) applied a DF model to the Stripa research mine in Sweden.

DUAL POROSITY

If the rock matrix containing the fracture network has significant primary permeability, a dual porosity (DP) model (Fig. 12.6d) may be used. In this conceptual model, flow through the fractures is accompanied by exchange of water and solute to and from the surrounding porous rock matrix. Obviously, the fracture network as well as the properties of the porous blocks must be described prior to modeling. Aquifer tests may indicate whether a system behaves as a dual porosity system and results can be interpreted to estimate hydraulic conductivities (e.g., Streltsova-Adams, 1978). Exchange between the fracture network and the porous blocks is represented by a term that describes the rate of mass transfer (Huyakorn et al., 1983; Huyakorn and Pinder, 1983; Dykhuizen, 1990). Glover (1987) reported an application of a finite element DP model to simulate solute transport in oil shale.

SUMMARY

Although most modeling studies of fractured systems use the EPM approach, it is evident that there are a number of important practical problems for which this conceptual model is inappropriate. There are a few DF and DP codes in the public domain, including SWIFT II (Reeves et al., 1986a,b). However, all three of the conceptual models discussed above require assumptions that vastly oversimplify the actual fractured system. Some researchers have experimented with models that combine approaches to achieve a more realistic conceptual model. For example, Schwartz and Smith (1988) used a DF model to deduce the transport pattern in an REV of a fractured system and then developed stochastic expressions that reproduced the transport pattern in a regional EPM model. Harrison et al. (1991) discussed a simulation of fractured clay that incorporated flow through individual fractures, the fracture network, and the surrounding porous medium. Results illustrate the complexity of the head distribution that can be expected in fractured systems (Fig. 12.7).

KARST SYSTEMS

Groundwater systems in limestone, dolomite, marble, evaporites, and other soluble rocks may behave as typical fracture flow systems as described above or

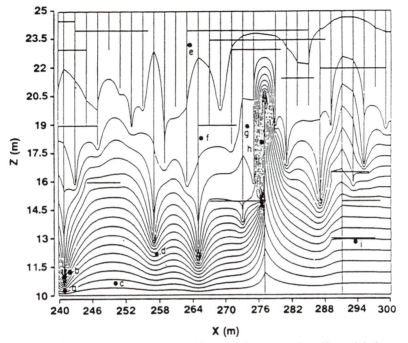

Fig. 12.7 Head distribution produced by a two-dimensional profile model of groundwater flow in a discretely fractured clayey deposit (Harrison et al., 1991). Fractures are represented by horizontal and vertical lines. Equipotential lines are solid curved lines. The head generally decreases with depth as water flows to an underlying aquifer. Vertical exaggeration is 3.33.

may be influenced by conduit flow. In soluble rocks, fracture apertures and primary porosity are enlarged with time, creating a system with paths of higher hydraulic conductivity (Ford and Williams, 1989). In general, soluble rock with high primary porosity develops a diffuse flow system with few karst features, whereas rock with little primary porosity develops a conduit flow system with karst (Ford and Williams, 1989).

Modeling is complicated by the difficulty in characterizing hydrogeologic properties of karst including conduit geometry. Three modeling approaches have been used to simulate groundwater flow in karst.

1. Groundwater flow in the porous and fractured rock is assumed to be governed by Darcy's law so that one of the three conceptual models used for other fractured rocks, as discussed above, can be applied. Thrailkill (1974, 1989) and Yusun and Ji (1988) described the application of a DF model to simulate flow in karst formations having pipes or conduits.
2. For mature karst, a "black box" approach is used, whereby functions are developed to reproduce input and output responses, namely recharge and springflow. Typically, springflow is used to represent an integrated

response of the groundwater system to recharge or stress. Analyses of discharge hydrographs, recharge, and geochemistry of the recharge and spring water are used to develop equations that describe discharge from the system. These equations may or may not include standard aquifer parameters. Examples of this type of modeling include studies by Dreiss (1989a,b), Simpson (1988), and Aiguang et al. (1988).

3. Another approach uses the aquifer response functions developed in item 2 to derive equivalent porous media parameters for use in EPM models (Pulido-Bosch and Padilla, 1988).

It is clear from the above discussion that it is possible to develop models of flow in fractured rock, including karst. However, all of the modeling techniques discussed above require detailed information about the nature of the fracture or conduit network. Field techniques to collect these data have not been perfected. Predictions made with these models will continue to be influenced by large uncertainties and require field validation.

12.7   Density-Dependent Flow of Miscible Fluids

Standard flow models assume that the density of groundwater is constant and approximately equal to 1.0 gm/cm^3. (Water at 4°C has a density of 0.999973 gm/cm^3.) This assumption is valid for water with low concentrations of total dissolved solids (TDS) and/or temperatures in the range of most shallow aquifers. Simulation of flow involving water with high TDS or higher or lower than normal temperatures requires that the effects of density be included in the model. For example, salt water has a TDS of roughly 35,000 mg/l, giving it a density of 1.025 gm/cm^3. Water heated to 50°C has a density of 0.988047 gm/cm^3. These small changes in density may cause significant changes in the flow field.

Examples of density-dependent flow involving miscible fluids include seepage of some types of landfill leachate, saltwater intrusion, waste injection into saline aquifers, heat storage in aquifers, brine disposal, freshwater storage in saline aquifers, and liquid-phase geothermal flow. Problems involving miscible fluids are complex because the density of the solution is dependent on solute concentration and temperature. Thus, it may be necessary to solve three models—flow, solute transport, and heat transport. The governing equation for flow must be written in terms of pressure (p) and intrinsic permeability (k) because total head (h) and hydraulic conductivity (K) are functions of density (ρ): $h = z + p/\rho g$ and $K = k\rho g/u$, where g is the constant of acceleration, u is the dynamic viscosity, and z is the elevation head.

Numerical models that incorporate density effects first solve a flow model based on Eqn. 12.7. If the density of the fluid changes as a result of changes in TDS or temperature, it may be necessary to solve a solute transport model

(Section 12.5) and/or a heat transport model. The heat transport equation is similar to the advection-dispersion equation (Eqn. 12.11) and may be written as follows:

$$\frac{\partial}{\partial x_i}\left(D_{ij}\frac{\partial T}{\partial x_j}\right) - \rho C_W \frac{\partial(q_i T)}{\partial x_i} = \rho C_s \frac{\partial T}{\partial t} \qquad (12.14)$$

where T is temperature, q_i is specific discharge, ρ is density, C_W is the heat capacity of water, and C_s is the heat capacity of the solute. The dispersion coefficient, D_{ij}, now a function of thermal conductivity, is calculated from Eqn. 12.9 by replacing D_d by the thermal conductivity of the porous medium. When simulating heat transport, the dynamic viscosity (u) in the flow equation (Eqn. 12.7) may vary spatially with temperature.

Models that simulate density-dependent flow require initial pressure and density distributions. At the beginning of a time step, these initial values are used to generate the first approximation of the flow field. The resulting head values are input to the transport models, which redistribute solute (Eqn. 12.11) and/or temperature (Eqn. 12.14). A new density distribution is calculated from the transport results, ending the first iteration of the first time step. The second iteration begins with the substitution of the newly calculated densities into the flow model. Iteration is continued until closure is attained. This process is repeated for all time steps.

There are many examples of density-dependent modeling in the literature. Many of these are applications to saltwater intrusion problems (e.g., Andersen et al., 1988; Bush, 1988; Voss and Souza, 1987; Huyakorn et al., 1987). Hickey (1989) simulated the injection of liquid waste into a saltwater formation using the finite difference model SWIP. Herbert et al. (1988) used a finite element code to simulate flow in the vicinity of a salt dome. Frind (1982) used finite elements to simulate transient density-dependent transport of leachate.

Public-domain software for density-dependent flow includes the USGS codes SUTRA, MOCDENSE, VARDEN, and HST3D. SUTRA (Voss, 1984b) combines finite elements and integrated finite differences to simulate two-dimensional problems under saturated and partially saturated conditions. Applications of SUTRA include Voss and Souza (1987) and Bush (1988). MOCDENSE (Sanford and Konikow, 1985) is a two-dimensional finite difference model for saturated conditions developed from the flow and solute transport code USGS MOC (Konikow and Bredehoeft, 1978). The code assumes that density and viscosity are functions of concentration only and that constituents are chemically conservative. VARDEN is described by Kuiper (1983, 1985) and Kontis and Mandle (1988). HST3D (Kipp, 1987) is a relatively new three-dimensional finite difference model that simulates groundwater flow, heat, and solute transport. The finite element code CFEST (Gupta et al., 1987), includes density and temperature effects on miscible solute transport in three dimensions.

12.8   Other Models

A few other types of specialty models that are sometimes used in groundwater work are discussed briefly below.

Compartment models of the unsaturated zone, such as SESOIL (Bonazoun-tas and Wagner, 1984) and PRZM (Carsel et al., 1984), are sometimes used to route contaminants through the soil zone to the water table. See Pennell et al. (1990) for a discussion of these codes.

Optimization or management models link codes for flow and/or solute transport to an optimization code that includes decision variables related to economics (Lefkoff and Gorelick, 1987). These models may address questions related to optimal placement of wells for water supply or aquifer remediation and problems of policy evaluation and water allocation.

Geochemical codes calculate the concentrations of ions in solution in water at chemical equilibrium by solving a mass balance problem. Some of these codes perform reaction path modeling whereby the chemical evolution of groundwater is followed along a flow line as material is dissolved or a compound is added from an external source. There have been a few attempts to incorporate geochemical codes into contaminant transport modeling (Mangold and Tsang, 1991; Abriola, 1987), but their application to practical problems is still limited. The FASTCHEM package of codes (Hostetler et al., 1988) developed for the Electric Power Research Institute is an example of this type of coupled model.

The purpose of this chapter is to point out the need to use specialized models when simulating complex groundwater systems. The material in this book provides a first step toward acquiring modeling expertise. Given the current interest in groundwater contamination problems, we can expect that data collection and field characterization will become more sophisticated. This will lead to better understanding of the physical processes simulated by the models discussed in this chapter and eventually to better models. Development of better models will encourage the use of more complex models in practical applications. Yet, the need for analysis of groundwater flow and advective transport will remain paramount.

"Once more unto the breach, dear friends, once more..."
—*Henry V, Act III*

Box 12.1
Calculating Streamfunctions with a Flow Code _____

A standard groundwater flow code can be used to generate a flow net by first calculating heads and then, in a second simulation, streamfunctions. A flow net is produced by superimposing equipotential lines and streamfunction contours, as in Fig. 12.3. We saw in Section 12.2 (eqn. 12.3) that the streamfunction simulation requires substituting $1/K_y$ for K_x and $1/K_x$ for K_y in the flow code. Boundary conditions must also be formulated in terms of the streamfunction. This is easily done once the relations between head-defined and streamfunction-defined boundaries are understood. Details on the way in which boundaries conditions are transferred from the head problem to the streamfunction problem are given by Bramlett

and Borden (1990) and Fogg and Senger (1985). The procedure is summarized briefly below.

Specified head (type 1) boundaries translate to derivative (type 2) streamfunction boundaries. Type 2 streamfunction boundary values, F, are calculated from the head values as follows:

$$F_i = h_{i-1/2} - h_{i+1/2} \qquad (1)$$

That is, the streamfunction value for node i is calculated by taking the difference between heads at points midway between nodes i and

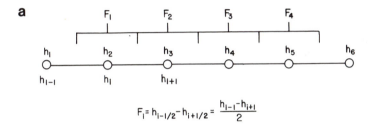

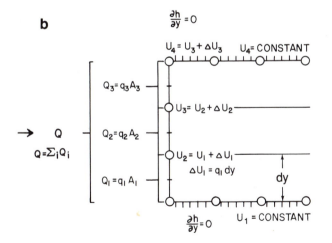

Fig. 1 Definition of streamfunction boundary conditions.
(a) Type 2 (derivative) streamfunction boundary conditions are defined from the head solution.
(b) Type 1 (specified value) streamfunction boundary conditions are defined by computing boundary fluxes. A no-flow boundary in the head problem is a constant streamfunction boundary.

$i - 1$ and between i and $i + 1$ (Fig. 1a). Normally, values at the midway points are computed by assuming a linear variation in head. Differences in head should be taken consistently in either a clockwise or counterclockwise direction. These values are then entered into a flow code as volumetric flux values via injection or pumping wells.

Flux (type 2) boundaries in the head problem become specified value (type 1) boundary conditions in the streamfunction solution. No-flow boundaries in the head problem are simulated as constant-value streamfunction boundaries; usually one such boundary is set equal to zero. Changes in flux across the boundary are calculated from the head solution and used to assign specified streamfunction values. Assignment begins at a known streamfunction value, usually a reference value of zero; the boundary values are then calculated as $U_{i+1} = U_i + \Delta U_i$ where ΔU_i equals Q_i (Fig. 1b).

Streamfunctions are continuous functions; thus, point sources and sinks, e.g., wells and infiltration ponds, cause discontinuities in the domain. They can be represented only by "cutting a hole" in the model domain to connect the discontinuity to a boundary. The flux is then distributed along the cut. See Frind and Mantanga (1985) for details of this procedure.

Finite element codes and finite difference codes that use mesh-centered grids can be used directly to generate streamfunctions by simple manipulation of input data as described above. An example of the streamfunction solution for a steady-state flow problem (Box 2.1, Fig. 1) with a mesh-centered grid (Box 4.2, Fig. 1a) is presented in Fig. 2. The same mesh-centered grid can be used for both head and streamfunction generation. The head solution is shown in Box 2.1, Fig. 2b and Box 4.2, Fig. 2a. The flow through the system as calculated from the water balance of the head solution is 28.8 m²/day.

STREAMFUNCTIONS (m²/d)

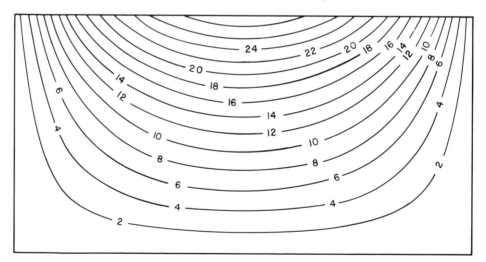

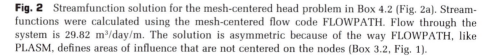

Fig. 2 Streamfunction solution for the mesh-centered head problem in Box 4.2 (Fig. 2a). Streamfunctions were calculated using the mesh-centered flow code FLOWPATH. Flow through the system is 29.82 m³/day/m. The solution is asymmetric because of the way FLOWPATH, like PLASM, defines areas of influence that are not centered on the nodes (Box 3.2, Fig. 1).

Streamfunctions (m^2/day)

0.000	13.203	21.271	26.234	28.955	29.821	28.941	26.204	21.221	13.125	0.000
0.000	8.240	14.413	18.546	20.897	21.655	20.884	18.520	14.371	8.180	0.000
0.000	5.215	9.477	12.512	14.300	14.885	14.290	12.492	9.446	5.174	0.000
0.000	3.107	5.747	7.696	8.873	9.262	8.866	7.683	5.727	3.081	0.000
0.000	1.457	2.716	3.662	4.241	4.434	4.238	3.656	2.706	1.445	0.000
0.000	0.000	0.000	0.000	0.000	0.000	0.000	0.000	0.000	0.000	0.000

Fig. 2 *(continued)*

FLOWPATH (Franz and Guiguer, 1990), a mesh-centered finite difference flow code, was used to generate the streamfunction values shown in Fig. 2. The flow through the system as calculated by FLOWPATH can be read from the streamfunction value in column 6 of the top row and is equal to 29.8 m^2/day.

In a finite element model or a finite difference model with a mesh-centered grid, both type 1 and type 2 boundaries always fall on the node. When this is the case there is no diffi-culty in transferring between type 1 head and type 2 streamfunction boundaries and vice versa. In a block-centered grid, however, specified head boundaries fall on the node whereas flux boundaries fall on the edge of the block, causing the problem domain to change size when shifting from the head to the streamfunction problem. For this reason, use of a block-centered code is not recommended for flow net construction.

Box 12.2
Initial Conditions for Unsaturated Flow Models

The following are some options for selecting initial conditions for unsaturated flow models.

1. *Static steady state* corresponds to a steady-state moisture profile where $\psi = -z$ or $h = 0$, assuming that $z = 0$ at the water table and W^* in Eqn. 12.6 is zero. The static steady-state moisture profile typically has high tensions in the upper part of the profile if the soil is thick, owing to the condition that $\psi = -z$. Such high tensions are unrealistic for humid zone conditions but may be ap-propriate for arid zone settings. High tensions may cause numerical instabili-ties in some models. The code VS2D (Lappala et al., 1987) was designed to handle highly nonlinear problems of unsaturated flow.

2. *Dynamic steady state* corresponds to a steady-state moisture profile with infil-tration occurring at a constant rate. The dynamic steady-state condition pro-duces an initial moisture profile that is relatively dry at the top of the column if

the infiltration rate is low and a relatively wet profile if the infiltration rate is high. A dynamic steady-state profile that is relatively wet (at "field capacity") over most of the soil column could be produced, for example.

3. *Non–steady state* corresponds to a head distribution from a previous simulation that has not reached steady state. The dynamic steady-state profile (item 2 above) may be too wet or too dry for the purposes of the simulation. If this is the case, a non–steady state model–generated moisture profile may be used as an

initial condition. The simulation that generates the initial condition may use a dynamic steady-state initial condition. The model-generated initial profile might be selected after a period of no infiltration during which most of the drainage from the column has occurred. The influence of the initial condition diminishes as the simulation progresses, so that errors associated with selecting the initial condition decrease with simulated time. Stoertz et al. (1991) showed examples of different types of initial conditions.

Problems

12.1 Use a flow model to calculate streamfunctions for Problems 6.3a and 6.3b.
- **(a)** Calculate streamfunctions under homogeneous and isotropic conditions with T = 140 m²/day.
- **(b)** Calculate streamfunctions again after inserting the low hydraulic conductivity zone.

12.2 Use a flow model to calculate streamfunctions and the equipotential distribution for the problem shown in Fig. P12.1.
- **(a)** Assume the aquifer is isotropic and homogeneous with a hydraulic conductivity of 10 m/day. Use the results to construct a flow net.
- **(b)** Assume the aquifer is anisotropic and homogeneous with K_x = 10 m/day and K_z = 1 m/day. Use the results to construct a flow net.

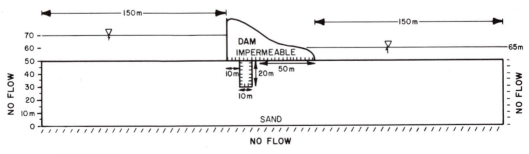

Fig. P12.1

12.3 Read the paper by Berkowitz et al. (1988) on the use of equivalent porous media to simulate groundwater flow in fractured media. List the assumptions required to use this approach. Contrast this approach with that of Harrison et al. (1991) and Schwartz and Smith (1988).

12.4 Solution of the advection-dispersion equation to simulate solute transport requires careful attention to the Peclet number (P_e) (Huyakorn and Pinder, 1983).

    (a) Describe the problems associated with grid design when using finite differences or finite element techniques to solve the advection-dispersion equation.

    (b) Examine documentation of transport models that use particle tracking methods, e.g., RNDWALK (Prickett et al., 1981) and USGS MOC (Konikow and Bredehoeft, 1978), to solve the advection-dispersion equation. Are these models similarly restricted by the magnitude of the Peclet number? Explain your answer.

12.5 The discovery of aquifers contaminated with NAPLs has focused attention on the development of methods for aquifer remediation. Simulation of the movement of the free phase unreacted NAPL requires a multiphase flow model. Review the work of Kueper and Frind (1991a) and Faust et al. (1989). List the field data required to simulate NAPL migration.

REFERENCES

Aboufirassi, M., and M.A. Marino, 1984, Cokriging of aquifer transmissivities from field measurements of transmissivity and specific capacity, Math. Geology 16(1), pp. 19–35.

Abriola, L.M., 1984, *Multiphase Migration of Organic Compounds in a Porous Medium*, Springer-Verlag, 232 p.

Abriola, L.M., 1987, Modeling contaminant transport in the subsurface: an interdisciplinary challenge, Reviews of Geophysics 25(2), pp. 125–134.

Abriola, L.M., and G.F. Pinder, 1985a, A multiphase approach to the modeling of porous media contamination by organic compounds. 1. Equation Development, Water Resources Research 21(1), pp. 11–18.

Abriola, L.M., and G.F. Pinder, 1985b, A multiphase approach to the modeling of porous media contamination by organic compounds. 2. Numerical simulation, Water Resources Research 21(1), pp. 19–26.

Ahlfeld, D.P., J.M. Mulvey, and G.F. Pinder, 1988, Contaminated groundwater remediation design using simulation, optimization and sensitivity theory. 2. Analysis of a field site, Water Resources Research 24(3), pp. 442–452.

Aiguang, C., L. Hanchen, X. Juanming, and S. Jinli, 1988, Stochastic modelling of the karst spring of Xin'an Shanxi province, *In: Karst Hydrogeology and Karst Environment Protection*, Proc. of the IAH 21st Congress, Grulin, China, Geological Publ. House, Beijing, pp. 621–628.

Allen, M.B. III, 1985, Numerical modelling of multiphase flow in porous media, Water Resources 8(4), pp. 162–187.

Alley, W.M., and P.A. Emery, 1986, Groundwater model of the Blue River Basin, Nebraska—twenty years later, Journal of Hydrology 85, pp. 225–250.

Andersen, P.F., C.R. Faust, and J.W. Mercer, 1984, Analysis of conceptual designs for remedial measures at Lipari landfill, New Jersey, Ground Water 22(2), pp. 176–190.

Andersen, P.F., J.W. Mercer, and H.O. White, Jr., 1988, Numerical modeling of salt-water intrusion at Hallandale, Florida, Ground Water 26(5), pp. 619–630.

Anderson, M.P., 1979, Using models to simulate the movement of contaminants through groundwater flow systems, CRC Critical Reviews in Environmental Control 9(2), pp. 97–156.

Anderson, M.P., 1989, Hydrogeologic facies models to delineate large-scale spatial trends in glacial and glaciofluvial sediments, Geol. Soc. of Amer. Bull., 101, pp. 501–511.

Anderson, M.P., 1991, Aquifer heterogeneity—a geological perspective, *In: Parameter Identification and Estimation for Aquifer and Reservoir Characterization*, Nat'l. Water Well Assoc., Columbus, Ohio, pp. 3–22.

Anderson, M.P., and J.A. Munter, 1981, Seasonal reversals of groundwater flow around lakes and the relevance to stagnation points and lake budgets, Water Resources Research 17(4), pp. 1139–1150.

Anderson, M.P., D.S. Ward, E.G. Lappala, and T.A. Prickett, 1992, Computer models for subsurface water, *In: Handbook of Hydrology* (D. Maidment, editor), McGraw-Hill, Chap. 22.

Andrews, C.B., and M.P. Anderson, 1978, Impact of a power plant on the groundwater system of a wetland, Ground Water 16(2), pp. 105–111.

Appel, C.A., 1976, A note on computing finite difference interblock transmissivities, Water Resources Research 12(3), pp. 561–563.

Aral, M.M., 1989, *Ground Water Modeling in Multilayer Aquifers: steady flow*, Lewis Publishers, 114 p.

Aucott, W.R., 1988, The predevelopment ground-water flow system and hydrologic characteristics of the Coastal Plain aquifers of South Carolina, USGS, Water Resources Investigation Report 86-4347, 65 p.

Baehr, A.L., and M.Y. Corapcioglu, 1987, A compositional multiphase model for groundwater contamination by petroleum products. 2. Numerical solution, Water Resources Research, 23(1), pp. 201–213.

Bailey, Z.C., 1988, Preliminary evaluation of ground-water flow in Bear Creek Valley, the Oak Ridge Reservation, Tennessee, USGS, Water Resources Investigation Report 88-4010, 12 p.

Bair, E.S., R.A. Sheets, and S.M. Eberts, 1990, Particle-tracking analysis of flow paths and travel times from hypothetical spill sites within the capture area of a wellfield, Ground Water 28(6), pp. 884–892.

Barwell, V.K., and D.R. Lee, 1981, Determination of horizontal-to-vertical hydraulic conductivity from seepage measurement on lakebeds, Water Resources Research 17(3), pp. 565–570.

Baski, H.A., 1979, Ground-water computer models—intellectual toys, Ground Water 17(2), pp. 177–179.

Bear, J., 1972, *Dynamics of Fluids in Porous Media*, Amer. Elsevier, N.Y., 764 p.

Bear, J., and A. Verruijt, 1987, *Modeling Groundwater Flow and Pollution*, D. Reidel Publishing Co., 408 p.

Beljin, M.S., 1987, Representation of individual wells in two-dimensional ground water modeling, *In: Solving Ground Water Problems with Models*, National Water Well Association, Dublin, Ohio, pp. 340–351.

Belmans, C., J.G. Wesseling, and R.A. Feddes, 1983, Simulation model of a cropped soil: SWATRE, Journal of Hydrology 63, pp. 271–286.

Bennett, G.D., A.L. Kontis, and S.P. Larson, 1982, Representation of multiaquifer well effects in three-dimensional ground-water flow simulation, Ground Water 20(3), pp. 334–341.

Berkowitz, B., J. Bear, and C. Braester, 1988, Continuum models for contaminant transport in fractured porous formations, Water Resources Research 24(8), pp. 1225–1236.

Blandford, T.N., and P.S. Huyakorn, 1990, WHPA: a modular semi-analytical model for the delineation of wellhead protection areas, U.S. EPA, Office of Ground-Water Protection.

Bloomsburg, G.L., and R.E. Rinker, 1983, Ground-water modeling with an interactive computer, Ground Water 21(2), pp. 208–211.

Bonazountas, M., and J.M. Wagner, 1984, SESOIL: a seasonal soil compartment model, U.S. Environmental Protection Agency, Washington, D.C., 562 p.

Boonstra, J., and N.A. deRidder, 1981, *Numerical Modelling of Groundwater Basins—a User-Oriented Manual*, International Institute for Land Reclamation and Improvement, The Netherlands, 226 p.

Bouma, J., F.G. Baker, and P.L.M. Veneman, 1974, Measurements of water movement in soil pedons above the water table, Wisc. Geol. Nat. Hist. Surv. 27, 114 p.

Bouwer, H., 1989, The Bouwer and Rice slug test—an update, Ground Water 27(3), pp. 304–309.

Bouwer, H., and R.C. Rice, 1976, A slug test for determining hydraulic conductivity of unconfined aquifers with completely or partially penetrating wells, Water Resources Research 12(3), pp. 423–428.

Bradbury, K.R., 1982, Hydrogeologic relationships between Green Bay of Lake Michigan and onshore aquifers in Door County, Wisc, Ph.D. thesis, University of Wisconsin-Madison, 287 p.

Brahana, J.V., and T.O. Mesko, 1988, Hydrogeology and preliminary assessment of regional flow in the upper Cretaceous and adjacent aquifers in the northern Mississippi embayment, USGS, Water Resources Investigation Report 87-4000, 65 p.

Bramlett, W., and R.C. Borden, 1990, Numerical generation of flow nets—the FLOWNS model, Ground Water 28(6), pp. 946–950.

Bredehoeft, J.D., and G.F. Pinder, 1970, Digital analysis of areal flow in multiaquifer groundwater systems: a quasi three-dimensional model, Water Resources Research 6(3), pp. 883–888.

Bredehoeft, J.D., P. Betzinski, C.C. Villanueva, G. de Marsily, A.A. Konoplyantsev, and J.U. Uzoma, 1982, *Ground-Water Models. Vol I. Concepts, Problems, and Methods of Analysis with Examples of Their Application*, The Unesco Press, 235 p.

Bredehoeft, J.D., C.E. Neuzil, and P.C.D. Milly, 1983, Regional flow in the Dakota Aquifer: a study of the role of confining layers, USGS, Water Supply Paper 2237, 45 p.

Brown, J.G., and J.H. Eychaner, 1988, Simulation of five groundwater withdrawal projections for the Black Mesa area, Navajo and Hopi Indian Reservations, Arizona, USGS Water-Resources Investigations Report 88-4000, 52 p.

Buckles, D.R., and K.R. Watts, 1988, Geohydrology, water quality and preliminary simulations of ground-water flow of the alluvial aquifer in the upper Black Squirrel Creek basin, El Paso County, Colorado, USGS, Water Resources Investigation Report 88-4017, 49 p.

Bush, P.W., 1988, Simulation of saltwater movement in the Floridan aquifer system, Hilton Head Island, South Carolina. USGS, Water Supply Paper 2331, 19 p.

Bush, P.W., and R.H. Johnston, 1986, Floridan regional aquifer-system study, In: Regional Aquifer-System Analysis Program of the U.S. Geol. Survey: Summary of Projects, 1978–84 (R.J. Sun, ed.), USGS Circ. 1002, pp. 17–29.

Buxton, H., and T.E. Reilly, 1986, A technique for analysis of ground-water systems at regional and subregional scales applied on Long Island, New York, USGS, Water Supply Paper 2310, pp. 129–142.

Buxton, H.T., T.E. Reilly, D.W. Pollock, and D.A. Smolensky, 1990, Particle tracking analysis of recharge areas on Long Island, New York, Ground Water 29(1), pp. 63–71.

Cacas, M.C., E. Ledoux, G. de Marsily, B. Tillie, A. Barbreau, E. Durand, B. Feuga, and P. Peaudecerf, 1990a, Modeling fracture flow with a stochastic discrete fracture network: calibration and validation. 1. The flow model, Water Resources Research 26(3), pp. 479–489.

Cacas, M.C., E. Ledoux, G. de Marsily, B. Tillie, A. Barbreau, E. Durand, B. Feuga, and P. Peaudecerf, 1990b, Modeling fracture flow with a stochastic discrete fracture network: calibration and validation. 2. The transport model, Water Resources Research 26(3), pp. 491–500.

Carrera, J., 1988, State of the art of the inverse problem applied to the flow and solute transport equations, In: Groundwater Flow and Quality Modelling (E. Custodio et al., eds.),D. Reidel Publ. Co., pp. 549–583.

Carrera, J., and S.P. Neuman, 1984, Adjoint state finite element estimation of aquifer parameters under steady-state and transient conditions, In: Proceedings of the 5th International Conference on Finite Elements in Water Resources, Springer-Verlag, N.Y.

Carrera, J., and S.P. Neuman, 1986a, Estimation of aquifer parameters under transient and steady state conditions. 1. Maximum likelihood method incorporating prior information, Water Resources Research 22(2), pp. 199–210.

Carrera, J., and S.P. Neuman, 1986b, Estimation of aquifer parameters under transient and steady state conditions. 2. Uniqueness, stability and pollution algorithms, Water Resources Research 22(2), pp. 211–227.

Carrera, J., and S.P. Neuman, 1986c, Estimation of aquifer parameters under transient and steady state conditions. 3. Application to synthetic and field data, Water Resources Research 22(2), pp. 228–242.

Carrera, J., G.R. Walter, M. Kuhn, H. Bentley, and J. Fabryka-Martin, 1984, Three-dimensional modeling of saline pond leakage calibrated by IN-VERT-3, a quasi three-dimensional, transient, parameter estimation program, In: Practical Applications of Ground Water Models, Nat'l. Water Well Assoc., Columbus, Ohio, pp. 547–569.

Carsel, R.F., R.L. Jones, J.L. Hansen, R.L. Lamb, and M.P. Anderson, 1988, A simulation procedure for groundwater quality assessments of pesticides, Journal of Contaminant Hydrology 2(1), pp. 125–138.

Carsel, R.F., C.N. Smith, L.A. Mulkey, J.D. Dean, and P. Jowise, 1984, User's manual for the pesticide root zone model (PRZM), Release 1, U.S. Environmental Protection Agency, Washington, D.C., U.S. EPA-60013-84-109, 216 p.

Cederberg, G.A., R.L. Street, and J.O. Leckie, 1985, A groundwater mass-transport and equilibrium chemistry model for multicomponent systems, Water Resources Research 21(8), pp. 1095–1104.

Cedergren, H.R., 1967, Seepage, Drainage, and Flow Nets, Wiley and Sons, 534 p.

Celia, M.A., E.T. Bouloutas, and R.L. Zarba, 1990, A general mass-conservative numerical solution for the unsaturated flow equation, Water Resources Research 26(7), pp. 1483–1496.

Chapelle, F.H., 1986, A solute-transport simulation of brackish-water intrusion near Baltimore, Maryland, Ground Water 24(3), pp. 304–311.

Chargaff, E., 1978, Heraclitean Fire: Sketches from a Life before Nature, The Rockefeller University Press, N.Y., 252 p.

Cherkauer, D., and J. McBride, 1988, A remotely-operated seepage meter for use in large lakes and rivers, Ground Water 26(2), pp. 165–177.

Christian, J.T., 1980, Flow nets by the finite element method, Ground Water 18(2), pp. 178–181.

Clarke, D., 1987, Microcomputer programs for groundwater studies, In: Developments in Water Science 30, Elsevier.

Clifton, P.M., and S.P. Neuman, 1982, Effects of kriging and inverse modeling on conditional simulation of the Avra Valley aquifer in southern Arizona, Water Resources Research 18(4), pp. 1215–1234.

Cohen, P., O.L. Franke, and B.L. Foxworthy, 1968, An atlas of Long Island's Water Resources, N.Y. Water Resources Commission, Bulletin 62, 117 p.

Connell, J.F., and Z.C. Bailey, 1989, Statistical and simulation analysis for hydraulic-conductivity data for Bear Creek and Melton Valleys, Oak Ridge Reservation, Tennessee, USGS, Water-Resources Investigations Report 89-4062, 49 p.

Cooley, R.L., 1977, A method of estimating parameters and assessing reliability for models of steady state groundwater flow. 1. Theory and numerical properties, Water Resources Research 13(2), pp. 318–324.

Cooley, R.L., 1979, A method of estimating parameters and assessing reliability for models of steady state groundwater flow, Water Resources Research 15(3), pp. 603–617.

Cooley, R.L., 1982, Incorporation of prior information on parameters into non-linear regression groundwater flow models. 1. Theory, Water Resources Research 18(4), pp. 965–976.

Cooley, R.L., L.F. Konikow, and R.L. Naff, 1986, Non-linear-regression ground-water flow modeling of a deep regional aquifer system, Water Resources Research, 22(13), pp. 1759–1778.

Cooley, R.L., and R.L. Naff, 1990, Regression modeling of ground-water flow, USGS, Techniques of Water-Resources Investigations 03-B4, 232 p.

Cooper, H.H., Jr., F.A. Kohout, H.R. Henry, and R.E. Glover, 1964, Sea water in coastal aquifers, USGS, Water Supply Paper 1613-C, 84 p.

Corapcioglu, M.Y., and A.L. Baehr, 1987, A compositional multiphase model for groundwater contamination by petroleum products. 1. Theoretical considerations, Water Resources Research, 23(1), pp. 191–200.

Corapcioglu, M.Y., and M.A. Hossain, 1990, Ground-water contamination by high-density immiscible hydrocarbon slugs in gravity-driven gravel aquifers, Ground Water 28(3), pp. 403–412.

Czarnecki, J.B., and R.K. Waddell, 1984, Finite-element simulation of ground-water flow in the vicinity of Yucca Mountain, Nevada-California, USGS, Water-Resources Investigations Report 84-4349, 38 p.

Dagan, G., 1989, *Flow and Transport in Porous Formations*, Springer-Verlag, 465 p.

Dagan, G., and V. Nguyen, 1989, A comparison of travel time and concentration approaches to modeling transport by groundwater, Journal of Contaminant Hydrology 4(1), pp. 79–92.

Danskin, W.R., 1988, Preliminary evaluation of the hydrogeologic system in Owens Valley, California, USGS, Water-Resources Investigations Report 88-4003, 76 p.

Darr, R.E., 1979, Ground-water computer models—practical tools, Ground Water 17(2), pp. 174–176.

d'Astous, A.R., W.W. Ruland, J.F.G. Bruce, J.A. Cherry, and R.W. Gillham, 1989, Fracture effects in the shallow groundwater zone in weathered Sarnia area clay, Canadian Geotechnical Journal 26, pp. 43–56.

Davies-Smith, A., E.L. Bolke, and C.A. Collins, 1988, Geohydrology and digital simulation of the ground-water flow system in the Umatilla Plateau and Horse Heaven Hills area, Oregon and Washington, USGS, Water-Resources Investigations Report 87-4268, 72 p.

Davis, L.A., and S.P. Neuman, 1983, Documentation and User's Guide: UN-SAT2 Variably Saturated Flow Model, U.S. Nuclear Regulatory Commission, NUREG/CR-33907, 200 p.

Davis, S.N., and R.J.M. De Wiest, 1966, *Hydrogeology,* John Wiley and Sons, New York, 463 p.

de Lima, V., and J.C. Olimpio, 1989, Hydrogeology and simulation of groundwater flow at superfund-site wells G and H, Woburn, Massachusetts, USGS, Water-Resources Investigations Report 89-4059, 39 p.

de Marsily, G., 1986, *Quantitative Hydrogeology*, Academic Press, 440 p.

Demond, A.H., and P.V. Roberts, 1987, An examination of relative permeability relations for two-phase flow in porous media, Water Resources Bulletin 23(4), pp. 617–628.

Desbarats, A.J., 1990, Macrodispersion in sand-shale sequences, Water Resources Research 26(1), pp. 153–164.

Domenico, P.A., 1972, *Concepts and Models in Groundwater Hydrology*, McGraw-Hill, N.Y., 405 p.

Domenico, P.A., and F.W. Schwartz, 1990, *Physical and Chemical Hydrogeology*, John Wiley and Sons, N.Y., 824 p.

Dreiss, S.J., 1989a, Regional scale transport in a karst aquifer. 1. Component separation of spring flow hydrographs, Water Resources Research 25(1), pp. 117–125.

Dreiss, S.J., 1989b, Regional scale transport in a karst aquifer. 2. Linear systems and time moment analysis, Water Resources Research 25(1), pp. 126–134.

Duffield, G.M., D.R. Buss, and D.E. Stephenson, 1990, Velocity prediction errors related to flow model calibration uncertainty, *In: Calibration and Reliability in Groundwater Modelling* (K. Kovar, ed.), IAHS Publ. 195, pp. 397–406.

Dunlap, L.E., and J.M. Spinazola, 1984, Interpolating water-table altitudes in west-central Kansas using kriging techniques, U.S. Water Supply Paper 2238, 19 p.

Durbin, T.J., and C. Berenbrock, 1985, Three-dimensional simulation of free-surface aquifers by finite-element method, USGS, Water Supply Paper 2270, pp. 51–67.

Dverstorp, B., and J. Andersson, 1989, Application of the discrete fracture network concept with field data: possibilities of model calibration and validation, Water Resources Research 25(3), pp. 540–550.

Dykhuizen, R.C., 1990, A new coupling term for dual-porosity models, Water Resources Research 26(2), pp. 351–356.

Eberts, S.M., and E.S. Bair, 1990, Simulated effects of quarry dewatering near a municipal well field, Ground Water 28(1), pp. 37–47.

El-Kadi, A.I., 1988, Applying the USGS mass-transport model (MOC) to remedial actions by recovery wells, Ground Water 26(3), pp. 281–288.

Emmons, P.J., 1988, A digital simulation of the glacial-aquifer system in Sanborn and parts of Beadle, Miner, Hanson, Davison, and Gerauld Counties, South Dakota, USGS, Water-Resources Investigations Report 87-4082, 59 p.

England, L.A., and R.A. Freeze, 1988, Finite-element simulation of long-term transient regional ground-water flow, Ground Water 26(3), pp. 298–308.

Faust, C.R., L.R. Silka, and J.W. Mercer, 1981, Computer modeling and groundwater protection, Ground Water 19(4), pp. 362–365.

Faust, C.R., J.H. Guswa, and J.W. Mercer, 1989, Simulation of three-dimensional flow of immiscible fluids within and below the unsaturated zone, Water Resources Research 25(2), pp. 2449–2464.

Faust, C.R., P.N. Sims, C.P. Spalding, and P.F. Andersen, 1990, FTWORK: a three-dimensional groundwater flow and solute transport code (U), Westinghouse Savannah River Co., WSRC-RP-89-1085, 173 p.

Feinstein, D.T., and M.P. Anderson, 1987, Recharge to and potential for contamination of an aquifer system in northeastern Wisconsin, Wisconsin Water Resources Center, Technical Report 87-01, 112 p.

Fetter, C.W., 1988, *Applied Hydrogeology* (2nd edition), Merrill Publishing Co., Columbus, Ohio, 592 p.

Fipps, G., R.W. Skaggs, and J.L. Nieber, 1986, Drains as a boundary condition in finite elements, Water Resources Research 22(11), pp. 1613–1621.

Fogg, G.E., 1986, Groundwater flow and sand body interconnectedness in a thick, multiple-aquifer system, Water Resources Research 22(5), pp. 679–694.

Fogg, G.E., and R.K. Senger, 1985, Automatic generation of flow nets with conventional ground-water modeling algorithms, Ground Water 23(3), pp. 336–344.

Ford, D., and P. Williams, 1989, *Karst Geomorphology and Hydrology*, Unwin Hyman, London, 601 p.

Forster, C., and L. Smith, 1988, Groundwater flow systems in mountainous terrain. 1. Numerical modeling technique, pp. 999–1010; 2. Controlling factors, pp. 1011–1023, Water Resources Research 24(7).

Franke, O.L., and T.E. Reilly, 1987, The effects of boundary conditions on the steady-state response of three hypothetical ground-water systems; results and implications of numerical experiments, USGS, Water Supply Paper 2315, 19 p.

Franke, O.L., T.E. Reilly, and G.D. Bennett, 1987, Definition of boundary and initial conditions in the analysis of saturated ground-water flow systems; an introduction, USGS, Techniques of Water-Resources Investigations 03-B5, 15 p.

Franks, B.J., 1988, Hydrogeology and flow of water in a sand and gravel aquifer contaminated by wood-preserving compounds, Pensacola, Florida, USGS, Water-Resources Investigations Report 87-4260, 72 p.

Franz, T., and N. Guiguer, 1990, FLOWPATH, two-dimensional horizontal aquifer simulation model, Waterloo Hydrogeologic Software, Waterloo, Ontario, 74 p.

Freeze, R.A., 1969, The mechanism of natural groundwater recharge and discharge. 1. One-dimensional, vertical, unsteady, unsaturated flow above a recharging or discharging ground-water flow system, Water Resources Research 5(1), pp. 153–171.

Freeze, R.A., 1971, Three-dimensional, transient, saturated-unsaturated flow in a groundwater basin, Water Resources Research 7(2), pp. 347–366.

Freeze, R.A., 1975, A stochastic-conceptual analysis of one-dimensional groundwater flow in nonuniform homogeneous media, Water Resources Research 11, pp. 725–741.

Freeze, R.A., and J.A. Cherry, 1979, *Groundwater*, Prentice-Hall, 604 p.

Freeze, R.A., and P.A. Witherspoon, 1967, Theoretical analysis of regional groundwater flow: 2. Effect of water-table configuration and subsurface permeability variation, Water Resources Research 3, pp. 623–634.

Freeze, R.A., J. Massmann, L. Smith, T. Sperling, and B. James, 1990, Hydrogeological decision analysis: 1. A framework, Ground Water 28(5), pp. 738–766.

Freshley, M.D., and M.J. Graham, 1988, Estimation of ground-water travel time at the Hanford site: description, post work, and future needs, PNL-6328, UC-70, Battelle Memorial Institute, 57 p.

Freyberg, D.L., 1988, An exercise in ground-water model calibration and prediction, Ground Water 26(3), pp. 350–360.

Friedman, R., C. Ansell, S. Diamond, and Y.Y. Haimes, 1984, The use of models for water resources management, planning and policy, Water Resources Research 20(7), pp. 793–802.

Frind, E.O., 1979, Exact aquitard response functions for multiple aquifer mechanics, Advances in Water Resources 2, pp. 77–82.

Frind, E.O., 1982, Simulation of long-term transient density-dependent transport in ground water. Advances in Water Resources 5, pp. 73–88.

Frind, E.O., and B.B. Matanga, 1985, The dual formulation of flow for contaminant transport modeling, 1. Review of theory and accuracy aspects, Water Resources Research 21(2), pp. 159–169.

Frind, E.O., G.B. Matanga, and J.A. Cherry, 1985, The dual formulation of flow for contaminant transport modeling, 2. The Borden aquifer, Water Resources Research 21(2), pp. 170–182.

Frind, E.O., A. Sudicky, and S.L. Schellenberg, 1988, Micro-scale modelling in the study of plume evolution in heterogeneous media, *In: Groundwater Flow and Quality Modelling* (Custodio, E., A. Gurgui, J.P.L. Ferriera, eds.), D. Reidel Publishing Co., Boston, pp. 439–462.

Frind, E.O., and M.J. Verge, 1978, Three-dimensional modeling of groundwater flow systems, Water Resources Research 14(5), pp. 844–856.

Gale, J.E., 1982, Assessing the permeability characteristics of fractured rock, Geol. Soc. Amer. Spec. Paper 189, pp. 163–181.

Gambolati, G., F. Toffolo, and F. Uliana, 1984, Groundwater response under an electronuclear plant to a river flood wave analyzed by a nonlinear finite element model, Water Resources Research 20(7), pp. 903–913.

Garabedian, S.P., 1986, Application of a parameter-estimation technique to modeling the regional aquifer underlying the Eastern Snake River plain, Idaho, USGS, Water Resources Investigation Report 2278, 60 p.

Gelhar, L.W., 1986, Stochastic subsurface hydrology from theory to applications, Water Resources Research (Supplement) 22(9), pp. 135S–145S.

Gelhar, L.W., 1991, *Stochastic Subsurface Hydrology*, Prentice-Hall, in press.

Gelhar, L.W., A. Mantoglou, C. Welter, and K.R. Rehfeldt, 1985, A review of field-scale physical solute transport processes in saturated and unsaturated porous media, EPRI EA-4190, Electric Power Research Institute, Palo Alto, CA, 116 p.

GeoTrans, Incorporated, 1987, STLINE: Version 1.9 user's manual, GeoTrans, Inc., Herndon, VA, 29 p.

GeoTrans, Incorporated, 1988, SEFTRAN: A simple and efficient two-dimensional groundwater flow and transport model, Version 2.7, Herndon, VA, 142 p.

Gerhart, J.M., 1984, A model of regional ground-water flow in secondary-permeability terrane, Ground Water 22(2), pp. 168–175.

Gerhart, S.M., and G.J. Lazorchick, 1988, Evaluation of the ground-water resources of the Lower Susquehanna River Basin, Pennsylvania and Maryland, USGS, Water Supply Paper 2284, 128 p.

Gilham, R.W., and R.N. Farvolden, 1974, Sensitivity analysis of input parameters in numerical modeling of steady state regional groundwater flow, Water Resources Research 10(3), pp. 529–538.

Gingarten, A.C., 1982, Flow test evaluation of fractured reservoirs, Geol. Soc. of Amer. Spec. Paper 189, pp. 237–263.

Glover, K.C., 1987, A dual porosity model for simulating solute transport in oil shale, Water Resources Investigations Report 86-4047, 88 p.

Glover, K.C., 1988, A finite-element model for simulating hydraulic interchange of surface and ground water, USGS, Water-Resources Investigations Report 86-4319, 90 p.

Gomez-Hernandez, J.J., and S.M. Gorelick, 1989, Effective groundwater model parameter values: influence of spatial variability of hydraulic conductivity, leakance, and recharge, Water Resources Research 25(3), pp. 405–419.

Goode, D.J., 1990, Particle velocity interpolation in block-centered finite difference groundwater flow models, Water Resources Research 26(5), pp. 925–940.

Goode, D.J., and L.F. Konikow, 1989, Modification of a method-of-characteristics solute-transport model to incorporate decay and equilibrium-controlled sorption or ion exchange, USGS, Water-Resources Investigations Report 89-4030, 65 p.

Goode, D.J., and L.F. Konikow, 1990a, Reevaluation of large-scale dispersivities for a waste chloride plume: effects of transient flow, In: *Calibration and Reliability in Groundwater Modelling*, IAHS Publ. 195, pp. 417–426.

Goode, D.J., and L.F. Konikow, 1990b, Apparent dispersion in transient groundwater flow, Water Resources Research 26(10), pp. 2339–2352.

Gray, W.G., 1984, Comparison of finite difference and finite element methods, In: *Fundamentals of Transport Phenomena in Porous Media* (J. Bear and M.Y. Corapcioglu, eds.) Martinus Nijhoff, NATO ASI Series, Series F, No. 82, pp. 899–952.

Groschen, G.E., 1985, Simulated effects of projected pumping on the availability of freshwater in the Evangeline aquifer in an area southwest of Corpus Christi, Texas, USGS, Water Resources Investigation Report 85-4182, 103 p.

Grove, D.B., and K.G. Stollenwerk, 1987, Chemical reactions simulated by ground-water-quality models, Water Resources Bulletin 23(4), pp. 601–615.

Gupta, A.D., and P.R.O. Shrestha, 1986, Contaminant movement under pump-age-recharge condition in steady ground-water flow system, Ground Water 24(3), pp. 342–350.

Gupta, S.K., C.R. Cole, C.T. Kincaid, and A.M. Monti, 1987, Coupled fluid energy and solute transport (CFEST) model: formulation and user's manual, BMI/ONWI-660, Office of Nuclear Waste Isolation, Battelle Memorial Institute, Columbus, Ohio, 466 p.

Guswa, J.H., and D.R. Le Blanc, 1985, Digital models of ground-water flow in the Cape Cod aquifer system, Massachusetts, USGS, Water Supply Paper 2209, 64 p.

Hamilton, D.A., 1982, *Groundwater Modeling: Selection, Testing and Use*, Volume 1, Michigan Dept. of National Resources, 199 p.

Hamilton, P.A., and J.D. Larson, 1988, Hydrogeology and analysis of the ground-water flow system in the coastal plain of southeastern Virginia, USGS, Water-Resources Investigations Report 87-4240, 175 p.

Hardt, W.R., and J.R. Freckleton, 1987, Aquifer response to recharge and pumping, San Bernardino ground-water basin, California, USGS, Water-Resources Investigations Report 86-4140, 69 p.

Hardt, W.F., and C.B. Hutchinson, 1978, Model aids planners in predicting rising ground-water levels in San Bernadino, CA, Ground Water 16(6), pp. 424–431.

Harrison, B., E.A. Sudicky, and J.A. Cherry, 1991, Numerical analysis of solute migration through fractured clayey deposits in the underlying aquifers, Water Resources Research, in press.

Hearne, G.A., 1985, Mathematical model of the Tesuque aquifer system near Pojoaque, New Mexico, USGS, Water Supply Paper 2205, 75 p.

Heath, R.C., 1983, Basic ground-water hydrology, USGS, Water Supply Paper 2220, 84 p.

Helgesen, J.O., S.P. Larson, and A.C. Razem, 1982, Model modifications for simulation of flow through stratified rocks in eastern Ohio, U.S. Geol. Survey, Water-Resources Investigations Report 82-4019, 109 p.

Hemker, C.J., and H. van Elburg, 1987, Micro-Fem, Version 2.0 User's Manual, Hemker and van Elburg, Amsterdam, The Netherlands, 152 p.

Hendry, M.J., and G.D. Buckland, 1990, Causes of soil salinization. 1. A basin in southern Alberta, Canada, Ground Water 28(3), pp. 385–393.

Hendry, M.J., J.A. Cherry, and E.I. Wallich, 1986, Origin and distribution of sulfate in a fractured till in southern Alberta, Canada, Water Resources Research 22(1), pp. 45–61.

Hensel, B.R., R.C. Berg, and R.A. Griffin, 1990, Numerical estimates of potential for groundwater contamination from land burial of municipal wastes in Illinois, HWRIC RR035, Illinois Hazardous Waste Research and Information Center, 84 p.

Herbert, A.W., C.P. Jackson, and D.A. Lever, 1988, Coupled groundwater flow and solute transport with fluid density strongly dependent upon concentration, Water Resources Research 24(10), pp. 1781–1795.

Herzog, B.L., and W.J. Morse, 1984, A comparison of laboratory and field determined values of hydraulic conductivity at a waste disposal site, Proceedings of the Seventh Annual Madison Waste Conference, pp. 30–52.

Hickey, J.J., 1989, Circular convection during subsurface injection of liquid waste. St. Petersburg, Florida, Water Resources Research 25(7), pp. 1481–1495.

Hill, M.C., 1990a, MODFLOWP: A computer program for estimating parameters of a transient, three-dimensional ground-water flow model using nonlinear regression. USGS, Open-File Report, 317 p.

Hill, M.C., 1990b, Preconditioned conjugate-gradient 2 (PCG2), a computer program for solving ground-water flow equations. USGS, Water-Resources Investigations Report 90-4048, 43 p.

Hillman, G.R., and J.P. Verscharen, 1988, Simulation of the effects of forest cover, and its removal, on subsurface water, Water Resources Research 24(2), pp. 305–314.

Hochmuth, D.P., and D.K. Sunada, 1985, Ground-water model of two-phase immiscible flow in coarse material, Ground Water 23(5), pp. 617–626.

Holmes, K.J., W. Chu, and D.R. Erickson, 1989, Automated calibration of a contaminant transport model for a shallow sand aquifer, Ground Water 27(4), pp. 501–508.

Hopmans, J.W., and E. Guttierez-Rave, 1988, Calibration of a root water uptake model in spatially variable soils, Journal of Hydrology 103, pp. 53–65.

Hornberger, G.M.J., J. Ebert, and I. Remson, 1970, Numerical solution of the Boussinesq equation for aquifer-stream interaction, Water Resources Research 6(2), pp. 601–608.

Hossain, M.A., and M.Y. Corapcioglu, 1988, Modifying the USGS solute transport computer model to predict high-density hydrocarbon migration, Ground Water 26(6), pp. 717–723.

Hostetler, C.J., R.L. Erikson, J.S. Fruchter, and C.T. Kincaid, 1988, Overview of FASTCHEM™ code package: application to chemical transport problems, EQ-5870-CCM, vol. 1, Electric Power Research Institute, Palo Alto, CA.

Hotchkiss, W.R., and J.F. Leavings, 1986, Hydrogeology and simulation of water flow in strata above the Bearpaw Shale and equivalents of eastern Montana and northeastern Wyoming, USGS, Water-Resources Investigations Report 85-4281, 72 p.

Hsieh, P.A., S.P. Neuman, G.K. Stiles, and E.S. Simpson, 1985, Field determination of three-dimensional hydraulic conductivity tensor of anisotropic media. 2. Methodology and application to fractured rocks, Water Resources Research 21(11), pp. 1667–1676.

Huyakorn, P.S., and G.F. Pinder, 1983, *Computational Methods in Subsurface Flow*, Academic Press, 473 p.

Huyakorn, P.S., B.H. Lester, and C.R. Faust, 1983, Finite element techniques for modeling groundwater flow in fractured aquifers, Water Resources Research 19(4), pp. 1019–1035.

Huyakorn, P.S., M.J. Ungs, L.A. Mulkey, and E.A. Sudicky, 1987, A three-dimensional analytical method for predicting leachate migration, Ground Water 25(5), pp. 588–598.

Hvorslev, M.J., 1951, Time-lag and soil permeability in groundwater observations, Waterways Experiment Station Bulletin 36, 50 p.

HydroGeoLogic, Inc., 1989, EPACMS: Composite model for simulating leachate migration from surface impoundments and Monte Carlo uncertainty analysis. Prepared for U.S. EPA, Office of Solid Waste, 144 p.

Istok, J., 1989, *Groundwater Modeling by the Finite Element Method*, Amer. Geophy. Union, Water Resources Monograph 13, 495 p.

Jackson, R.E., and K.J. Inch, 1989, The in-situ adsorption of ^{90}Sr in a sand aquifer at the Chalk River Nuclear Laboratories, Journal of Contaminant Hydrology 4(1), pp. 27–50.

Jamieson, G.R., and R.A. Freeze, 1983, Determining hydraulic conductivity distributions in a mountainous area using mathematical modeling, Ground Water 21(2), pp. 168–177.

Javandel, I., C. Doughty, and C.F. Tsang, 1984, *Groundwater Transport: Handbook of Mathematical Models*, Water Resources Monograph 10, Amer. Geophy. Union, Washington, D.C., 228 p.

Johnson, M.J., D.J. Londquist, J. Laudon, and J.T. Mitten, 1988, Geohydrology and mathematical simulation of the Pajaro Valley aquifer system, Santa Cruz and Monterey Counties, California, USGS, Water-Resources Investigations Report 87-4281, 62 p.

Jorgensen, D.G., 1981, Geohydrologic models of the Houston District, Texas, Ground Water 19(4), pp. 418–428.

Jorgensen, D.G., 1989a, Accounting for intracell flow in models with emphasis on water-table recharge and stream-aquifer interaction—Part 1, problems and concepts, Water Resources Research 25(4), pp. 669–676.

Jorgensen, D.G., 1989b, Accounting for intracell flow in models with emphasis on water-table recharge and stream-aquifer interaction—Part 2, a procedure, Water Resources Research 25(4), pp. 677–684.

Journel, A.G., and C.J. Huijbregts, 1978, *Mining Geostatistics*, Academic Press, 600 p.

Kaluarachchi, J.J., and J.C. Parker, 1989, An efficient finite element method for modeling multiphase flow, Water Resources Research 23, pp. 43–54.

Kaluarachchi, J.J., and J.C. Parker, 1990, Modeling multicomponent organic chemical transport in three-fluid-phase porous media, Journal of Contaminant Hydrology 5(4), pp. 349–374.

Karanjac, J., M. Altankaynak, and G. Oval, 1977, Mathematical model of Uluova Plain, Turkey—a training and management tool, Ground Water 15(5), pp. 348–357.

Kauffmann, C., W. Kinzelbach, and J.J. Fried, 1990, Simultaneous calibration of flow and transport models and optimization of remediation measures, *In: Calibration and Reliability in Groundwater Modelling* (K. Kovar, ed.), IAHS Publ. 195, pp. 159–170.

Keidser, A. D., Rosbjerg, K. Hogh Jensen, and K. Bitsch, 1990, A joint kriging and zonation approach to inverse groundwater modeling, *In: Calibration and Reliability in Groundwater Modelling* (K. Kovar, ed.), IAHS Publ. 195, pp. 171–184.

Keller, C.K., G. van der Kamp, and J.A. Cherry, 1989, A multiscale study of the permeability of a thick clayey till, Water Resources Research 25(11), pp. 2299–2318.

Kendy, E., and K.R. Bradbury, 1988, Hydrogeology of the Wisconsin River Valley in Marathon County, Wisconsin, Wisconsin Geological and Natural History Survey, Information Circular 64, 66 p.

Kincaid, C.T., 1988, FASTCHEM™ package, V.3: User's guide to the ETUBE pathline and streamtube database code, EPRI EA-5870-CCM, Electric Power Research Institute.

Kinzelbach, W., 1986, *Groundwater Modelling: An Introduction with Sample Programs in BASIC*, Developments in Water Science, 25, Elsevier, 334 p.

Kipp, K.L., Jr., 1987, HST3D; a computer code for simulation of heat and solute transport in three-dimensional ground-water flow systems, USGS, Water Resources Investigations Report 86-4095, 517 p.

Knott, J.T., and J.C. Olimpio, 1986, Estimation of recharge rates to the sand and gravel aquifer using environmental tritium, Nantucket Island, Massachusetts, USGS, Water Supply Paper 2297, 26 p.

Koch, D.L., and T.A. Prickett, 1989, User's manual for RD3D, a three-dimensional mass transport random walk model attachment to the USGS MODFLOW three-dimensional flow model, Joint Engineering Technologies Assoc. and T.A. Prickett & Assoc. Scientific Publ. 3, Ellicott City, MD, 102 p.

Konikow, L.F., 1977, Modeling chloride movements in the alluvial aquifer at the Rocky Mountain Arsenal, Colorado, USGS, Water Supply Paper 2044, 43 p.

Konikow, L.F., 1978, Calibration of ground-water models, In: *Verification of Mathematical and Physical Models in Hydraulic Engineering*, American Society of Civil Engineers, N.Y., pp. 87–93.

Konikow, L.F., 1986, Predictive accuracy of a ground-water model—lessons from a postaudit, Ground Water 24(2), pp. 173–184.

Konikow, L.F., and J.D. Bredehoeft, 1974, Modeling flow and chemical quality changes in an irrigated stream-aquifer system, Water Resources Research 10(3), pp. 546–562.

Konikow, L.F., and J.D. Bredehoeft, 1978, Computer model of two-dimensional solute transport and dispersion in ground water, Techniques of Water-Resources Investigations Book 7, Chap. C2, USGS, 90 p.

Konikow, L.F., and M. Person, 1985, Assessment of long-term salinity changes in an irrigated stream-aquifer system, Water Resources Research 21(11), pp. 1611–1624.

Kontis, A.L., and R.J. Mandle, 1988, Modification of a three-dimensional ground-water flow model to account for variable water density and effects of multiaquifer wells, USGS, Water-Resources Investigations Report 87–4265, 78 p.

Krabbenhoft, D.P., and M.P. Anderson, 1986, Use of a groundwater model for hypothesis testing, Ground Water 24(1), pp. 49–55.

Krabbenhoft, D.P., M.P. Anderson, and C.J. Bowser, 1990, Estimating ground-water exchange with lakes. 2. Calibration of a three-dimensional, solute transport model to a stable isotope plume, Water Resources Research 26(10), pp. 2455–2462.

Krishnamurthi, N., D.K. Sunada, and R.A. Longenbaugh, 1977, Mathematical modeling of natural groundwater recharge, Water Resources Research 13(4), pp. 720–724.

Krohelski, J.T., 1986, Hydrogeology and ground-water use and quality, Brown County, Wisconsin, Wisconsin Geological and Natural History Survey, Information Circular 57, 42 p.

Kueper, B.H., and E.O. Frind, 1991a, The behavior of dense, nonaqueous phase liquids in heterogeneous porous media. 1. Model development, verification and validation, Water Resources Research 27(6), pp. 1049–1057.

Kueper, B.H., and E.O. Frind, 1991b, The behavior of dense, nonaqueous phase liquids in heterogeneous porous media. 2. Parameter measurement and model application in a spatially correlated random field, Water Resources Research 27(6), pp. 1059–1070.

Kuiper, L.K., 1983, A numerical procedure for the solution of the steady state variable density ground water flow equation, Water Resources Research 19(1), pp. 234–240.

Kuiper, L.K., 1985, Documentation of a numerical code for the simulation of variable density ground-water flow in three dimensions, USGS, Water-Resources Investigations Report 84-4302, 98 p.

Kuppusamy, T., J. Sheng, J.C. Parker, and R.J. Lenhard, 1987, Finite-element analysis of multiphase immiscible flow through soils, Water Resources Research 23(4), pp. 625–631.

Land, L.F., 1977, Utilizing a digital model to determine the hydraulic properties of a layered aquifer, Ground Water 15(2), pp. 153–159.

Lapham, W.W., 1989, Use of temperature profiles beneath streams to determine rates of vertical ground-water flow and vertical hydraulic conductivity, USGS, Water Supply Paper 2337, 35 p.

Lappala, E.G., R.W. Healy, and E.P. Weeks, 1987, Documentation of computer program VS2D to solve the equations of fluid flow in variably saturated porous media, USGS, Water-Resources Investigations Report 83-4099, 184 p.

Leahy, P.P., 1982, A three-dimensional ground-water-flow model modified to reduce computer memory requirements and better simulate confining-bed and aquifer pinchouts, USGS, Water-Resources Investigations Report 82-4023, 59 p.

Leake, S.A., 1977, Simulation of flow from an aquifer to a partially penetrating trench, USGS, Journal of Research 5(5), pp. 535–540.

Leake, S.A., 1990, Interbed storage changes and compaction in models of regional groundwater flow, Water Resources Research 26(9), pp. 1939–1950.

Lee, D.D., 1977, A device for measuring seepage flux in lakes and estuaries, Limnology and Oceanography 2, pp. 140–147.

Lee, R.W., and D.J. Strickland, 1988, Geochemistry of groundwater in Tertiary and Cretaceous sediments of the South Eastern Coastal Plain in eastern Georgia, South Carolina and south eastern North Carolina, Water Resources Research 24(2), pp. 291–304.

Lefkoff, L.J., and S.M. Gorelick, 1987, AQMAN: Linear and quadratic programming matrix generator using two-dimensional ground-water flow simulation for aquifer management modeling, U.S. Geol. Surv., Water Resources Investigations Report 87-4061, 164 p.

Leismann, H.M., and E.O. Frind, 1989, A symmetric-matrix time integration scheme for the efficient solution of advection-dispersion problems, Water Resources Research 25(6), pp. 1133–1140.

Lewis, B.D., and F.S. Goldstein, 1982, Evaluation of a predictive ground-water solute-transport model at the Idaho National Engineering Laboratory, Idaho, USGS, Water-Resources Investigations Report 82–85, 71 p.

Lewis, F.M., C.I. Voss, and J. Rubin, 1986, Numerical simulation of advective-dispersive multisolute transport with sorption, ion exchange and equilibrium chemistry, USGS, Water-Resources Investigations Report 86-4022, 165 p.

Liggett, J.A., 1987, Advances in the boundary integral equation method in subsurface flow, Water Resources Bull. 23(4), pp. 637–651.

Liggett, J.A., and P.L.-F. Liu, 1983, *The Boundary Integral Equation Method for Porous Media Flow*, Allen and Unwin, 255 p.

Lin, C.L., 1972, Digital simulation of the Boussinesq equation for a water table aquifer, Water Resources Research 8(3), pp. 691–698.

Lindholm, G.F., 1986, Smoke River Plain regional aquifer-system study, *In:* Regional Aquifer-System Analysis Program of the U.S. Geological Survey: Summary of Projects 1978–84 (R.J. Sun, ed.), USGS Circular 1002, pp. 88–106.

Lines, G.C., 1976, Digital model to predict effects of pumping from the Arikaree aquifer in the Dwyer area, southeastern Wyoming. USGS, Water-Resources Investigations/Open-File Report 8-76, 24 p.

Long, J.C., and D.M. Billaux, 1987, From field data to fracture network modeling: an example incorporating spatial structure, Water Resources Research 23(7), pp. 1201–1216.

Long, J.C., P. Bilmour, and P.A. Witherspoon, 1985, A model for steady fluid flow in random three-dimensional networks of disc-shaped fractures, Water Resources Research 21(8), pp. 1105–1115.

Long, J.C., J.S. Remer, C.R. Wilson, and P.A. Witherspooon, 1982, Porous media equivalents for networks of discontinuous fractures, Water Resources Research 18(3), pp. 645–658.

Luckey, R.R., and D.M. Stephens, 1987, Effect of grid size on digital simulation of ground-water flow in the southern High Plains of Texas and New Mexico, USGS, Water-Resources Investigations Report 87-4085, 32 p.

Luckey, R.R., E.D. Gutentag, F.J. Heimes, and J.B. Weeks, 1986, Digital simulation of ground-water flow in the High Plains aquifer in parts of Colorado, Kansas, Nebraska, New Mexico, Oklahoma, South Dakota, Texas, and Wyoming, USGS, Professional Paper 1400-D, 57 p.

Luckey, R.R., E.D. Gutentag, F.J. Heimes, and J.B. Weeks, 1988, Effects of future ground-water pumpage on the High Plains aquifer in parts of Colorado, Kansas, Nebraska, New Mexico, Oklahoma, South Dakota, Texas, and Wyoming, USGS, Professional Paper 1400-E, pp. E1–E44.

McBride, M.S., and H.O. Pfannkuch, 1975, The distribution of seepage within lakebeds, USGS, Journal of Research 3(5), pp. 505–512.

McDonald, M.G., 1984, Development of a multi-aquifer well option for a modular ground water flow model, Proc. Practical Applications of Ground Water Models, National Water Well Assoc., Dublin, Ohio, pp. 786–796.

McDonald, M.G., and A.W. Harbaugh, 1988, A modular three-dimensional finite-difference ground-water flow model, Techniques of Water-Resources Investigations 06-A1, USGS, 576 p.

McLaughlin, D., and W.K. Johnson, 1987, Comparison of three groundwater modeling studies, Journal of Water Resources Planning and Management 113(3), pp. 405–421.

McLeod, R.S., 1975, A digital-computer model for estimating hydrologic changes in the aquifer system in Dane County, Wisconsin, Wisconsin Geologic and Natural History Survey, Information Circular 30, 40 p.

Maclay, R.W., and L.F. Land, 1988, Simulation of flow in the Edwards aquifer, San Antonio region, Texas, and refinement of storage and flow concepts, USGS, Water Supply Paper 2336-A, pp. A1–A48.

Makepeace, S.V., 1989, Simulation of groundwater flow in a coarse-grained alluvial aquifer in the Jocko Valley, Flathead Indian Reservation, Montana, MS thesis, University of Montana, 155 p.

Mandle, R.J., and A.L. Kontis, 1986, Directions and rates of ground-water movement in the vicinity of Kesterson Reservoir, San Joaquin Valley, California, USGS, Water-Resources Investigations Report 86-4196, 57 p.

Mangold, D.C., and C.-F. Tsang, 1991, A summary of subsurface hydrological and hydrochemical models, Reviews of Geophysics 29(1), pp. 51–79.

Mantoglou, A., and J. Wilson, 1982, The turning bands method for simulation of random fields using line generation by a spectral method, Water Resources Research 18, pp. 1379–1394.

Martin, M.M., 1984, Simulated ground-water flow in the Potomac aquifers, New Castle County, Delaware, USGS, Water-Resources Investigations Report 84-4007, 85 p.

Masch, F.D., and K.L. Denny, 1966, Grain size distribution and its effect on the permeability of unconsolidated sands, Water Resources Research 2(4), pp. 665–677.

Maslia, M.L., and L.R. Hayes, 1988, Hydrogeology and simulated effects of ground-water development of the Floridan aquifer system, southwest Georgia, northwest Flordia, and southernmost Alabama, USGS, Professional Paper 1403-H, pp. H1–H71.

Maslia, M.L., and R.H. Johnston, 1984, Use of a digital model to evaluate hydrogeologic controls on groundwater flow in a fractured rock aquifer at Niagara Falls, New York, USA, Journal of Hydrology 75, pp. 167–194.

Maslia, M.L., and R.B. Randolph, 1987, Methods and computer program documentation for determining anisotropic transmissivity tensor components of two-dimensional ground-water flow, USGS, Water Supply Paper 2308, 46 p.

Massmann, J., and R.A. Freeze, 1989, Updating random hydraulic conductivity fields: a two-step procedure, Water Resources Research 25(7), pp. 1763–1765.

Maxey, G.B., 1964, Hydrostratigraphic units, Journal of Hydrology 2, pp. 124–129.

Medina, A., J. Carrera, and G. Galarza, 1990, Inverse modelling of coupled flow and solute transport problems, In: Calibration and Reliability in Groundwater Modelling (K. Kovar, ed.), IAHS Publ. 195, pp. 185–194.

Mendoza, C.A., and E.O. Frind, 1990, Advective-dispersive transport of dense organic vapors in the unsaturated zone. 1. Model development, Water Resources Research 26(3), pp. 379–387.

Mendoza, C.A., and E.O. Frind, 1990, Advective-dispersive transport of dense organic vapors in the unsaturated zone. 2. Sensitivity analysis, Water Resources Research 26(3), pp. 388–397.

Meng, A.A., III, and J.F. Harsh, 1988, Hydrogeologic framework of the Virginia Coastal Plain, USGS, Professional Paper 1404-C, pp. C1–C82.

Menke, W., 1989, Geophysical Data Analysis: Discrete Inverse Theory, Academic Press, Inc., 289 p.

Mercer, J.W., S.D. Thomas, and B. Ross, 1982, Parameters and variables appearing in repository siting models, NUREG/CR-3066, U.S. Nuclear Regulatory Commission, 244 p.

Mercer, J.W., L.R. Silka, and C.R. Faust, 1983, Modeling ground-water flow at Love Canal, New York, Journal of Environmental Engineering 109(4), pp. 924–942.

Miller, D., and L. Benson, 1983, Simulation of solute transport in a chemically reactive heterogeneous system: model development and application, Water Resources Research 19(2), pp. 381–391.

Miller, R.T., and C.I. Voss, 1987, Finite-difference grid for a doublet well in an anisotropic aquifer, Ground Water 24(4), pp. 490–496.

Mitten, H.T., G.C. Lines, C. Berenbrock, and T.J. Durbin, 1988, Water resources of Borrego Valley and vicinity, San Diego County, California: Phase 2—Development of a ground-water flow model, USGS, Water-Resources Investigations Report 87-4199, 27 p.

Moore, J.E., 1979, Contributions of ground-water modeling to planning, Journal of Hydrology 43, pp. 121–128.

Morgan, D.S., 1988, Geohydrology and numerical model analysis of ground-water flow in the Goose Lake Basin, Oregon and California, USGS, Water-Resources Investigations Report 87-4058, 92 p.

Morris, D.A., and Johnson, A.I., 1967, Summary of hydrologic and physical properties of rock and soil materials as analyzed by the Hydrologic Laboratory of the U.S. Geological Survey 1948–1960, USGS, Water Supply Paper 1839-D.

Morrissey, D.J., 1989, Estimation of the recharge area contributing water to a pumped well in a glacial-drift, river-valley aquifer, USGS, Water Supply Paper 2338, 41 p.

Muldoon, M.A., K.R. Bradbury, D.M. Mickelson, and J.W. Attig, 1988, Hydrogeologic and geotechnical properties of Pleistocene materials in north-central Wisconsin, Wisconsin Water Resources Center, Technical Report 88-03, 58 p.

Narasimhan, T.N., and P.A. Witherspoon, 1978, Numerical model for saturated-unsaturated flow in deformable porous media. 3. Applications, Water Resources Research 14(6), pp. 1017–1034.

National Research Council, 1990, Ground Water Models: Scientific and Regulatory Applications, National Academy Press, 303 p.

Nelson, R.W., 1978, Evaluating the environmental consequences of ground-water contamination, Parts 1–4, Water Resources Research 14(3), pp. 409–450.

Neuman, S.P., 1973, Saturated-unsaturated seepage by finite elements, American Society of Civil Engineers, Journal of the Hydraulics Division 99(12), pp. 2233–2250.

Neuman, S.P., 1975, Analysis of pumping test data from anisotropic unconfined aquifers considering delayed gravity response, Water Resources Research 11(2), pp. 329–342.

Neuman, S.P., 1976, User's guide for FREESURF I, Dept. of Hydrology and Water Resources, Univ. of Arizona, Tuscon, Arizona, 22 p.

Neuman, S.P., 1979, Perspective on "delayed yield," Water Resources Research 15(4), pp. 899–908.

Neuman, S.P., 1980, Adjoint-state finite element equations for parameter estimation, 3rd International Conference on Finite Elements in Water Resources, University of Mississippi, Oxford.

Neuman, S.P., and E.L. Jacobson, 1984, Analysis of non-intrinsic spatial variability by residual kriging with application to regional groundwater levels, Mathematical Geology 16, pp. 499–521.

Neuman, S.P., G.E. Fogg, and E.A. Jacobson, 1980, A statistical approach to the inverse problem of aquifer hydrology. 2. Case study, Water Resources Research 16(1), pp. 32–58.

Neuman, S.P., G.R. Walter, H.W. Bentley, J.J. Ward, and D.D. Gonzalez, 1984, Determination of horizontal aquifer anisotropy with three wells, Ground Water 2(1), pp. 66–72.

Neuman, S.P., and P.A. Witherspoon, 1969, Theory of flow in a confined two aquifer system, Water Resources Research 5(4), pp. 803–816.

Neuman, S.P., and P.A. Witherspoon, 1970, Finite element method of analyzing steady seepage with a free surface, Water Resources Research 6(3), pp. 889–897.

Neuman, S.P., and P.A. Witherspoon, 1971, Analysis of nonsteady flow with a free surface using the finite element method, Water Resources Research 7(3), pp. 611–623.

Newsom, J.M., and J.L. Wilson, 1988, Flow of ground water to a well near a stream—effect of ambient ground-water flow direction, Ground Water 26(6), pp. 703–711.

Oberlander, P.L., and R.W. Nelson, 1984, An idealized ground-water flow and chemical transport model (S-PATHS), Ground Water 22(4), pp. 441–449.

Olsthoorn, T.N., 1985, The power of the electronic worksheet: modeling without special programs, Ground Water 23(3), pp. 381–390.

Ophori, D.U., and J. Toth, 1989, Characterization of ground-water flow by field mapping and numerical simulation, Ross Creek Basin, Alberta, Canada, Ground Water 27(2), pp. 193–201.

Osborne, M., and J. Sykes, 1986, Numerical modeling of immiscible organic transport at the Hyde Park landfill, Water Resources Research 22(1), pp. 25–33.

Ousey, J.R., Jr., 1986, Modeling steady-state groundwater flow using microcomputer spreadsheets, Journal of Geological Education, 34, pp. 305–311.

Ozbilgin, M.M., and D.C. Dickerman, 1984, A modification of the finite-difference model for simulation of two dimensional ground-water flow to include surface–ground water relationships, USGS, Water-Resources Investigations Report 83-4251, 98 p.

Papadopulos, I.S., 1966, Nonsteady flow to multiaquifer wells, Journal of Geophysical Research 71(20), pp. 4791–4797.

Parker, J.C., 1989, Multiphase flow and transport in porous media, Reviews of Geophysics 27(3), pp. 311–328.

Patrick, L.D., T.P. Brabets, R.L. Glass, 1989, Simulation of ground-water flow at Anchorage, Alaska, 1955–83, USGS, Water-Resources Investigations Report 88-4139, 41 p.

Peck, A., S.M. Gorelick, G. de Marsily, S. Foster, and V. Kovalevsky, 1988, Consequences of spatial variability in aquifer properties and data limitation for groundwater modelling practice, International Association of Hydrological Sciences, Publication 175, 272 p.

Pennell, K.D., A.G. Hornsby, R.E. Jessup, and P.S.C. Rao, 1990, Evaluation of five simulation models for predicting aldicarb and bromide behavior under field conditions, Water Resources Research 26(11), pp. 2679–2694.

Pennequin, D.F.E., 1983, Ground-water flow model determinations of areal hydrology and geology, Ground Water 21(5), pp. 552–557.

Peralta, R.C., P.W. Dutram, A.W. Peralta, and A. Yazdanian, 1986, Saturated thickness for drought and litigation protection, Ground Water 24(3), pp. 357–364.

Pernik, M., 1987, Sensitivity analysis of multilayer, finite-difference model of the Southeastern Coastal Plain regional aquifer system: Mississippi, Alabama, Georgia, and South Carolina, USGS, Water-Resources Investigations Report 87-4108, 53 p.

Person, M., and L.F. Konikow, 1986, Recalibration and predictive reliability of a solute transport model of an irrigated stream-aquifer system, Journal of Hydrology 87, pp. 145–165.

Peters, J.G., 1987, Description and comparison of selected models for hydrologic analysis of ground-water flow, St. Joseph River Basin, Indiana, USGS, Water-Resources Investigations Report 86-4199, 125 p.

Philip, J.R., 1980, Field heterogeneity: some basic issues, Water Resources Research 16(2), pp. 443–448.

Phillips, S.W., 1987, Hydrogeology, degradation of ground-water quality, and simulation of infiltration from the Delaware River into the Potomac aquifers, northern Delaware, USGS, Water-Resources Investigations Report 87-4185, 86 p.

Pickens, J.F., and R.W. Gillham, 1980, Finite element analysis of solute transport under hysteretic unsaturated flow conditions, Water Resources Research 16(6), pp. 1071–1078.

Piggott, A.R., and D. Elsworth, 1989, Physical and numerical studies of a fracture system model, Water Resources Research 25(3), pp. 457–462.

Pikul M.F., R.L. Street, and I. Remson, 1974, A numerical model based on coupled one-dimensional Richards and Boussinesq equations, Water Resources Research 10(2), pp. 295–302.

Pinder, G.F., and L.M. Abriola, 1986, On the simulation of nonaqueous phase organic compounds in the subsurface, Water Resources Research 22(9), pp. 1095–1195.

Pinder, G.F., and W.G. Gray, 1976, Is there a difference in the finite element method? Water Resources Research 12(1), pp. 105–107.

Pinder, G.F., and W.G. Gray, 1977, *Finite Element Simulation in Surface and Subsurface Hydrology*, Academic Press, 295 p.

Pinder, G.F., and C. Voss, 1979, AQUIFEM, a finite element model for aquifer evaluation (documentation), Dept. of Water Resources Engineering, Royal Institute of Technology, Stockholm, Sweden, Report 7911; TRITA-VAT-3806.

Pollock, D.W., 1988, Semianalytical computation of path lines for finite-difference models, Ground Water 26(6), pp. 743–750.

Pollock, D.W., 1989, Documentation of computer programs to complete and display pathlines using results from the U.S. Geological Survey modular three-dimensional finite-difference ground-water model, USGS, Open File Report 89-381, 81 p.

Potter, S.T., and W.J. Gburek, 1986, Simulation of the seepage face—limitations of a one-dimensional approach, Journal of Hydrology 87, pp. 379–394.

Potter, S.T., and W.J. Gburek, 1987, Seepage face simulation using PLASM, Ground Water 25(6), pp. 722–732.

Prickett, T.A., 1967, Designing pumped well characteristics into electric analog models, Ground Water 5, pp. 38–46.

Prickett, T.A., 1979, Ground-water computer models—state of the art, Ground Water 17(2), pp. 167–173.

Prickett, T.A., and C.G. Lonnquist, 1971, Selected Digital Computer Techniques for Groundwater Resource Evaluation, Illinois State Water Survey, Bulletin 55, 62 p.

Prickett, T.A., T.G. Naymik, and C.G. Lonnquist, 1981, A "Random-Walk" Solute Transport Model for Selected Groundwater Quality Evaluations, Illinois State Water Survey, Bulletin 65, 103 p.

Prince, K.R., O.L. Franke, and T.E. Reilly, 1988, Quantitative assessment of the shallow ground-water flow system associated with Connetquot Brook, Long Island, New York, USGS, Water Supply Paper 2309, 28 p.

Pritchett, J.W., and S.K. Garg, 1980, Determination of effective well block radii for numerical reservoir simulations, Water Resources Research 16(4), pp. 665–674.

Prudic, D.E., 1989, Documentation of a computer program to simulate stream-aquifer relations using a modular, finite difference ground-water flow model, USGS, Open-File Report 88-729, 113 p.

Pucci, A.A., and J.E. Murashige, 1987, Applications of universal kriging to an aquifer study in New Jersey, Ground Water 25(6), pp. 672–678.

Pulido-Bosch, A., and A. Padilla, 1988, Some considerations about the simulation of karstic aquifers In: Karst Hydrogeology and Karst Environment Protection, Proc. IAH 21st Congress, Guilin, China, Geological Publ. House, Beijing, pp. 583–588.

Quinones-Aponte, V., 1989, Horizontal anisotropy of the principal ground-water flow zone in the Salinas alluvial fan, Puerto Rico, Ground Water 27(4), pp. 491–500.

Rao, B.K., and D.L. Hathaway, 1989, A three-dimensional mixing cell solute transport model and its application, Ground Water 27(4), pp. 509–516.

Reeves, M., D.S. Ward, N.D. Johns, and R.M. Cranwell, 1986a, Theory and Implementation for SWIFT II, the Sandia Waste-Isolation Flow and Transport Model for Fractured Media, Release 4.84, NUREG/CR-3328, SAND83-1159, Sandia National Laboratories, NM, 189 p.

Reeves, M., D.S. Ward, N.D. Johns, and R.M. Cranwell, 1986b, Data input guide for SWIFT II, the Sandia Waste-Isolation Flow and Transport Model for Fractured Media, Release 4.84, NUREG/CR-3162, SAND 83-0242, Sandia National Laboratories, NM, 146 p.

Rehfeldt, K.R., 1988, Prediction of macrodispersivity in heterogeneous aquifers, Ph.D. thesis, Dept. of Civil Engineering, MIT, 233 p.

Reilly, T.E., 1984, A Galerkin finite-element flow model to predict the transient response of a radially symmetric aquifer, USGS, Water Supply Paper 2198, 33 p.

Remson, I., S.M. Gorelick, and J.F. Fliegner, 1980, Computer models in groundwater exploration, Ground Water 18(5), pp. 447–451.

Remson, I., G.M. Hornberger, and F.J. Molz, 1971, *Numerical Methods in Subsurface Hydrology*, Wiley-Interscience, 389 p.

Reynolds, R.J., 1987, Hydrogeology of the surficial outwash aquifer at Cortland, Cortland County, New York, USGS, Water Resources Investigation Report 85-4090, 43 p.

Risser, D.W., 1988, Simulated water-level and water-quality changes in the Bolson-Fill aquifer, post headquarters area, White Sands Missile Range, New Mexico, U.S. Geol. Survey, Water Resources Investigation Report 87-4152, 71 p.

Robertson, J.B., 1974, Digital modeling of radioactive and chemical waste transport in the Snake River Plain aquifer at the National Reactor Testing Station, Idaho, USGS, Open File Report ID0–22054, 41 p.

Robertson, W.D., and J.A. Cherry, 1989, Tritium as an indicator of recharge and dispersion in a groundwater system in central Ontario, Water Resources Research 25(6), pp. 1097–1110.

Robertson, W.D., J.A. Cherry, and S.L. Schiff, 1989, Atmospheric sulfur deposition 1950–1985 inferred from sulfate in groundwater, Water Resources Research 25(6), pp. 1111–1124.

Robson, S.G., 1974, Feasibility of digital water quality modeling illustrated by application at Barstow, CA, USGS, Water-Resources Investigations Report 46-73, 66 p.

Robson, S.G., 1978, Application of digital profile modeling techniques to ground-water solute transport at Barstow, California, USGS, Water Supply Paper 2050, 28 p.

Rogers, P.P., 1983, Book review of: Analyzing natural systems: analysis for regional residuals—environmental quality management, EOS 64(25), p. 419.

Rumbaugh, J.O., III, and G.M. Duffield, 1989, Model Cad™: Computer-aided design for ground-water modeling, Version 1.0, Geraghty & Miller, Inc., Reston, VA., 52 p.

Rushton, K.R., and S.C. Redshaw, 1979, *Seepage and Groundwater Flow*, John Wiley and Sons, N.Y., 339 p.

Rushton, K.R., and L.A. Wedderburn, 1973, Starting conditions for aquifer simulations, Ground Water 11(1), pp. 37–42.

Sampler, J., J. Carrera, G. Galarza, and A. Medina, 1990, Application of an automatic calibration technique to modelling an alluvial aquifer, *In: Calibration and Reliability in Groundwater Modelling* (K. Kovar, ed.), IAHS Publ. 195, pp. 87–96.

Sanford, W.E., and L.F. Konikow, 1985, A two-constituent solute-transport model for ground water having variable density, USGS, Techniques of Water-Resources Investigations 85-4279, 88 p.

Sanford, W.E., and L.F. Konikow, 1989, Simulation of calcite dissolution and porosity changes in salt water mixing zones in coastal aquifers, Water Resources Research 25(4), pp. 655–667.

Sauveplane, C., 1984, Pumping test analysis in fractured aquifer formations: state of the art and some perspectives, *In: Groundwater Hydraulics* (J. Rosensheim and G.D. Bennett, eds.), American Geophysical Union, Washington, D.C., pp. 171–206.

Schafer-Perini, A., J.L. Wilson, M. Perini, 1991, Efficient and accurate front tracking for two-dimensional groundwater flow models. Water Resources Research 27, in press.

Schmelling, S.G., and R.R. Ross, 1989, Contaminant transport in fractured media: models for decision makers, EPA/540/4-89/004, 8 p.

Schwartz, F.W., and L. Smith, 1988, A continuum approach for modeling mass transport in fractured media, Water Resources Research 24(8) pp. 1360–1372.

Seaber, P.R., 1988, Hydrostratigraphic units, *In: Hydrogeology* (W. Back, J.S. Rosenshein, and P.R. Seaber, eds.), *The Geology of North America*, v. 0-2, Geol. Soc. Amer., pp. 9–14.

Senger, R.K., and G.E. Fogg, 1987, Regional underpressuring in deep brine aquifers, Palo Duro Basin, Texas. 1. Effects of hydrostratigraphy and topography, Water Resources Research 23(8), pp. 1481–1493.

Senger, R.K., and G.E. Fogg, 1990, Stream functions and equivalent fresh-water heads for modeling regional flow of variable-density ground water. 2. Application and implications for modeling strategy, Water Resources Research 26(9), pp. 2097–2106.

Senger, R.K., C.W. Kreitler, and G.E. Fogg, 1987, Regional underpressuring in deep brine aquifers, Palo Duro Basin, Texas. 2. The effect of Cenozoic basin development, Water Resources Research 23(8), pp. 1494–1504.

Shafer, J.M., 1987, Reverse pathline calculation of time related capture zones in nonuniform flow, Ground Water 25(3), pp. 283–289.

Shafer, J.M., 1990, GWPATH, Version 4.0, J.M. Shafer, Champaign, IL.

Siegel, D.I., 1988, The recharge-discharge function of wetlands near Juneau, Alaska: Part I. Hydrogeological investigations, Ground Water 26(4), pp. 427–434.

Silliman, S.E., and A.L. Wright, 1988, Stochastic analysis of paths of high hydraulic conductivity in porous media, Water Resources Research 24(11), pp. 1901–1910.

Simmers, J., 1988, *Estimation of Natural Groundwater Recharge*, NATO ASI Series C: V. 222, D. Reidel Publishing Company, 510 p.

Simpson, E.S., 1988, The discrete state compartment model and its applications to flow through karstic aquifers, *In: Karst Hydrogeology and Karst Protection*, Proc. IAH 21st Congress, Guilin, China, Geological Publ. House, Beijing, pp. 671–676.

Sims, P.N., P.F. Andersen, D.E. Stephenson, and C.R. Faust, 1989, Testing and benchmarking of a three-dimensional groundwater flow and solute transport model, *In: Solving Ground Water Problems with Models*, IGWMC, Indianapolis, pp. 821–841.

Smith, L., and R.A. Freeze, 1979, Stochastic analysis of steady state groundwater flow in a bounded domain. 2. Two-dimensional simulations, Water Resources Research 15, pp. 1543–1559.

Smith, L., and F.W. Schwartz, 1984, An analysis of fracture geometry on mass transport in fractured media, Water Resources Research 20(9), pp. 1241–1252.

Sophocleus, M., and C.A. Perry, 1985, Experimental studies in natural groundwater-recharge dynamics: the analysis of observed recharge events, Journal of Hydrology 81, pp. 297–332.

Stark, J.R., and M.G. McDonald, 1980, Ground water of coal deposits, Bay County, Michigan, USGS, Open-File Report 80-591, p. 36.

Steenhuis, T.S., C.D. Jackson, S.K.J. Kung, and W. Brutsaert, 1985, Measurement of groundwater recharge on eastern Long Island, New York, U.S.A., Journal of Hydrology 79, pp. 145–169.

Stephens, D.B., 1983, Groundwater flow and implications for groundwater contamination north of Prewitt, New Mexico, USA, Journal of Hydrology 61, pp. 391–408.

Stephens, D.B., and R. Knowlton, Jr., 1986, Soil water movement and recharge through sand at a semiarid site in New Mexico, Water Resources Research 22(6), pp. 881–889.

Stoertz, M.W., M.P. Anderson, and K.R. Bradbury, 1991, Field investigation and numerical studies of groundwater recharge through unsaturated sand: A methodology applied to central Wisconsin, Wisconsin Geological and Natural History Survey, Circular 71, 52 p.

Stoertz, M.W., and K.R. Bradbury, 1989, Mapping recharge areas using a ground-water flow model—a case study, Ground Water 27(2), pp. 220–229.

Strack, O.D.L., 1987, The analytic element method for regional groundwater modeling, *In: Solving Ground Water Problems with Models*, National Water Well Assoc., Columbus, Ohio, pp. 929–941.

Strack, O.D.L., 1988, *Groundwater Mechanics*, Prentice-Hall, Englewood Cliffs, N.J., 732 p.

Streltsova-Adams, T.D., 1978, Well hydraulics in heterogeneous aquifer formations, Advances in Hydroscience 11, pp. 357–418.

Sudicky, E.A., 1989, The Laplace transform technique: a time-continuous finite element theory and application to mass transport in groundwater, Water Resources Research 25(8), pp. 1833–1846.

Sudicky, E.A., R.W. Gillham, and E.O. Frind, 1985, Experimental investigation of solute transport in stratified porous media. 1. The nonreactive case, Water Resources Research 21(7), pp. 1035–1042.

Sudicky, E.A., R.W. Gillham, and E.O. Frind, 1985, Experimental investigation of solute transport in stratified porous media. 2. The reactive case, Water Resources Research 21(7), pp. 1043–1050.

Sun, R.J., 1986, Regional aquifer-system analysis program of the U.S. Geological Survey, Summary of Projects, 1978–84, USGS, Circular 1002, 264 p.

Swenson, F.A., 1968, New theory of recharge to the Artesian Basin of the Dakotas, Geol. Soc. of Amer. Bull., 79, pp. 163–182.

Sykes, J.F., J.L. Wilson, and R.W. Andrews, 1985, Sensitivity analysis for steady state groundwater flow using adjoint operators, Water Resources Research 21(3), pp. 359–371.

Tanaka, H.H., and J.R. Hollowell, 1966, Hydrology of the alluvium of the Arkansas River, Muskogee, Oklahoma, to Fort Smith, Arkansas, USGS, (Water Supply Paper 1809-T, 42 p.

Taylor, O.J., 1988, Predicted effects of underground mine flooding at Tract C-b in Piceance basin, northwestern Colorado, USGS, Water-Resources Investigations Report 87-4189, 16 p.

Theis, C.V., 1935, The relation between the lowering of the piezometric surface and the rate and duration of discharge of a well using groundwater storage, Trans. Amer. Geophys. Union 2, pp. 519–524.

Thomas, H.E., and L.B. Leopold, 1964, Groundwater in North America, Science 143, pp. 1001–1006.

Thomas, J.M., S.M. Carlton, and L.B. Hines, 1989, Ground-water hydrology and simulated effects of development in Smith Creek valley, a hydrologically closed basin in Lander County, Nevada, USGS, Profession Paper 1409-E, pp. E1–E57.

Thrailkill, J., 1974, Pipe flow models of a Kentucky limestone aquifer, Ground Water 12(4), pp. 202–205.

Thrailkill, J., 1989, Shallow conduit-flow carbonate aquifers: conceptual models and parameter evaluation, In: Recent Advances in Ground-Water Hydrology (J.E. Moore, A.A. Zaporozec, S.C. Csallany, and T.C. Varney, eds.), American Institute of Hydrology, pp. 153–159.

Todd, D.K., and J. Bear, 1959, River seepage investigations, Contrib. no. 20, Hydraulic Lab., Water Resource Center, Univ. of California, Berkeley.

Tompson, A.F.B., R. Ababou, and L.W. Gelhar, 1989, Implementation of the three-dimensional turning bands random field generator, Water Resources Research 25(10), pp. 2227–2243.

Toran, L., and K.R. Bradbury, 1988, Ground-water flow model of drawdown and recovery near an underground mine, Ground Water 26(6), pp. 724–733.

Toth, J., 1962, A theory of groundwater motion in small drainage basins in central Alberta, J. Geophys. Res. 67, pp. 4375–4387.

Toth, J., 1963, A theoretical analysis of groundwater flow in small drainage basins, J. Geophys. Res. 68, pp. 4795–4812.

Townley, L.R., 1990, AQUIFEM-N: a multi-layered finite element aquifer flow model, user's manual and description, CSIRO Division of Water Resources, Perth, Western Australia.

Townley, L.R., and M.R. Davidson, 1988, Definition of a capture zone for shallow water table lakes, Journal of Hydrology 104, pp. 53–76.

Townley, L.R., and J.L. Wilson, 1980, Description of and user's manual for a finite element aquifer flow model AQUIFEM-1, MIT Ralph M. Parsons Laboratory for Water Resources and Hydrodynamics, Technology Adaptation Program Report No. 79-3, 294 p.

Townley, L.R., and J.L. Wilson, 1983, Conditional second moment analysis of groundwater flow: the cumulative effects of transmissivity and head measurements, In: Papers of the International Conference on Groundwater and Man, Australian Government Publ. Service, Canberra.

Townley, L.R., and J.L. Wilson, 1985, Computationally efficient algorithms for parameter estimation and uncertainty propagation in numerical models of groundwater flow, Water Resources Research 21(12), pp. 1851–1860.

Trapp, H., Jr., and L.H. Geiger, 1986, Three-dimensional steady-state simulation of flow in the sand-and-gravel aquifer, southern Escambia County, Florida, USGS, Water-Resources Investigations Report 85-4278, 149 p.

Trescott, P.C., G.F. Pinder, and S.P. Larson, 1976, Finite-difference model for aquifer simulation in two dimensions with results of numerical experiments, USGS, Techniques of Water-Resources Investigations, Book 7, 116 p.

Tsang, Y.W., and C.F. Tsang, 1987, Channel model of flow through fractured media, Water Resources Research 23(3), pp. 467–479.

Tsang, Y.W., and C.F. Tsang, 1988, Flow and tracer transport in fractured media: a variable aperture channel model and its properties, Water Resoures Research 24(12), pp. 2049–2060.

Tucciarelli, Y., 1989, A semiautomatic mesh generation algorithm for three-dimensional groundwater tetrahedral finite element models, Water Resources Research 25(3), pp. 573–576.

Tyler, S.W., and S.W. Wheatcraft, 1990, Fractal processes in soil water retention, Water Resources Research 26(5), pp. 1047–1054.

U.S. Environmental Protection Agency, 1987, Guidelines for delineation of wellhead protection areas, U.S. EPA–Office of Ground-Water Protection.

U.S. Geological Survey, 1976, Digital-computer model of the sandstone aquifer in southeastern Wisconsin, SE Wisconsin Regional Planning Commission Technical Report 16, 42 p.

Urish, D.W., and M.M. Ozbilgin, 1989, The coastal ground-water boundary, Ground Water 27(3), pp. 310–315.

van der Heijde, P.K.M., A.I. El-Kadi, S.A. Williams, 1988, Groundwater modeling: An overview and status report. U.S. EPA, EPA/600/2-89/028, 242 p.

Van Everdingen, R.D., 1963, Groundwater flow-diagrams in sections with exaggerated vertical scale, Geological Survey of Canada Paper 63-27, 21 p.

Voss, C.I., 1984a, AQUIFEM-SALT: a finite-element model for aquifers containing a seawater interface, USGS, Water-Resources Investigations Report 84-4263, 37 p.

Voss, C.I., 1984b, A finite-element simulation model for saturated-unsaturated, fluid-density–dependent groundwater flow with energy transport or chemically-reactive single-species solute transport, USGS, Water-Resources Investigations Report 84-4369, 409 p.

Voss, C.I., and W.R. Souza, 1987, Variable density flow and solute transport simulation of regional aquifers containing a narrow freshwater–saltwater transition zone. Water Resources Research 23(10), pp. 1851–1866.

Waddell, R.K., 1982, Two-dimensional, steady-state model of ground-water flow, Nevada Test Site and vicinity, Nevada-California, USGS, Water-Resources Investigations 82-4085, 72 p.

Walton, W.C., 1987, *Groundwater Pumping Tests: Design and Analysis*, Lewis Publishers, Chelsea, MI, 201 p.

Walton, W., 1989, *Numerical Groundwater Modeling: Flow and Contaminant Migration*, Lewis Publishers, 272 p.

Wang, C.P., and R.E. Williams, 1984, Aquifer testing, mathematical modeling, and regulatory risk, Ground Water 22(3), pp. 285–296.

Wang, H.F., and M.P. Anderson, 1977, Finite differences and finite elements as weighted residual solutions to Laplace's equation, *In: Finite Elements in Water Resources* (W.G. Gray, G.F. Pinder, and C.A. Brebbia, eds.), Pentech Press, London, pp. 2.167–2.178.

Wang, H.F., and M.P. Anderson, 1982, *Introduction to Groundwater Modeling: Finite Difference and Finite Element Methods*, W.H. Freeman, 256 p.

Ward, D.S., D.R. Buss, J.W. Mercer, and S.S. Hughes, 1987, Evaluation of a groundwater corrective action at the Chem-Dyne hazardous waste site using a telescopic mesh refinement modeling approach, Water Resources Research 23(4), pp. 603–617.

Watson, K.K., 1986, Numerical analysis of natural recharge to an unconfined aquifer, *In: Conjunctive Water Use* (S.M. Gorelick, ed.), IAHS Publication 156, pp. 323–333.

Watts, K.R., 1989, Potential hydrologic effects of ground-water withdrawals from the Dakota aquifer, southwestern Kansas, USGS, Water Supply Paper 2304, 47 p.

Weeks, E.P., 1969, Determining the ratio of horizontal to vertical permeability by aquifer-test analysis, Water Resources Research 5(1), pp. 196–214.

Weeks, J.B., and R.J. Sun, 1987, Regional aquifer-system analysis program of the U.S. Geological Survey—bibliography, 1978–1986, USGS, Water-Resources Investigations Report 87-4138.

Weiss, J.A., and A.K. Williamson, 1985, Subdivision of thick sedimentary units into layers for simulation of ground-water flow, Ground Water 23(6), pp. 767–774.

Wexler, E.J., and P.E. Maus, 1988, Ground-water flow and solute transport at a municipal landfill site on Long Island, New York. Part 2. Simulation of ground-water flow, USGS, Water-Resources Investigations Report, 86-4106, 44 p.

White, I., 1988, Comment on: "A natural gradient experiment on solute transport in a sand aquifer: spatial variability of hydraulic conductivity and its role in the dispersion process" by E.A. Sudicky, Water Resources Research 24(6), pp. 892–894.

Williams, T.A., and A.K. Williamson, 1989, Estimating water-table altitudes for regional ground-water flow modeling, U.S. Gulf Coast, Ground Water 27(3), pp. 333–340.

Wilson, J.L., and D.A. Hamilton, 1978, Influence of strip mines on regional ground-water flow, Journal of the Hydraulics Division, ASCE, 104(HY9), pp. 1213–1223.

Wilson, J.L., L.R. Townley, and A. Sa da Costa, 1979, Mathematical development and verification of a finite element aquifer flow model: AQUIFEM-1, TAP Report 79-2, Massachusetts Institute of Technology, Cambridge, MA, 114 p.

Winter, T.C., 1976, Numerical simulation of the interaction of lakes and ground water, USGS, Professional Paper 1001, 45 p.

Winter, T.C., 1983, The interaction of lakes with variably saturated porous media, Water Resources Research 19(5), pp. 1203–1218.

Witherspoon, P.A., J.C.S. Long, E.L. Majer, and L.R. Myer, 1987, A new seismic-hydraulic approach to modeling flow in fractured rocks, In: *Solving Ground-Water Problems with Models*, National Water Well Association, Dublin, Ohio, pp. 793–826.

Woessner, W.W., and M.P. Anderson, 1990, Setting calibration targets and assessing model calibration—room for improvement: an example from North America, In: *Calibration and Reliability in Groundwater Modelling* (K. Kovar, ed.), IAHS Pub. 195, pp. 279–290.

Wong, K-F.V., G.T. Yeh, and E.C. Davis, 1987, Predictive application of an ORNL geohydrology model, Ground Water 25(3), pp. 342–345.

Yager, R.M., 1986, Simulation of ground-water flow and infiltration from the Susquehanna River to a shallow aquifer at Kirkwood and Conklin, Boone County, New York, USGS, Water-Resources Investigations Report 86-4123, 70 p.

Yager, R.M., 1987, Simulation of ground-water flow near the nuclear-fuel reprocessing facility at the Western New York Nuclear Service Center, Cattaraugus County, New York, USGS Water Resources Investigation Report 85-4308, 58 p.

Yazicigil, H., and L.A.V. Sendlein, 1981, Management of ground water contaminated by aromatic hydrocarbons in the aquifer supplying Ames, Iowa, Ground Water 19(6), pp. 648–665.

Yeh, G.T., 1987, 3D-FEMWATER: a three-dimensional finite element model of water flow through saturated-unsaturated media, Oak Ridge National Laboratory, Publication No. 2904, 314 p.

Yeh, W., 1986, Review of parameter identification procedures in groundwater hydrology: the inverse problem, Water Resources Research 22(2), pp. 95–108.

Yusun, C. and B. Ji, 1988, The media and movement of karst water, *In: Karst Hydrogeology and Karst Environment Protection*, Proc. IAH 21st Congress, Guilin, China, Geological Publ. House, Beijing, pp. 555–564.

Zheng, C., 1989, PATH3D, S.S. Papadopulos & Assoc., Rockville, MD.

Zheng, C., 1990, MT3D, a modular three-dimensional transport model, S.S. Papadopulos & Assoc., Rockville, MD.

Zheng, C., G.D. Bennett, and C.B. Andrews, 1991, Analysis of ground water remedial alternatives at a Superfund site. Ground Water 29, in press.

Zheng, C., K.R. Bradbury, and M.P. Anderson, 1988a, Role of interceptor ditches in limiting the spread of contaminants in ground water, Ground Water 26(6), pp. 734–742.

Zheng, C., H.F. Wang, M.P. Anderson, and K.R. Bradbury, 1988b, Analysis of interceptor ditches for control of groundwater pollution, Journal of Hydrology 98, pp. 67–81.

Zucker, M.B., I. Remson, J. Ebert, and E. Aguado, 1973, Hydrologic studies using the Boussinesq equation with a recharge term, Water Resources Research 9(3), pp. 586–592.

INDEX